A Student's Guide to Einstein's Major Papers

A Student's Guide
to Einstein's Major Papers

Robert E. Kennedy
Department of Physics, Creighton University

OXFORD
UNIVERSITY PRESS

OXFORD

UNIVERSITY PRESS

Great Clarendon Street, Oxford OX2 6DP

Oxford University Press is a department of the University of Oxford.
It furthers the University's objective of excellence in research, scholarship,
and education by publishing worldwide in

Oxford New York

Auckland Cape Town Dar es Salaam Hong Kong Karachi
Kuala Lumpur Madrid Melbourne Mexico City Nairobi
New Delhi Shanghai Taipei Toronto

With offices in

Argentina Austria Brazil Chile Czech Republic France Greece
Guatemala Hungary Italy Japan Poland Portugal Singapore
South Korea Switzerland Thailand Turkey Ukraine Vietnam

Oxford is a registered trade mark of Oxford University Press
in the UK and in certain other countries

Published in the United States
by Oxford University Press Inc., New York

British Library Cataloguing in Publication Data
Data available

Library of Congress Cataloging in Publication Data
Data available

Typeset by SPI Publisher Services, Pondicherry, India
Printed and bound by
CPI Group (UK) Ltd, Croydon, CR0 4YY

ISBN 978–0–19–969403–7

3 5 7 9 10 8 6 4 2

To my wife, Mary
for her continued support and constant quiet encouragement

Contents

Acknowledgments   xiii

Introduction   xv

1 Setting the Stage for 1905   **1**

1.1 Overview   1

1.2 Historical Background   2

    1.2.1 600 BC to AD 200: The Contribution of the
Early Greeks   2

    1.2.2 The 1600s: The Contribution of Galileo and Newton   6

    1.2.3 The 1800s: The Contribution of Maxwell and Lorentz   13

    1.2.4 The Worldview in 1900   15

1.3 Albert Einstein   15

    1.3.1 The Pre-College Years   15

    1.3.2 The College Years   17

    1.3.3 From College to 1905   19

1.4 Discussion and Comments   20

1.5 Appendices   21

    1.5.1 Science Today   21

    1.5.2 Newton's Law of Gravitation from Kepler's Laws   25

1.6 Notes   27

1.7 Bibliography   31

2 Radiation and the Quanta   **33**

2.1 Historical Background   33

    2.1.1 Thermodynamics and Entropy   33

    2.1.2 Blackbody Radiation   34

    2.1.3 Max Planck's Derivation of the Radiation Density   38

2.2 Albert Einstein's Paper, "On a Heuristic Point of View
Concerning the Production and Transformation of Light"   39

    2.2.1 On a Difficulty Encountered in the Theory of
"Blackbody Radiation"   40

    2.2.2 On Planck's Determination of the Elementary Quanta   41

    2.2.3 On the Entropy of Radiation   42

    2.2.4 Limiting Law for the Entropy of Monochromatic
Radiation at Low Radiation Density   43

2.2.5 Molecular-Theoretical Investigation of the
       Dependence of the Entropy of Gases and Dilute
       Solutions on the Volume   43
2.2.6 Interpretation of the Expression for the Dependence
       of the Entropy of Monochromatic Radiation on
       Volume According to Boltzmann's Principle   44
2.2.7 On Stokes' Rule   45
2.2.8 On the Generation of Cathode Rays by Illumination
       of Solid Bodies   46
2.2.9 On the Ionization of Gases by Ultraviolet Light   47
2.3 Discussion and Comments   47
2.4 Appendices   49
2.4.1 Entropy and Irreversibility   49
2.4.2 Planck's derivation of $\rho\left(\nu, T\right)$ 50
2.4.3 Wien's Expression for Entropy   51
2.5 Notes   52
2.6 Bibliography   55

3 The Atom and Brownian Motion   56
3.1 Historical Background   56
3.1.1 The Atom   57
3.1.2 Brownian Motion   60
3.1.3 The Worldview in 1900   60
3.2 Albert Einstein's Paper, "A New Determination
    of Molecular Dimensions"   62
3.2.1 On the Influence on the Motion of a Liquid
       Exercised by a Very Small Sphere Suspended in It   63
3.2.2 Calculation of the Coefficient of Viscosity of a Liquid
       in Which Very Many Irregularly Distributed Small
       Spheres are Suspended   66
3.2.3 On the Volume of a Dissolved Substance Whose
       Molecular Volume is Large Compared to that
       of the Solvent   67
3.2.4 On the Diffusion of an Undissociated Substance
       in a Liquid Solution   68
3.2.5 Determination of the Molecular Dimensions with the
       Help of the Relations Obtained   69
3.3 Albert Einstein's Paper, "On the Movement of Small
    Particles Suspended in Stationary Liquids Required by the
    Molecular-Kinetic Theory of Heat"   70
3.3.1 On the Osmotic Pressure Attributable to Suspended
       Particles   71
3.3.2 Osmotic Pressure from the Standpoint of the
       Molecular-Kinetic Theory of Heat   72
3.3.3 Theory of Diffusion of Small Suspended Spheres   73
3.3.4 On the Random Motion of Particles Suspended in a
       Liquid and Their Relation to Diffusion   75

    3.3.5 Formula for the Mean Displacement of Suspended
        Particles. A New Method of Determining the True
        Size of Atoms   76

3.4 Discussion and Comments   76

3.5 Appendices   78

    3.5.1 Derivation of the Expressions for u, v, and w 78
    3.5.2 Derivation of the Expression for $W = $ Energy per
        Unit Time Converted into Heat   85
    3.5.3 Derivation of the Coefficient of Viscosity of a Liquid
        in Which Very Many Irregularly Distributed Spheres
        are Suspended   90
    3.5.4 Determination of the Volume of a Dissolved Substance   93
    3.5.5 Derivation of the Expression for Entropy   93
    3.5.6 Derivation of $B = JV^{*n}$ 95
    3.5.7 Derivation of $\nu = f(x,t)$ 96
    3.5.8 Derivation of $\langle x^2 \rangle$ 98

3.6 Notes   98

3.7 Bibliography   103

4 The Special Theory of Relativity   **105**

4.1 Historical Background   105

    4.1.1 The Relativity of Galileo Galilei and of Isaac Newton   105
    4.1.2 The Lorentz Transformations (from Lorentz)   108

4.2 Albert Einstein's Paper, "On the Electrodynamics of
    Moving Bodies"   113

    4.2.1 Definition of Simultaneity   115
    4.2.2 On the Relativity of Lengths and Times   116
    4.2.3 Theory of Transformation of Coordinates and Time
        from a System at Rest to a System in Uniform
        Translational Motion Relative to It   118
    4.2.4 The Physical Meaning of the Equations Obtained
        Concerning Moving Rigid Bodies and Moving Clocks   120
    4.2.5 The Addition Theorem of Velocities   121
    4.2.6 Transformation of the Maxwell–Hertz Equations for
        Empty Space. On the Nature of the Electromotive
        Forces that Arise upon Motion in a Magnetic Field   122
    4.2.7 Theory of Doppler's Principle and of Aberration   124
    4.2.8 Transformation of the Energy of Light Rays. Theory
        of the Radiation Pressure Exerted on Perfect Mirrors   126
    4.2.9 Transformation of the Maxwell–Hertz Equations
        when Convection Currents Are Taken into
        Consideration   128
    4.2.10 Dynamics of the (Slowly Accelerated) Electron   128

4.3 Albert Einstein's Paper, "Does the Inertia of a Body
    Depend Upon Its Energy Content?"   129

4.4 Discussion and Comments   131

4.5 Appendices   133
    4.5.1 Lorentz and the Transformed Maxwell Equations   133
    4.5.2 Derivation of the Lorentz Transformation Equations   140
    4.5.3 The Electromagnetic Field Transformations   146
    4.5.4 The Doppler Principle   150
    4.5.5 The Electrodynamic Lorentz Force   153
4.6 Notes   155
4.7 Bibliography   159

5 The General Theory of Relativity   **161**
5.1 Historical Background   161
    5.1.1 Lingering Questions   161
    5.1.2 Generalizing the Special Theory of Relativity   163
    5.1.3 The Equivalence of a Gravitational Field and an
          Accelerated Reference Frame   164
    5.1.4 The Timeline from 1905 to 1916   167
5.2 Albert Einstein's Paper, "The Foundation of the General
    Theory of Relativity"   171

**Part A: "Fundamental Considerations on the Postulate
of Relativity"**   **171**
    5.2.1 Observations on the Special Theory of Relativity   171
    5.2.2 The Need for an Extension of the Postulate
          of Relativity   172
    5.2.3 The Space-Time Continuum. Requirement of
          General Covariance for the Equations Expressing
          General Laws of Nature   174
    5.2.4 The Relation of the Four Coordinates to
          Measurement in Space and Time   175

**Part B: "Mathematical Aids to the Formulation of
Generally Covariant Equations"**   **178**
    5.2.5 Contravariant and Covariant Four-Vectors   179
    5.2.6 Tensors of the Second and Higher Ranks   181
    5.2.7 Multiplication of Tensors   182
    5.2.8 Some Aspects of the Fundamental Tensor $g_{\mu\nu}$ 183
    5.2.9 The Equation of the Geodetic Line. The Motion
          of a Particle   186
    5.2.10 The Formation of Tensors by Differentiation   187
    5.2.11 Some Cases of Special Importance   188
    5.2.12 The Riemann–Christoffel Tensor   191

Part C: "Theory of the Gravitational Field"   **192**
    5.2.13 Equations of Motion of a Material Point in the
          Gravitational Field. Expression for the
          Field-Components of Gravitation   192

5.2.14 The Field Equations of Gravitation in the Absence
of Matter — 193

5.2.15 The Hamiltonian Function for the Gravitational
Field. Laws of Momentum and Energy — 194

5.2.16 The General Form of the Field Equations
of Gravitation — 196

5.2.17 The Laws of Conservation in the General Case — 198

5.2.18 The Laws of Momentum and Energy for Matter,
as a Consequence of the Field Equations — 198

Part D: "Material Phenomena" — **199**

5.2.19 Euler's Equations for a Frictionless Adiabatic Fluid — 199

5.2.20 Maxwell's Electromagnetic Field Equations for
Free Space — 200

Part E: — **205**

5.2.21 Newton's Theory as a First Approximation — 205

5.2.22 The Behaviour of Rods and Clocks in the Static
Gravitational Field. Bending of Light Rays. Motion
of the Perihelion of a Planetary Orbit — 208

5.3 Discussion and Comments — 213

5.3.1 Verification of the General Theory of Relativity — 213

5.3.2 Beyond the General Theory of Relativity:
Cosmology and the Unified Field Theory — 216

5.4 Appendices — 223

5.4.1 Multiplication of Tensors — 223

5.4.2 Some Aspects of the Fundamental Tensor $g_{\mu\nu}$ — 224

5.4.3 The Equation of the Geodetic Line — 225

5.4.4 The Formation of Tensors by Differentiation — 229

5.4.5 Some Cases of Special Importance — 232

5.4.6 The Riemann–Christoffel Tensor — 239

5.4.7 The Hamiltonian Function for the
Gravitational Field — 241

5.4.8 Calculation of the Bending of Starlight — 249

5.4.9 Calculation of the Precession of the Perihelion
of Mercury — 249

5.4.10 The Bending of Starlight Experiment — 253

5.4.11 Newton's Bucket — 254

5.5 Notes — 255

5.6 Bibliography — 262

6 Einstein and Quantum Mechanics — **265**

6.1 Historical Background — 265

6.2 The Evolution of Quantum Mechanics — 267

6.2.1 The Theory of Specific Heat (1906) — 267

6.2.2 The Dual Nature of Radiation (1909) — 268

6.2.3 The Bohr Atom (1913)   269

6.2.4 Spontaneous and Induced Transitions (1916)   271

6.2.5 The Compton Scattering Experiment (1923)   271

6.2.6 Bose–Einstein Statistics (1924)   272

6.2.7 Einstein, de Broglie (1924), and Schrödinger (1926)   275

6.2.8 Einstein and Bohr (1927, 1930)   277

6.3 Discussion and Comments   281

6.4 Appendices   282

6.4.1 The Specific Heat of Dulong and Petit   282

6.4.2 The Commutator of P and Q 282

6.5 Notes   283

6.6 Bibliography   287

7 Epilogue   **290**

7.1 The Inflexible Boundary Condition   290

7.2 Notes   293

7.3 Bibliography   295

Index   **297**

Acknowledgments

I would like to thank my many friends and colleagues at Creighton University and at the University of Notre Dame who have contributed to the development of this book. At Creighton University both the Graduate School and the College of Arts and Sciences provided support for the development of this manuscript, while my colleagues (both within and outside of the Department of Physics) continue to send me new material on Einstein. In particular, I thank Menachem Mor for the many fruitful discussions on the historical perspectives and on the Jewish context regarding Albert Einstein, and a special thank you to Paul Nienkamp who "re-created" Lorentz' development of the Lorentz transformations. I am indebted to the Department of Physics at the University of Notre Dame where I spent three sabbatical leaves working with their faculty. Of particular help were the conversations with James Cushing that interested me in the history and philosophy of science and his comments on many of the scientific aspects of Einstein's work. I want especially to thank Don Howard for the many stimulating discussions on the material included herein and for his encouragement to pursue this project. Thank you to Oxford University Press (Sonke Adlung, April Warman, and Clare Charles). I extend a special thank you to the reviewers of the manuscript who included in their comments a number of insights, some of which have been included in the text (Sections 5.1.4 and 5.2.20). In closing, I extend a particular thank you to my wife Mary, and my children (and children-in-law) Bob, Erin and Peter, Chris and Michelle, Mary Shannon, Mike and Amy for their support.

Introduction

Our understanding of nature underwent a revolution in the early twentieth century – from the classical physics of Galileo, Newton, and Maxwell to the modern physics of relativity and quantum mechanics. The dominant figure in this revolutionary change was Albert Einstein.

In 1905, Einstein produced breakthrough work in three distinct areas of physics: on the size and the effects of atoms; on the quantization of the electromagnetic field; and on the special theory of relativity. In 1916, he produced a fourth breakthrough work, the general theory of relativity. Einstein's scientific work is the main focus of this book. The book sets many of his major works into their historical context, with an emphasis on the pathbreaking works of 1905 and 1916. It also develops the detail of his papers, taking the reader through the mathematics to help the reader discover the simplicity and insightfulness of his ideas and to grasp what was so "revolutionary" about his work.

As with any revolution, the story told after the fact is not always an accurate portrayal of the events and their relation to one another at the time of the revolution. Following Einstein's work in 1905, more efficient and more convenient ways were found to reach the same results but, in such revisions, many of the original insights were lost. Today, many people hold historically incorrect views of Einstein's papers, mainly regarding the insights and reasoning that led to the results. For example:

- The quantum paper was not written to explain the photoelectric effect, rather, it was written to explain the Wien region of blackbody radiation;
- The Brownian motion paper was not written to explain Brownian motion, Einstein was not even certain his work would pertain to Brownian motion;
- The relativity paper was not written to explain the Michelson–Morley experiment, etc.

By working through Einstein's original papers, the reader will gain a better appreciation for Einstein's revolutionary insights as well as a historically more accurate picture of them.

Just as a person cannot hope to appreciate the significance of the American Revolution without some knowledge of the American colonies before 1776, one cannot hope to appreciate the significance of the scientific revolution of the early 1900s without some knowledge of the state of science at that time. In order to help the reader appreciate the deep impact of Einstein's work, chapter one briefly lists some key concepts and issues in the history and philosophy of science, together with some recommendations for further reading for the interested student. To complete setting the context for 1905, chapter one concludes with a discussion of several of the factors in Einstein's life that contributed to his worldview, ranging from his early childhood, through the German and Swiss school systems, his marriage to Mileva Marić, and to his position at the patent office.

Chapters two through five discuss the four major works of Einstein, one per chapter. As the general theory of relativity became the base for the development of cosmology and unified field theories, an overview of Einstein's contribution to these fields is included at the end of the chapter on the general theory of relativity. Despite the perception that Einstein was constantly fighting the advances of quantum mechanics, from 1905 to 1924 he stood virtually alone in defense of the idea that the quantum is a real constituent of the electromagnetic field. This was in opposition to Planck's idea that it was merely the exchange of electromagnetic energy between radiation and matter that was quantized.[1] It was not until the mid 1920s that Einstein became the strong dissenter from the conventional interpretation of quantum mechanics, the role he played famously in the Bohr–Einstein debates.[2] Einstein's contributions to the development of quantum mechanics are discussed in chapter six.

To remove one hindrance to reading the original papers, the notation and phrasing have been updated: the electric and magnetic fields of today were previously referred to as electric and magnetic forces; the speed of light is denoted c, not V as in Einstein's original papers; the mathematical cross product $\vec{A} \times \vec{B}$ was written as $\vec{A} \cdot \vec{B}$; etc.

Obviously not everything Einstein did can be put into one book with any detail. For example, *The Collected Papers of Albert Einstein*[3] was, as of 2011, a 12-volume set of Einstein's papers and correspondence – and this included his papers only through the early 1920s!

It is assumed that the reader has a copy of Einstein's original papers for reference. They are available from a number of sources. The most complete source is *The Collected Papers of Albert Einstein*. With each volume is a companion English translation volume, containing translations of papers that were not in English in the original volume. The volumes of the original writings contain a number of essays and editorial comments that are quite informative, but they are not included in the companion translation volumes. These essays and editorial comments provide a very good introduction to the various topics and Einstein's contribution. The serious reader is encouraged to access these editorial comments to gain a fuller, and a more complete, picture of Einstein's contributions. Nearly all of the references to the writings of Einstein are to this source, listed as (for example) CPAE1, p. 123 (*The Collected Papers of Albert Einstein*, Volume 1, page 123), listing also the companion English translation volume immediately following as CPAE2 ET, p. 456 (*The Collected Papers of Albert Einstein*, Volume 2, English translation, page 456). All five of Einstein's 1905 papers, with a good introduction to each of them by John Stachel, can be found in *Einstein's Miraculous Year*.[4] A collection of all of Einstein's papers in the volumes of *Annalen der Physik* can be found in the Wiley publication, *Einstein's Annalen Papers*, by Jürgen Renn (the papers are in the original, not in translation).[5] Renn's book has a nice introductory essay for each of the four major areas of Einstein's work. The Dover publication, *Albert Einstein: Investigations on the Theory of Brownian Motion*, contains the two 1905 papers on the atom: "A New Determination of Molecular Dimensions" and "On the Movement of Small Particles Suspended in Stationary Liquids Required by the Molecular-Kinetic Theory of Heat."[6] Another Dover publication, *Principle of Relativity*,[7] contains the two special theory of relativity papers of 1905 and the general theory of relativity paper of 1915, as well as the cosmology paper of 1917.

The selection and presentation of the material included in the book, unavoidably, will reflect the bias of the author. To minimize the impact of that bias, and to avoid misrepresentations of the source material, extensive use of quotations has been made. The extensive citation of sources, also, is intended to aid the reader interested in pursuing further a particular item. At the end of each chapter, the sources are referenced in detail and a summary of the literature used in the preparation of the chapter is included in the bibliography for that chapter.

A Synopsis of the Purpose of Each Chapter

1   Setting the Stage for 1905

This chapter attempts to give the reader some awareness of the evolution of scientific thought from the early Greek natural philosophers (Pythagoras, Plato, Aristotle, etc.) through the work of Galileo, Newton, and Maxwell to the ideas of Einstein. Its purpose is to provide a brief overview, not to provide a detailed picture of the history and philosophy of physical science.

The first portion of the chapter is a brief history of physical science, highlighting selected events in our evolving understanding of the universe we inhabit, from the motion of the heavens to an understanding of its basic constituents. The focus is on the ideas leading to the works of Einstein: the universe is orderly and understandable; mathematics describes this underlying order; new and better data lead to the revision of previous ideas; and our advancing understanding of nature generally leads to a more unified framework for understanding nature. At the beginning of each of the science chapters, additional material on the history of the topic is presented. The second portion of chapter one looks at the events in Einstein's life prior to 1905, from his childhood years through the German school system, through college, his marriage to Mileva Marić, and to his position in the patent office. These are the years and the events leading to the *annus mirabilis* of 1905.

2   Radiation and the Quanta

Chapter two details the paper, "On a Heuristic Point of View Concerning the Production and Transmission of Light,"[8] one of the 1905 *annus mirabilis* papers. This is often referred to as the "photoelectric effect" paper. However, Einstein used the photoelectric effect as but one of three possible examples at the end of the paper. His focus in the paper is not on the photoelectric effect but, rather, on a thermodynamic treatment of the Wien region of the blackbody radiation, showing that the expression for the entropy of the radiation can be made identical to the expression for the entropy of an ideal gas of non-interacting particles.

3   The Atom and Brownian Motion

Chapter three details the two papers, "A New Determination of Molecular Dimensions"[9] and "On the Movement of Small Particles Suspended in Stationary Liquids Required by the Molecular-Kinetic Theory of Heat."[10] The first of these is the work of Einstein's doctoral dissertation. The second is often referred to as the "Brownian motion" paper, although Einstein himself was not certain his results pertained to Brownian motion. His goal was to find further evidence for the atomic hypothesis. Einstein's "proof" of the reality of atoms is the subject of chapter three.

4   The Special Theory of Relativity

Chapter four details the papers, "On the Electrodynamics of Moving Bodies"[11] and "Does the Inertia of a Body Depend on its Energy Content?"[12] the fourth and fifth of the 1905 *annus mirabilis* papers. The first of these is the special theory of relativity. Beginning with a discussion of clocks running synchronously, Einstein derives the Lorentz transformations for position and time and, subsequently, using the Lorentz transformations he derives the transformations for the electric and magnetic fields. The second of these papers is very short, essentially an addendum to the first paper, in which the famous relation $E = mc^2$ is obtained.

5   The General Theory of Relativity

Chapter five details the paper, "The Foundation of the General Theory of Relativity,"[13] published in 1916. This paper builds on concerns left to be answered from the special theory of relativity of 1905: Why should the theory of relativity be restricted to uniform velocities? Why do inertial mass and gravitational mass have the same value? Why do all objects, regardless of their composition, fall with the same acceleration in a given gravitational field? From considerations such as these came the realization that the effects of gravity and those of an accelerating reference frame are equivalent and, eventually, that gravity is expressible as a property of space itself, but of a four-dimensional space that has curvature and is non-Euclidean. This chapter concludes with a discussion of the tests of the general theory of relativity and its application in cosmology and the unified field theory.

6   Einstein and Quantum Mechanics

Beyond the "photoelectric effect" paper of 1905, Einstein made a number of major contributions to quantum mechanics: the anomalous low specific heat of certain materials at low temperature; defense of the quantum as a constituent of the electromagnetic field; the wave–particle dual nature of radiation; Bose–Einstein statistics; the meaning of quantum mechanics. Each of these developments is introduced, plus Einstein's work with de Broglie and Schrödinger, and the "debates" with Bohr.

7   Epilogue

The Epilogue is a summary of Einstein's insistent focus on "the inflexible boundary condition of agreeing with physical reality,"[14] and how this was the source of his insights, the guide for the development of his theories, and the verification of the correctness of his ideas. For his ideas on the quantum, he looked to the photoelectric effect; for the atom to Brownian motion; for the special theory of relativity to the constancy of the speed of light; for the general theory of relativity to the precession of the perihelion of Mercury; and for cosmology to the known structure of the universe. For the unified field theory he had no such physical phenomena to guide him.

This book looks not only to detail the major works of Albert Einstein, it also attempts to set Einstein's work into a historical and philosophical context. Perhaps a disclaimer, a "truth in advertising" is appropriate. My training is as a physicist and as a teacher of physics, not as a philosopher or historian of science. I am interested in broadening the view of our science students to realize and appreciate the historical development of science and its philosophical underpinnings. In the history and philosophy of physics there is much folklore and even some revisionist history. Trying as I might to avoid these, there are surely some places where I have succumbed. Trained historians and philosophers of science undoubtedly might have some uneasiness about some of what I have said. For these I apologize, but trust the reader to whom this book is aimed will appreciate the historical and philosophical context that is included.

Notes

1. Pais, Abraham, *Subtle is the Lord*, Oxford University Press, New York, 1982, p. 357.
2. Pais, Abraham, *Subtle is the Lord*, p. 358.

3. *The Collected Papers of Albert Einstein*, [CPAE], Princeton University Press, Princeton, NJ, 1989, Volume 1. Subsequent volumes in succeeding years.

4. Stachel, John, editor, *Einstein's Miraculous Year*, Princeton University Press, Princeton, NJ, 1998.

5. Renn, Jürgen, editor, *Einstein's Annalen Papers*, Wiley-VCH, Weinheim, Germany, 2005.

6. Fürth, R., editor, *Investigations on the Theory of Brownian Movement*, Dover Publications, New York, 1956.

7. Lorentz, H. A., Einstein, A., Minkowski, H., and Weyl, H., *The Principle of Relativity*, Dover Publications, New York, 1952.

8. Einstein, Albert, On a Heuristic Point of View Concerning the Production and Transformation of Light, *Annalen der Physik* 17 (1905), pp. 132–148; Stachel, John, editor, *The Collected Papers of Albert Einstein*, Volume 2, [CPAE2], Princeton University Press, Princeton, NJ, 1989, pp. 150–166; English translation by Anna Beck, [CPAE2 ET], pp. 86–103. The original text contains a number of editorial comments and introductory comments (pp. 134–148) that are quite informative.

9. Einstein, Albert, A New Determination of Molecular Dimensions, Dissertation, University of Zurich, 1905; Stachel, John, editor, *The Collected Papers of Albert Einstein*, Volume 2, [CPAE2], Princeton University Press, Princeton, NJ, 1989, pp. 183–202; English translation by Anna Beck, [CPAE2 ET], pp. 104–122. The original text contains a number of editorial comments and introductory comments (pp. 170–182) that are quite informative.

10. Einstein, Albert, On the Movement of Small Particles Suspended in Stationary Liquids Required by the Molecular-Kinetic Theory of Heat, *Annalen der Physik* 17 (1905), 549–560; [CPAE2, pp. 223–235; CPAE2 ET, pp. 123–134]. The original text contains a number of editorial comments and introductory comments (pp. 206–222) that are quite informative.

11. Einstein, Albert, On the Electrodynamics of Moving Bodies, *Annalen der Physik* 17 (1905), 891–921; Stachel, John, editor, *The Collected Papers of Albert Einstein*, Volume 2, [CPAE2], Princeton University Press, Princeton, NJ, 1989, pp. 275–306; English translation by Anna Beck, [CPAE2 ET], pp. 140–171. The original text contains a number of editorial comments and introductory comments (pp. 253–274) that are quite informative.

12. Einstein, Albert, Does the Inertia of a Body Depend Upon Its Energy Content? *Annalen der Physik* 18 (1905), 639–641; [CPAE2, pp. 311–314; CPAE2 ET, pp. 172–174].

13. Einstein, Albert, The Foundation of the General Theory of Relativity, 20 March, 1916, *Annalen der Physik* 49 (1916), 769–822; Kox, A. J., Klein, Martin, J., and Schulmann, Robert, editors, *The Collected Papers of Albert Einstein*, Volume 6, [CPAE6], Princeton University Press, Princeton, NJ, 1996, pp. 283–339; English translation by Alfred Engel, [CPAE6 ET], Princeton University Press, Princeton, NJ, 1997, pp. 146–200.

14. Cushing, James T., *Philosophical Concepts in Physics*, Cambridge University Press, Cambridge, 1998, p. 360.

Bibliography

Cushing, James T., *Philosophical Concepts in Physics*, Cambridge University Press, Cambridge, 1998.

Fürth, R., editor, *Investigations on the Theory of Brownian Movement*, Dover Publications, New York, 1956.

Kox, A. J., Klein, Martin J., and Schulmann, Robert, editors, *The Collected Papers of Albert Einstein*, Volume 6, [CPAE6], Princeton University Press, Princeton, NJ, 1996; English translation by Alfred Engel, [CPAE6 ET, 1997].

Lorentz, H. A., Einstein, A., Minkowski, H., and Weyl, H., *The Principle of Relativity*, Dover Publications, New York, 1952.

Pais, Abraham, *Subtle is the Lord*, Oxford University Press, New York, 1982

Renn, Jürgen, editor, *Einstein's Annalen Papers*, Wiley-VCH, Weinheim, Germany, 2005.

Stachel, John, editor, *The Collected Papers of Albert Einstein*, Volume 1, [CPAE1], Princeton University Press, Princeton, NJ, 1987; English translation by Anna Beck, [CPAE1 ET].

Stachel, John, editor, *The Collected Papers of Albert Einstein*, Volume 2, [CPAE2], Princeton University Press, Princeton, NJ, 1989; English translation by Anna Beck, [CPAE2 ET].

Stachel, John, editor, *Einstein's Miraculous Year*, Princeton University Press, Princeton, NJ, 1998.

Setting the Stage for 1905

1

1.1 Overview

In the early 1900s, our understanding of the world underwent a revolution from the classical physics of Galileo, Newton, and Maxwell to the modern physics of relativity and quantum mechanics. For his role in this revolution, Albert Einstein is justifiably placed with the giants of science – with Galileo, Newton, and Maxwell.

Just as a person cannot hope to appreciate the significance of the American Revolution without some knowledge of the American colonies before 1776, and of the people playing major roles in it, one cannot hope to appreciate the significance of the scientific revolution of the early 1900s without some knowledge of the state of science before 1905, and of the people playing major roles in it. In his 1905 papers, Albert Einstein built not only on the state of science as it had evolved over the centuries but also on events in his personal life that shaped his worldview. This chapter presents a context into which Einstein's work can be placed, leading to a fuller appreciation of his contribution to scientific thought and to a better understanding of the events that influenced his remarkable achievements.

One of the characteristics that sets physical science apart from mathematics is the demand of agreement with the physical world. As stated by James T. Cushing, "One major difference between the 'games' played by theoretical physicists and those played by pure mathematicians is that, aside from meeting the demands of internal consistency and mathematical rigor, a physical model must also meet the inflexible boundary condition of agreeing with physical reality."[1] It is, as we shall see, this inflexible boundary condition of agreement with physical reality that led to many of Einstein's insights and provided verification of, or corrective guidance for, his theories.

The science of today is built upon the ideas of those who went before, starting with the ancient Greek thought that nature was orderly, and that this order could be expressed mathematically. This "order" is referred to as the "Laws of Nature." Major advances in describing these "Laws of Nature" were contributed by Galileo and Newton in the seventeenth century, and by Einstein in the twentieth century. (See Appendix 1.5.1 for a discussion of "The Logic of Science" and "Falsification in Science.")

1.1	Overview	1
1.2	Historical Background	2
1.3	Albert Einstein	15
1.4	Discussion and Comments	20
1.5	Appendices	21
1.6	Notes	27
1.7	Bibliography	31

1.2   Historical Background

1.2.1   600 BC to AD 200: The Contribution of the Early Greeks

Our present concept of science dates from about 600 BC, associated with the Greek philosopher Thales of Miletus (c. 600 BC). Thales was aware of Egyptian discoveries of regularities in the heavens and began to question the meaning of such regularity, searching for an underlying order or some organizing principle. He began to ask "why" there were regularities in the heavens, going beyond simply describing the regularities.

The early Greeks saw nature as a "well-ordered whole, as a structure whose parts are related to each other in some definite pattern."[2] To Pythagoras (c. 500 BC) this well-ordered structure was expressed in numbers, and in ratios of small whole numbers. Pythagoras saw the universe as an orderly, beautiful structure described in harmony and number. Numbers were the essence of physical reality. To the Pythagoreans the goal of science was to "reproduce nature by a system of mathematical entities and their inter-relations."[3] This legacy of the Pythagoreans is still seen today in the close connection between mathematics and the physical sciences.

Mathematics as the foundation of our universe was further developed by Plato (429 BC–348 BC). Plato viewed our physical world as imperfect representations of ideal mathematical forms (in geometry a mathematical line has no width, while a physical line has width, etc.). To Plato the things "perceived by us are only imperfect copies, imitations or reflections of ideal forms...that can only be approached by pure thought."[4] In Plato's view, if the soul before being united to the body had acquired direct knowledge of the ideal forms, this knowledge may still be present. This knowledge might be "recalled" more so if the mind is properly stimulated by mathematical reasoning than by "empirical examination by the senses of the imperfect image of this ideal reality...Empiricism may be useful as a stimulus or support for mathematico-physical thought...but if the truth is to be found, empiricism has to be abandoned at a certain moment..."[5]

In astronomy, Plato's aim was "to save the phenomena."[6] In Simplicius' *Commentary*

...Plato lays down the principle that the heavenly bodies' motion is circular, uniform, and regular. Thereupon he sets the mathematicians the following problem: What circular motions, uniform and perfectly regular, are to be admitted as hypotheses so that it might be possible to save the appearances presented by the planets?[7]

Duhem writes, "The object of astronomy is here defined with utmost clarity: astronomy is the science that so combines circular and uniform motions as to yield a resultant motion like that of the stars. When its geometric constructions have assigned each planet a path which conforms to its visible path, astronomy has attained its goal, because *its hypotheses have then saved the appearances.*"[8]

Eudoxus (c. 408 BC–c. 355 BC), a student of Plato and considered the greatest mathematician of his day, developed a geocentric (earth-centered) model of the universe that "saved the appearances." In Eudoxus' model, the earth was at the center of the universe with the stars circling around it, the moon in a small circle with the sun, the planets, and the fixed stars further out. See Figure 1.1.

To Aristotle all knowledge originates in sense perceptions, leading to a "fundamentally empirical attitude towards the phenomena of nature."[10] Aristotle's physics is based on the concept that all motion needs a mover to maintain the motion. The fundamental law of Aristotelian dynamics is, "[A] constant force imparts to the body on which it acts a uniform motion, the velocity of which is directly proportional to the force and inversely proportional to the weight of the body."[11]

Motion is distinguished between natural and enforced. Natural motion, such as the spontaneous falling of a stone, or the spontaneous rising of smoke, is associated with the qualities of heavy (gravia) and light (levia). Natural motion is in a straight line to its goal. Heavy objects move toward the center of the universe (earth being the heaviest, water

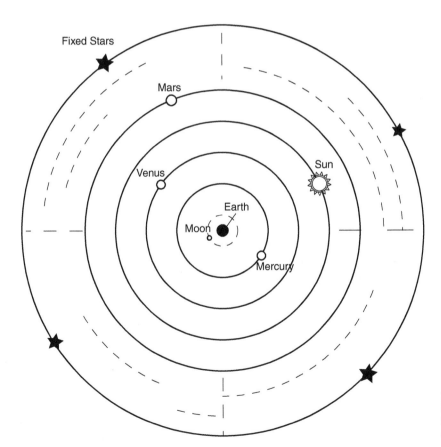

Fig. 1.1 The geocentric universe of Eudoxus.

Source: (Adapted from Zeilik, Michael, *Astronomy, The Evolving Universe.*[9])

less so) with light objects moving to the periphery of the universe (fire being the lightest, air less so).[12]

Each of the four elements (earth, water, fire, air) performs straight-line motion. (See Section 3.1.1 for a discussion of thought regarding the atom.) But the motion of the heavenly bodies is circular, indicating the heavens cannot be composed of the four terrestrial elements. The heavens were composed of a fifth element, called quintessence or the aether, that had neither gravity nor levity, could not transform into any of the terrestrial elements, and in which the natural motion was continuous and circular.[13]

In Aristotle's worldview, all bodies with "gravity" would move toward their proper place at the center of the universe. Earth being the heaviest would occupy the region closest to the center. With the earth situated at the center, other objects with "gravity" striving to reach the center of the universe would be seen as falling toward the earth. In Aristotle's worldview the earth, because it was the heaviest element, must be at the center of the universe.[14]

Aristotle was aware of other worldviews, such as the earth rotating on its axis, but rejected them as not fitting into his total worldview. To Plato, whose guiding principle is "to save the phenomena," a stationary earth with circular motion in the heavens or a rotating earth with the heavens stationary would be equally acceptable if they each predicted the motions of the stars with equal accuracy and precision. To Aristotle, the central location of the earth is necessary because of its heaviness, and it cannot be rotating since circular motion is not natural motion for "sublunar" elements. These views of Plato and of Aristotle exemplify what Duhem labels the formalistic and the realistic approaches. The formalistic approach [Plato] "considered the various geometrical models of planetary motions and of the construction of the cosmos as mathematical expedients.... [while] the realistic interpretation [Aristotle] of astronomical theory assigned physical reality to these geometrical patterns. Consistency then demanded that only those aspects of the patterns be retained which did not conflict with the physical, which meant commonsense reasoning."[15] The formalistic approach had two qualifications: (1) save the phenomena (good numerical results), and (2) the rule of greatest possible simplicity.[16]

Although starting from different bases, Aristotle's astronomy agreed with that of Plato and Eudoxus. "The axiom of the uniformity and circularity of the motions of the heavenly bodies, which Plato had formed on mathematical and religious grounds, had been supported by Aristotle with physical arguments and made an essential part of his world-system; enunciated unanimously by two such authoritative thinkers, it was bound to appear beyond all doubt. Nor did astronomers venture to deviate from this view before the beginning of the seventeenth century..."[17]

Plato's principle that the heavenly bodies motion is circular, and to "save the phenomena" was the guiding principle in astronomy for the next several centuries. But the growing accuracy of the empirical data caused refinements to the theories, all in accord with saving the phenomena and with combinations of uniform circular motion:

1. To explain why the summer half-year is longer than the winter half-year, the earth was moved a distance from the center of the motion of the sun, to a point called the eccentric.[18] See Figure 1.2.
2. To explain retrograde motion the epicycle was introduced. See Figures 1.3 and 1.4.

This, though, introduced a center for the natural circular motion of heavenly bodies that was other than the center of the universe. In Plato's formalistic approach this was readily accepted. In Aristotle's realistic approach "a natural circular motion cannot take place otherwise than round the immovable centre of the universe...in this case the earth.... The resulting conflict between Aristotelian physics and the astronomy that was to bear the name of Ptolemy...continued well into the Middle Ages..."[20]

3. The motion of a planet was to be uniform. To explain the non-uniform motion of a planet as seen from earth a second point, called the equant, was proposed, around which the motion of the planet would be uniform.[21] See Figure 1.5.

In the middle of the second century AD, Ptolemy of Alexandria (c. AD 90–168), a Greek astronomer, published *The Mathematical Syntaxis*, known also as the *Almagest*, a comprehensive summary of his work in astronomy, and of his predecessors. This compilation of all of the known work in astronomy into a single complete system to predict planetary motions became known as the Ptolemaic system. Sambursky comments, "In spite of the important results achieved by Ptolemy's theory, he was not really satisfied with this system and the complicated details which could not be reduced to the simplicity and unity of [the system of Eudoxus]."[22] As better information became known, the system became even more complicated, but it continued to save the phenomena.

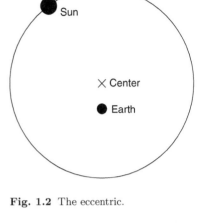

Fig. 1.2 The eccentric.

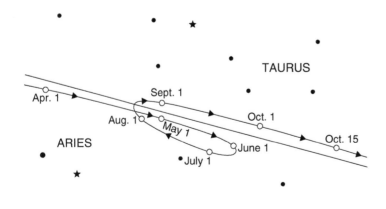

Fig. 1.3 Retrograde motion.

Source: (From McGrew, Timothy, *et al.*, *Philosophy of Science*.[19] Reproduced with permission.)

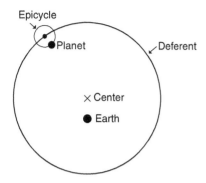

Fig. 1.4 The deferent and the epicycle.

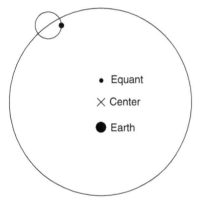

Fig. 1.5 The equant.

1.2.2   The 1600s: The Contribution of Galileo and Newton

1.2.2.1   Copernicus, Brahe, and Kepler

After the time of Ptolemy, interest in astronomy and science began to decline (coinciding with the decline of the Roman Empire and Greek civilization), to the point that knowledge from ancient Greece almost disappeared during the Dark Ages in Europe (roughly AD 400–900).

In the fourteenth century, to explain enforced motion, a concept called impetus was introduced by a group called the Paris Terminists. When a rock is thrown horizontally, what keeps it from falling vertically once it has left the hand of the thrower was troublesome for Aristotelian physics. The Paris Terminists said when the rock is thrown horizontally an "impetus is imparted to it, which causes the motion to continue after the body has been released."[23] The impetus was dependent on the quantity of matter in the body and on its velocity. Although expressed in vague terms by the Paris Terminists, the impetus can be considered the forerunner of the momentum we speak of today.[24]

In the spirit of Plato's formalistic approach, Nicolaus Copernicus made two fundamental changes to the Ptolemaic system: (1) he allowed the earth to be in motion, and (2) the motion must be uniform circular motion (no equants). Copernicus explained simply that the phenomena can be saved equally well by his new hypothesis or by the Ptolemaic system. Looking at the Copernican system, at first glance it appears as complicated as the Ptolemaic system, complete with epicycles and eccentrics (but no equants).[25] See Figures 1.6 and 1.7.

Somewhat surprisingly, reflecting on his theory in his later years, Copernicus "considered the greatest gain it had brought astronomy was not the changed position of the sun...but the elimination of the [equant]..."[28] It was not until after his death in May, 1543, that this work, *On the Revolutions of the Celestial Spheres*, was printed.

Better and more accurate astronomical data were needed to distinguish between the theories of Ptolemy and Copernicus. Tycho Brahe (1546–1601), a Danish astronomer, was able to measure the locations of the planets to two minutes of arc, down from the previously best-attainable ten minutes of arc. He collected these data over a twenty-year period, carefully recording all of his observations.[29] Brahe observed the Nova of 1572 and the comets of 1577, which he later showed to be "in the sphere of the fixed stars, and thus shattered the Aristotelian dogma of the immutability of the heavens."[30]

On the death of Brahe in 1601, Johannes Kepler (1571–1630), Brahe's assistant, was appointed his successor and took custody of Brahe's data. Kepler was a dedicated Copernican, one reason being Kepler's belief that "the sun should be at the center of the universe by virtue of its dignity and power, being a place where God would reside as prime mover."[31] Using Brahe's data to fit the orbit of Mars with combinations of circular motion, he "obtained agreement with Tycho's data to within eight minutes of arc..."[32] But Tycho Brahe's data was accurate to

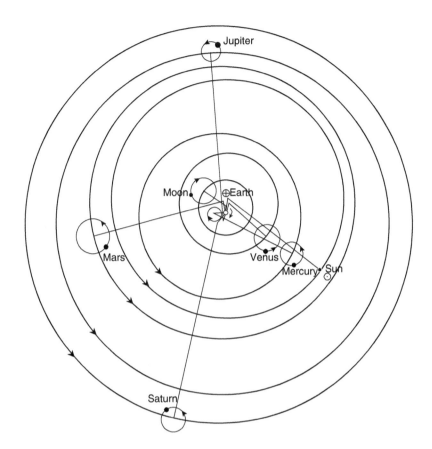

Fig. 1.6 The Ptolemaic system.

Source: (From Layzer, David, *Constructing the Universe.*[26] Reproduced with permission.)

within two minutes of arc. Because of the substantially more accurate data of Brahe, Kepler was led, eventually, to reject the necessity of circular motions and, after six years of labor on the orbit of Mars, Kepler came to the realization that an ellipse exactly fitted the data.[33] And the sun, befitting its "dignity and power," was located at one of the foci of the ellipse.[34]

Over a twenty-year period, Kepler found in Brahe's data what have come to be known as his three laws of planetary motion:[35]

1. *Kepler's first law*: the planets move on ellipses about the sun, with the sun at one focus. See Figure 1.8.
2. *Kepler's second law*: A radius vector drawn from the sun to the planet sweeps out equal areas in equal time. See Figure 1.9.
3. *Kepler's third law*: The ratio of the cube of the mean radius R of a planet's orbit to the square of its period τ is a fixed constant for all planets in the solar system.[38]

$$\frac{R^3}{\tau^2} = \text{constant}$$

See Figures 1.10 and 1.11.

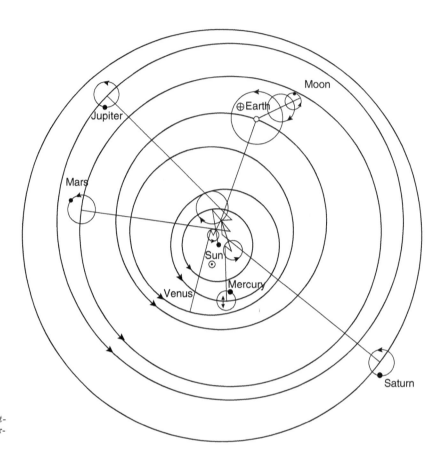

Fig. 1.7 The Copernican system.

Source: (From Layzer, David, *Constructing the Universe.*[27] Reproduced with permission.)

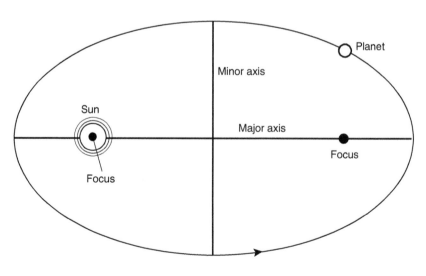

Fig. 1.8 Kepler's first law: Elliptical planetary orbits.

Source: (Adapted from Zeilik, Michael, *Astronomy, The Evolving Universe.*[36])

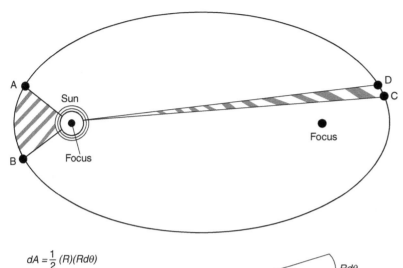

$dA = \frac{1}{2}(R)(Rd\theta)$

$\frac{dA}{dt} = \frac{1}{2}(R)(R\dot{\theta}) = \frac{1}{2}R^2\dot{\theta} = constant$

Fig. 1.9 Kepler's second law: The equal area rule for planetary orbits.

Source: (Adapted from Zeilik, Michael, *Astronomy, The Evolving Universe.*[37])

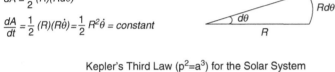

Fig. 1.10 Kepler's third law: $\frac{R^3}{\tau^2} = $ constant for planets around the sun.

Source: (From Zeilik, Michael, *Astronomy, The Evolving Universe.*[39] Reproduced with permission.)

Planet	Period (*p*, in years)	p^2	Distance (*a*, in AU)	a^3	p^2/a^3
Mercury	0.24	0.058	0.39	0.059	0.97
Venus	0.62	0.38	0.72	0.37	1.0
Earth	1.0	1.0	1.0	1.0	1.0
Mars	1.9	3.6	1.5	3.4	1.1
Jupiter	12.0	140.0	5.2	140.0	1.0
Saturn	29.0	840.0	9.5	860.0	0.98

Fig. 1.11 Table of Kepler's third law.

Source: (From Zeilik, Michael, *Astronomy, The Evolving Universe.*[40] Reproduced with permission.)

Following Brahe's discovery of the mutability of the heavenly sphere, Kepler's elliptical orbits were another dent in the perfection of the heavens (circular orbits).

1.2.2.2 Galileo Galilei

More than anyone, Galileo Galilei (1564–1642) is the central figure in the transition from Aristotelian physics to classical physics. In explaining projectile motion, his theories included the impetus theory of the Paris Terminists, rather than the ideas of Aristotle. On the structure of the universe, Galileo was a disciple of Copernicus rather than of Ptolemy. Having a strong mathematical perspective, Galileo was a Platonist, yet he believed the Copernican system was the physical truth about the universe.[41]

To understand the motion of falling bodies and projectile motion Galileo consciously restricted his investigation to a study of the motion itself, not the causes of the motion.[42] For Aristotle qualitative relations were predominant. Galileo changed the emphasis to quantitative relations. The causes of the motion would be left to others, after Galileo had more precisely determined the description of the motion. Experiments for Galileo were to verify relations he had obtained by mathematical reasoning, not to discover new phenomena.[43]

For free-fall, Galileo discovered a body is accelerated downward at a constant rate.[44] In explaining projectile motion, he showed that the vertical and horizontal components could be treated separately of one another, with the resultant combined motion being what we see. (Although this was a major advance, in reality the addition of velocity components had been used by the Greeks to explain the heavenly motions were composed of the motion of the deferent and epicycles.) Reflecting on projectile motion as seen by different observers, Galileo advanced the idea of the relative character of motion. From these reflections came a basic theory of relativity of motion that "the motion of a system of bodies relative to each other does not change if the whole system is subjected to a common motion."[45]

The law of inertia was at the foundation of Galileo's description of motion.[46] A smooth ball rolling down an inclined plane would accelerate, continually increasing its speed, while one projected up an incline would decelerate, continually slowing down. A ball rolling on a horizontal plane, therefore, would neither accelerate nor decelerate; it would continue at a constant speed in a straight line.[47] Galileo arrived at the conclusion that if no force acts on the body it continues at a constant velocity. This is Galileo's law of inertia. To Aristotle a constant force was required to maintain the constant velocity of a body. To Galileo a constant force provided a constant acceleration for a body. Galileo was dealing with motion absent all external influences in his treatment of impetus and inertia, closer to the world of Plato's ideal forms than to Aristotle's everyday world. But Galileo's ideas needed the refinement and clarification of the concept of force and the distinction between mass and weight. And gravity needed to be seen not as something intrinsic to the body but, rather, "as an external action exerted upon the body."[48]

Although Galileo did not invent the telescope, he was the first to recognize its value as a scientific instrument. "Turning it to the heavens" Galileo saw the surface of the moon was not smooth as Aristotelian physics claimed. He saw the moons of Jupiter, indicating a second center for motion (Jupiter).[49] To the dents in the Aristotelian worldview from Tycho Brahe (the heavens are mutable) and Kepler (motion in the heavens is not circular) Galileo adds two more – an example of imperfection in the heavens (the moon) and a second center of revolution (Jupiter). Later he would detect sunspots moving across the face of the sun, indicating also the mutability of the sun.

1.2.2.3 Isaac Newton

1666 is Isaac Newton's (1642–1727) *annus mirabilis* (actually a two-year span from 1665–1667). In 1665, the year Newton received his degree from Cambridge, the Great Plague of London (the outbreak of the Black Death in London in 1665–1666) had spread to Cambridge and the university was closed. He returned to the family home in the country, there inventing the branch of mathematics known as calculus, discovering many new ideas of light and optics, and laying the foundation for his work in mechanics and gravity. Although in the main discovered in 1666, his ideas on motion and gravity were not published until 1687 in the *Principia*,[50] his results of experiments and studies on light until 1704 in *Optics*,[51] and his results on calculus in 1736 in *On the Method of Series and Flucxions* (nine years after his death).[52]

Newton's major contribution to mechanics was to pull together what had been separate and fragmentary knowledge, and to assemble it into a systematic and consistent mathematical system.[53] Newton incorporated the impetus of Galileo and the Paris Terminists. He broke with the concept that one body affected the motion of another only through direct contact, introducing into his mechanics "action-at-a-distance" forces. Gravity became an external constant force that causes a constant acceleration; no longer is gravity an innate property of the body seeking to move the body to its natural place in the universe. The distinction between mass and weight is brought out of the background.[54] These are formalized as the axioms for mechanics, now known as Newton's three laws of motion:

I. *Newton's first law of motion*: Every body perseveres in its state of being at rest or of moving uniformly straight forward except as it is compelled to change its state by forces impressed.

II. *Newton's second law of motion*: A change in motion is proportional to the motive force impressed and takes place along the straight line in which that force is impressed.

III. *Newton's third law of motion*: To any action there is always an opposite and equal reaction; in other words, the actions of two bodies upon each other are always equal and always opposite in direction.[55]

Newton's laws of motion were valid in absolute space. But if valid in absolute space, it was determined they also were valid in any reference frame moving in uniform translational motion relative to absolute space. Although Newton believed in absolute space, he was unable to determine a way to distinguish the absolute space reference frame from the moving reference frame.[56] Newton raised this "inability" to a principle, today referred to as Newton's principle of relativity, "When bodies are enclosed in a given space, their motions in relation to one another are the same whether the space is at rest or whether it is moving uniformly straight forward without circular motion."[57]

In the *Principia*, Newton shows that [for circular motion] "if the periodic times are as the 3/2 powers of the radii...the centripetal forces will be inversely as the squares of the radii."[58] Generalizing this to an elliptical orbit, Newton shows the centripetal force of an object in an elliptical orbit "tending toward a focus of the ellipse...[is] inversely as the square of the distance..."[59] For the inverse square law, he then shows "that the squares of the periodic times in ellipses are as the cubes of the major axes [Kepler's third law]."[60] These are summarized as, "The forces by which the primary planets are continually drawn away from rectilinear motions and are maintained in their respective orbits are directed to the sun and are inversely as the squares of their distances from its center."[61] (See Appendix 1.5.2 for a present-day derivation of the inverse square force law for gravitation.)

If the sun attracts the planet with this force, by Newton's third law the planet must attract the sun with an equal force. The gravitational force is proportional to the masses of the two objects and inversely proportional to their separation.[62]

$$F_{gravity} = \frac{GM_1 M_2}{r^2}$$

where G is a constant (called the gravitational constant). The masses of the two objects, M_1 and M_2 in the equation, ensure consistency with Newton's three laws of motion and Galileo's finding that all objects fall with the same acceleration at the surface of the earth (assuming no air resistance).[63]

Having obtained the law of gravity for the earth about the sun, Newton checked the law of gravity for the moon in orbit around the earth. At the surface of the earth, the acceleration due to gravity is 9.8 m/s^2. The distance to the moon is about 60 times larger than the radius of the earth. Calculating the centripetal acceleration of the moon in orbit about the earth, $a = 2.74 \times 10^{-3}$m/s^2. If this is due to the gravitational attraction of the earth, the accelerations of gravity at the surface of the earth and at the orbit of the moon should differ by a factor of $(r_{moon'sorbit}/r_{earth})^2 = (60)^2 = 3600$. The ratio of accelerations is $(9.80/2.74 \times 10^{-3}) = 3560$. Newton concluded the force of gravity that kept the earth in orbit around the sun was the same force that kept the moon in orbit around the earth. The motions of planets around the sun, the motion of the moon about the earth, and the falling of objects near the surface of the earth could all be explained by a single force of gravity. It was a short jump to extend the law of gravity to every pair of material bodies in the universe.[64]

In comparison to the Aristotelian universe of separate laws for the heavens and for the earth, through his three laws of motion and the law of gravitation, Newton unified the description of the heavens and the earth: The same laws govern the motions in the heavens as govern motion on earth. From Newton's one law of gravitation come all three of Kepler's laws of planetary motion.

When two planets pass sufficiently close enough to one another the gravitational force between the planets, while small compared to that from the sun, is enough to slightly disturb their elliptical orbits. These predictions were in good agreement with observation. However, the planet Uranus, discovered in 1781, showed significant discrepancies between its observed motion and its predicted motion (even after accounting for the effects of other planets). Assuming the discrepancies were due to an unknown planet, scientists determined the necessary orbit of such an unknown planet and, in 1846, looking where directed, the planet Neptune was first seen. Similar discrepancies arose in the orbit of Neptune, leading to the prediction and eventual discovery, in 1930, of the planet Pluto.[65] As noted in Appendix 1.5.1.1 (The Logic of Science), locating these planets as predicted "verifies" Newton's law of gravity, but does not "prove" it. (We will see in Chapter 5 where it fell short in describing the orbit of the planet Mercury.)

In the case of Kepler, more accurate data by Brahe forced him to develop a new description of planetary orbits (elliptical orbits). For Galileo, his own more accurate data on falling bodies allowed him to determine that motion under only the influence of gravity was one of constant acceleration. Newton then used these more precise descriptions of motion by Kepler and Galileo, building on them to develop his laws of mechanics, including the law of universal gravitation.[66]

1.2.3 The 1800s: The Contribution of Maxwell and Lorentz

In the 1800s, a number of discoveries and advances were made that led to the modern physics of relativity and of quantum mechanics in the early 1900s. The introduction of quanta into our world picture is the material of Chapter 2 and the introduction of relativity is the material of Chapter 4. Just as Newton drew together information from those who preceded him as the foundation for his mechanics, so also did Einstein for his work in the early 1900s. The following is a summary of some of those items from the 1800s upon which Einstein was to draw. Others will be presented as needed in specific chapters.

1.2.3.1 The Aether and Electromagnetism

In the early 1800s (roughly 1800 to 1825), work by Thomas Young and Augustin Fresnel led to acceptance of the idea that light was a wave phenomenon. But there remained the question of what medium the waves were propagating in. Over the centuries, the aether of Aristotle (see Section 1.2.1) had been joined with an aether for gravity, an aether for electrostatics, and an aether for magnetism.[67] Since no medium was known for the propagation of light, the scientists postulated another aether – the luminiferous aether (sometimes spelled ether, sometimes spelled aether).[68]

Until the early 1800s, Electricity and Magnetism remained as separate and distinct areas of study. Then, in the period from 1820 to 1865, a number of discoveries and advances were made, linking electrical effects with magnetism and magnetic effects with electricity. In 1820, the Danish physicist Hans Christian Ørsted discovered that an electric current in a wire can deflect a compass needle. But the effect was present only for charges in motion, i.e., for an electric current. In 1831, the reverse effect of electromagnetic induction, i.e., magnetism giving rise to electric currents, was discovered by the English experimenter Michael Faraday, and shortly thereafter by the American Joseph Henry and the Russian H. F. E. Lenz.[69]

The discoveries of Ørsted, Faraday *et al.*, linking electrical effects with magnetism and magnetic effects with electricity, culminated in James Clerk Maxwell's paper of 1865, "A Dynamical Theory of the Electromagnetic Field."[70] In this paper, Maxwell presented the equations that have come to be known as Maxwell's equations. In today's notation, his mathematical equations describing the electromagnetic field are summarized as:[71]

$$\nabla.\boldsymbol{E} = 4\pi\rho$$

$$\nabla.\boldsymbol{B} = 0$$

$$\nabla \times \boldsymbol{E} = -\frac{1}{c}\frac{\partial \boldsymbol{B}}{\partial t}$$

$$\nabla \times \boldsymbol{B} = \frac{1}{c}\frac{\partial \boldsymbol{E}}{\partial t} + \frac{4\pi}{c}j$$

Not only were electricity and magnetism united as electromagnetism but, as Maxwell discovered, his equations showed that light also was included in his theory of electromagnetic waves. The aether of electricity and the aether of magnetism had been brought together as one electromagnetic aether and, in the electromagnetic aether, included also was the luminiferous aether of light.[72]

By 1900, the electromagnetic aether was an established part of scientific belief. In *The Theory of Electrons*,[73] Lorentz states his belief in the aether, "however different it may be from all ordinary matter."[74] Lorentz stated, further, that the aether always remains at rest [relative to absolute space].[75]

By 1875, Hendrik Lorentz had become convinced that Maxwell's theory needed to be "complemented by an electrical theory of matter which would show how the electromagnetic field of Maxwell interacts with matter."[76] To this end, Lorentz postulated the existence of an extremely small charged particle – the electron – a hypothetical unit of electrical charge.[77] This interaction between matter and the electromagnetic field is the Lorentz force. In today's notation, the Lorentz force is written as,[78]

$$\boldsymbol{F} = q\left(\boldsymbol{E} + \frac{\nu}{c} \times \boldsymbol{B}\right)$$

1.2.4   The Worldview in 1900

The Mechanical Worldview: By the late 1800's, the mechanics of Newton was well entrenched, and scientists were attempting to reduce their understanding of nature to a mechanical foundation.

The Electromagnetic Worldview: By 1900, because of the success of the electromagnetic theory, scientists were contemplating an electromagnetic foundation for mechanics.

The Energetics Worldview: Based on the success of thermodynamics in describing the world in terms of energy, scientists were contemplating the underlying structure of the world to be forms of energy and energy transformations.[79] (See Section 3.1.3 for further discussion of the Worldview around 1900.)

1.3   Albert Einstein

1.3.1   The Pre-College Years

Hermann Einstein (1847–1902), Albert's father, had shown an early ability in mathematics but, since his parents did not have the funds for him to pursue his studies at a university, Hermann became a merchant and, eventually, a partner in his cousin's featherbed company in Ulm, Wurttemberg. His mother, Pauline (Koch) (1858–1920), was a warm and caring person, and a talented pianist. She was from a family of means, bringing a breadth of culture and a love of literature and music to the marriage. Hermann and Pauline were married on August 8, 1876. Jewish traditions, such as a deep respect for learning, ran deep in the family but, as Clark says, Hermann and Pauline "were not merely Jews, but Jews who had fallen away.... [T]he essential root of the matter was lacking: the family did not attend the local synagogue. It did not deny itself bacon or ham ... " These were considered "ancient superstitions" by Hermann, as were most other Jewish traditions. Thus, Albert Einstein was nourished on a tradition that had broken with authority and sought independence. But the family tradition also included the Jewish tradition of self-help.[80]

On March 14, 1879, a son, Albert, was born to Hermann and Pauline Einstein in Ulm, Wurttemberg. Two years later, on November 18, 1881, their second child, a daughter, Marie (Maja), was born in Munich.

In 1880, Hermann and his brother Jakob, an electrical engineer, formed a company to manufacture electrical generating and transmission equipment and moved to Munich to set up the business. The business prospered and the family enjoyed a comfortable life there. As their home was on the grounds of the factory, Albert grew up in daily contact with electromechanical equipment.[81]

Normal childhood development proceeded slowly for the young Albert. As recalled by Einstein in his later years:

I sometimes ask myself how did it come that I was the one to develop the theory of relativity. The reason, I think, is that a normal adult never stops to think about problems of space and time. These are things he has thought of as a child. But my intellectual development was retarded, as a result of which I began to wonder about space and time only when I had already grown up. Naturally, I could go deeper into the problem than a child with normal abilities.[82]

When eight years old, Einstein entered the Luitpold Gymnasium. He did well in mathematics and the logically structured Latin, but not so well in Greek and modern foreign languages. When he was thirteen and scheduled to begin algebra and geometry he spent his summer vacation working through the proofs of the theorems by himself to see what he could understand on his own, often finding proofs that differed from those in his books.[83]

As an adult, looking back on his youth, Einstein comments that when he was about eleven years old, his religious feelings became so strong that he went through a period when he followed all of the religious precepts in detail, but which

...found an abrupt ending at the age of 12. Through the reading of popular scientific books I soon reached the conviction that much of the stories in the bible could not be true. The consequence was a positively fanatic [orgy of] freethinking coupled with the impression that youth is intentionally being deceived by the state through lies; it was a crushing impression. Suspicion against every kind of authority grew out of this experience, a skeptical attitude toward the convictions which were alive in any specific social environment – an attitude which has never again left me, even though later on, because of better insight into causal connections, it lost some of its original poignancy.[84]

If one could not trust religion, surely order and logic could be discovered in the world which

...exists independently of us human beings and which stands before us like a great, eternal riddle, at least partially accessible to our inspection and thinking. The contemplation of this world beckoned like a liberation, and I soon noticed that many a man whom I had learned to esteem and to admire had found inner freedom and security in devoted occupation with it.[85]

Commenting on another occasion, Einstein recalled,

At the age of 12 I experienced a second wonder of a totally different nature: in a little book dealing with Euclidian plane geometry, which came into my hands at the beginning of a school year. Here were assertions, as for example the intersection of the three altitudes of a triangle in one point, which – though by no means evident – could nevertheless be proved with such certainty that any doubt appeared to be out of the question. This lucidity and certainty made an indescribable impression on me. That the axiom had to be accepted unproved did not disturb me.[86]

When sixteen, Einstein began to wonder what a beam of light would be like if he could travel at the same speed as the beam of light:

If I pursue a beam of light with the velocity c (velocity of light in a vacuum), I should observe such a beam of light as a spatially oscillatory electromagnetic field at rest. However, there seems to be no such thing, whether on the basis of experience or according to Maxwell's equations. From the very beginning it appeared to me intuitively clear that, judged from the standpoint of such an observer, everything would have to happen according to the same laws as for an observer who, relative to the earth, was at rest. For how, otherwise, should the first observer know, i.e. be able to determine, that he is in a state of fast uniform motion? [87]

It would be another ten years, not until 1905, before he had the insights that would allow him to resolve this.

By 1894, the Einstein business had difficulty competing with the larger German companies. Hermann and Jakob closed the factory in Munich and transferred the business to Italy where the circumstances appeared more favorable. In 1895, when Albert's parents and sister, Maja, moved to Italy after the transfer of the business, Albert was left in Munich to finish his final year at the Gymnasium. In spring, 1895, without consulting his parents, Einstein left the gymnasium without acquiring his diploma. He refused to return to Munich and informed his parents he intended to give up his German citizenship. [88]

1.3.2 The College Years

With the encouragement of his father, Albert looked to continue his education, pursuing a program in engineering. But, without a certificate of graduation from the gymnasium, entry into the major universities of Europe was not possible, one exception being the Swiss Federal Polytechnic School (ETH) [89] in Zurich, Switzerland. The ETH, located in the German-speaking part of Switzerland, allowed success on an entrance examination in place of a certificate of graduation. In October 1895, Albert took the entrance examination, obtaining high marks in mathematics and science, but scoring low marks in languages and history and, overall, did not receive a passing mark. H. F. Weber, a professor in the physics section at the ETH, was so strongly impressed with Albert's performance on the scientific part of the examination that he gave Einstein permission to attend his lectures. However, the ETH advised Einstein to attend the cantonal secondary school in Aarau (20 miles west of Zurich), complete the work necessary for his diploma, and then he would be admitted. [90]

After one year in Aarau, in September 1896, Einstein took the "Matura," the graduation examination consisting of seven written examinations, and an oral examination. Of the nine candidates taking the examination, Einstein had the highest average over the written examinations. But, more importantly, passing the Matura allowed him to enroll in the ETH. [91] During his time at Aarau, he had decided his future would be as a physicist, not as an electrical engineer his family had envisioned.

In October 1896, Einstein enrolled at the ETH. During his years at the ETH he was supported on an allowance of 100 francs per month from his mother's family, the Kochs, of which he put away 20 francs each month to pay for his eventual application for Swiss citizenship. [92]

At the ETH, Einstein settled into the track for high-school physics and mathematics teachers, with an interest in theoretical physics. His program consisted primarily of physics, mathematics, and mathematical physics courses. He found most of the mathematics and mathematical physics courses irrelevant, believing not much more than calculus would be needed to pursue his interests in physics. (It was not until after 1907, as he was pursuing the general theory of relativity, that he realized the importance of advanced mathematics for physics.) The majority of his physics courses were taken from Professor Heinrich Friedrich Weber, fifteen in all.[93]

In his undergraduate days, Einstein was in the classes of Hermann Minkowski, an excellent mathematics teacher who later was to cast Einstein's special theory of relativity into a four-dimensional space. Marcel Grossman, a fellow student, was later to help Einstein secure a position at the patent office and collaborate with him for several years on the general theory of relativity. Mileva Marić, another fellow student, was to become Einstein's wife in 1903.

Although following some courses with intense interest, Einstein relied more on self-study than regular class attendance, studying the writings of the "masters of contemporary theoretical physics – above all, Hermann von Helmholtz, Gustav Robert Kirchhoff, Heinrich Hertz, Ludwig Boltzmann, Ernst Mach, and...James Clerk Maxwell."[94] He recalled Mach's book on mechanics as "a book which, with its critical attitude toward basic concepts and basic laws, made a deep and lasting impression on me." He had time for these, since, "In all there were only two examinations; for the rest one could do what one wanted...up to a few months before the examination." For examinations, Einstein would borrow the meticulously organized notes of Marcel Grossman.[95]

Heinrich Weber was a well-respected experimentalist, renowned in precision instrumentation and precision measurements.[96] He joined Hermann von Helmholtz in 1871 at the University of Berlin. Working in Helmholtz's laboratory, Weber produced two major papers on the specific heats of carbon, boron, and silicon, showing them to be noticeably smaller at low temperatures than predicted by the law of Dulong and Petit. (For nearly thirty years Weber's empirical findings would remain an anomaly – until Einstein, his former student, presented a new explanation. See Section 6.2.1.) In 1875, Weber accepted the position of professor of mathematical and technical physics at the Swiss Federal Polytechnic School (ETH) in Zurich. In 1887 and 1888, he published two additional major papers, both on light emission in incandescent solid bodies. The 1888 paper is of particular significance as it led to Wien's displacement law for the energy distribution in blackbody radiation (which, in turn, was used by Max Planck when he obtained the energy quantization relation, $E = h\nu$. See Section 2.1.3).[97]

Weber was considered by all an excellent teacher. Even Einstein acknowledged this, but was disappointed he did not teach the new and fascinating electromagnetic theory of Maxwell. Further, "He did not teach the foundations of physics, and he did not teach theoretical or mathematical physics."[98]

After two years of study, Einstein passed his qualifying examination and began concentrating on his physics. Of the fifteen courses Einstein took with Weber, five of them were laboratory courses. Weber's laboratory courses stressed the importance of measurement. Einstein talks of his enjoyment of the laboratory experience, that he was "...fascinated by the direct contact with experience..."[99]

From Weber's courses, Einstein received a solid introduction to physics, especially thermodynamics, electricity, and magnetism; a strong concern with precision measurement; and, likely, an introduction to blackbody radiation. From Weber's research came the work on specific heats, particularly the anomalous behavior of some materials at low temperatures. Moreover, Weber served as a mentor, from seeing Einstein's scientific potential even in the failed entrance examination, to encouraging and helping develop his scientific talents, to guiding his thesis for graduation.[100]

In July 1900, Einstein passed the final examination to complete the requirements for graduation. Upon graduation his allowance from the Koch family ended.

1.3.3 From College to 1905

At graduation, Einstein anticipated being offered an assistantship at the ETH to work with Weber. However, one was not forthcoming. As John Stachel writes, "An attitude of independence, considered excessive by his professors, may have played a role." Weber selected two mechanical engineers to work with him. When Einstein failed to secure a position at the ETH he began applying for assistantships at other institutions, but with no success.[101] After a succession of part-time jobs, through the intercession of his college friend, Marcel Grossman, in June 1902 he secured a permanent job at the Swiss patent office in Bern. Albert Einstein and Mileva Marić, were married on January 6, 1903. Their first son, Hans Albert, was born on May 14, 1904.[102]

Through an agreement between the ETH and the University of Zurich, students could complete their doctoral work at the ETH, but receive their degree from the University of Zurich (the ETH did not offer a doctoral degree). Einstein began his doctoral work at the ETH with Weber in the fall, 1900, working on it over the winter. He began work on thermoelectricity, but switched to molecular forces. By spring, 1901, Einstein and Weber had a falling-out and parted company. Einstein then turned to Alfred Kleiner at the University of Zurich to supervise his doctoral thesis. With Kleiner, Einstein again dealt with molecular forces, covering a wide range of topics related to kinetic theory. He officially submitted his dissertation to the University of Zurich on 23 November, 1901, and on 2 February, 1902, officially retracted it. Kleiner likely had suggested the dissertation be withdrawn because, at least in Einstein's view, Kleiner felt it contained "sharp criticism of Ludwig Boltzmann." Again, under the supervision of Kleiner, and on a topic different from the first two submissions of 1902, Einstein submitted his dissertation, "A new Determination of Molecular Dimensions," to the University of Zurich on July 20, 1905. It was accepted on July 24, 1905. As Abraham Pais describes it, "Einstein was now Herr Doktor."[103]

1.4   Discussion and Comments

During Einstein's days at the ETH, physicists were trying to unify all of physics under one "world-view," i.e., on a common foundation. Three such programs were in vogue: the mechanical worldview (Boltzmann, Hertz), the electromagnetic worldview (Lorentz, Abraham), and the energetic worldview (Ostwald, Duhem). (See Section 3.1.3 for further details of the three worldviews.) The general acceptance was a dualistic worldview, a mixture of the mechanical and electromagnetic worldviews.[104]

Albert Einstein undoubtedly entered the ETH with a bias toward atomism, and confirmed it through his study of Boltzmann's work while a student.[105] His formal introduction to thermodynamics was in Weber's course during his fourth year at the ETH, but with no mention of the current developments in the kinetic theory of Maxwell and Boltzmann.

From graduation in 1900 to 1905, Einstein published five papers, each of them indicating a belief in the kinetic theory of Boltzmann.[106]

> 1901 "Conclusions Drawn from the Phenomena of Capillarity"[107]
>
> 1902 "On the Thermodynamic Theory of the Difference in Potentials between Metals and Fully Dissociated Solutions of Their Salts and on an Electrical Method for Investigating Molecular Forces"[108]
>
> 1902 "Kinetic Theory of Thermal Equilibrium and of the Second Law of Thermodynamics"[109]
>
> 1903 "A Theory of the Foundations of Thermodynamics"[110]
>
> 1904 "On the General Molecular Theory of Heat"[111]

John Stachel describes the first two papers as "an investigation of the nature of molecular forces by means of the effect of such forces on various observable phenomena in liquids."[112] The third, fourth, and fifth papers are "devoted exclusively to the foundations of statistical physics..."[113] In these three papers Einstein reinterpreted the work of Ludwig Boltzmann on statistical mechanics, clarifying, adding, and/or putting on a firm foundation the distinction between microcanonical and canonical ensembles, the equipartition of energy, the equivalence of ensemble and time averages (the ergodic theorem), the concept of entropy, and the calculation of probabilities.[114]

Einstein was firmly committed to the atomic theory and to Boltzmann's approach to statistical physics.[115] This was the mechanical worldview, including also Newton's mechanical principle of relativity, and experiments were beginning to show the principle of relativity should be extended beyond mechanics to optical and electromagnetic phenomena.

However, the principle of relativity was in conflict with Maxwell's electromagnetic theory, which selected out a "rest frame," that of the aether. It was the success of Maxwell's theory, coupled with the inability to provide a mechanical underpinning for it, that left physicists with a dualistic mixture of mechanical and electromagnetic concepts. Albert Einstein's "striving for simplicity and unification led him to anticipate the ultimate elimination of this dualism." This led him to regard neither the mechanical worldview nor the electromagnetic worldview as unalterable.[116]

In his personal and scientific life, Einstein was neither a rebel nor a revolutionary seeking to overthrow authority. Rather, he had become free of any authority except the authority of reason.[117]

By 1905:

1. The mechanics of Newton was well established, including the principle of relativity, describing the motion of objects.
2. The electromagnetism of Maxwell was well established, describing the propagation of electromagnetic waves.
3. The aether was an established part of scientific belief. In *The Theory of Electrons*,[118] Lorentz states his belief in the aether, "however different it may be from all ordinary matter."[119] Lorentz stated, further, that the aether always remains at rest [relative to absolute space].[120]
4. The Lorentz transformations had already been obtained by Lorentz in 1904.
5. Atoms were accepted as a convenient means of visualizing systems, but their reality remained a point of controversy.

1.5 Appendices

1.5.1 Science Today

Science is sometimes viewed as marching inevitably and inexorably toward the ultimate truth, all scientists following the same "scientific method." But the ways of science are not monolithic. Hendrik Lorentz, from his own experience, cautions us that "one of the lessons which the history of science teaches us is ... that we must not too soon be satisfied with what we have achieved. The way of scientific progress is not a straight one which we can steadfastly pursue. We are continually seeking our course, now trying one path and then another, many times groping in the dark, and sometimes even retracing our steps."[121]

1.5.1.1 The Logic of Science

In the hypothetico-deductive model of science, one goes from observation to hypothesis to prediction to confirmation (or to refutation). However, confirmation does not verify the correctness of the hypothesis; it only tells us the hypothesis has not been refuted. This "logic of science" can be viewed in the following manner. If the proposition p always must have the consequence q, we write:[122]

$$p \Rightarrow q$$

For example, p could be the statement, "The earth is stationary in the center of the universe, with the sun revolving about the earth every 24 hours." The consequence q would be the sun appearing over the horizon each morning and setting each evening. This consequence q would verify (but not prove) the proposition p. On the other hand, if the sun does not appear over the horizon each morning and set each evening, this does "prove" that the proposition p was false, i.e., the absence of the consequence q proves that the proposition p was false.

But the converse is not true, i.e., the presence of q does not prove the truth of p:

$$q \not\Rightarrow p$$

For example, if $q = $ "the sun rises each morning and sets each evening," it could be due to $p = $ "the rotation of the earth on its axis." These can be written symbolically as:

$p \Rightarrow q$	If q appears it verifies (but does not prove) the truth of p.
$q \not\Rightarrow p$	The presence of q does not prove the truth of p.
$not(q) \Rightarrow not(p)$	If q does not appear, the statement p must be false.
$not(p) \not\Rightarrow not(q)$	If p is not present, it says nothing about q.

If q is present, it does not "prove" the validity of the statement p. However, if q is not present, it does "prove" that p is false.

No matter how many times, nor in how many circumstances, q verifies the proposition p, it takes but one falsification (q is absent) to falsify the proposition p. This is the way of science: we observe the world around us (the consequence q) and try to determine the "Laws of Nature" that govern it (the proposition p). For an example, see Section 1.2.2.3 (Isaac Newton) on Newton's law of gravity and Section 5.3.1.1 on the precession of the perihelion of the planet Mercury.

1.5.1.2 Experimental Falsification

Falsification is not a methodology peculiar to science; it is part of how we learn in the everyday world. The application of falsification to scientific knowledge is in the confirmation, or repudiation, of a scientific model in as much as an experiment agrees, or does not agree, with the model. In physics, Cushing writes,

One major difference between the 'games' played by theoretical physicists and those played by pure mathematicians is that, aside from meeting the demands of internal consistency and mathematical rigor, a physical model must ultimately also meet the inflexible boundary condition of agreeing with physical reality.[123]

In his 1968 book, *Conjectures and Refutations*, Karl Popper writes,

'When should a theory be ranked as scientific?' or 'Is there a criterion for the scientific character or status of a theory?'
The problem which troubled me at the time was neither, 'When is a theory true?' nor, 'When is a theory acceptable?' My problem was different. *I wished to distinguish between science and pseudo-science*; knowing very well that science often errs, and that pseudo-science may happen to stumble on the truth.
I knew, of course, the most widely accepted answer to my problem: that science is distinguished from pseudo-science – or from 'metaphysics' – by its *empirical method*, which is essentially *inductive*, proceeding from observation or experiment. But this did not satisfy me. On the contrary, I often formulated my problem as one of distinguishing between a genuinely empirical method and a non-empirical or even pseudo-empirical method – that is to say, a method which, although it appeals to observation and experiment does not come up to scientific standards. The latter method may be exemplified by astrology, with its stupendous mass of empirical evidence based on observation – on horoscopes and on biographies.
But as it was not the example of astrology which led me to my problem I should perhaps briefly describe the atmosphere in which my problem arose and the examples by which it was stimulated. After the collapse of the Austrian Empire there had been a revolution in Austria: the air was full of revolutionary slogans and ideas, and new and often wild theories. Among the theories which interested me Einstein's theory of relativity was no doubt by far the most important. Three others were Marx's theory of history, Freud's psychoanalysis, and Alfred Adler's so-called 'individual psychology.'[124]
I found that those of my friends who were admirers of Marx, Freud, and Adler were impressed by a number of points common to those theories, and by their apparent *explanatory power*. These theories appeared to be able to explain practically everything that happened within the fields to which they referred. The study of them seemed to have the effect of an intellectual conversion or revelation, opening your eyes to a new truth hidden from those not yet initiated. Once your eyes were thus opened you saw confirming instances everywhere: the world was full of <u>verifications</u> of the theory. Whatever happened always confirmed it. Thus its truth appeared manifest; and unbelievers were clearly people who did not want to see the manifest truth; who refused to see it, either because it was against their class interest, or because of their repressions which were still 'unanalyzed' and crying out for treatment.[125]
With Einstein's theory the situation was strikingly different. Take one typical instance – Einstein's prediction, just then confirmed by the findings of Eddington's expedition. Einstein's gravitational theory had led to the result that light must be attracted by heavy bodies (such as the sun), precisely as material bodies were attracted. As a consequence it could be calculated that light from a distant star whose apparent position was close to the sun would reach the earth from such a direction that the star would seem to be slightly

shifted away from the sun; or, in other words, that the stars close to the sun would look as if they had moved a little away from the sun, and from one another. This is a thing which cannot normally be observed since such stars are rendered invisible in daytime by the sun's overwhelming brightness; but during an eclipse it is possible to take photographs of them. If the same constellation is photographed at night one can measure the distances on the two photographs and check the predicted effect.

Now the impressive thing about this case is the *risk* involved in a prediction of this kind. If observation shows that the predicted effect is definitely absent, then the theory is simply refuted. The theory is *incompatible with certain possible results of observation* – in fact with results which everybody before Einstein would have expected. This is quite different from the situation I have previously described, when it turned out that the theories in question were compatible with the most divergent human behavior, so that it was practically impossible to describe any human behavior that might not be claimed to be a verification of these theories.[126]

One can sum up all this by saying that *the criterion of the scientific status of a theory is its falsifiability, or refutability, or testability.*[127]

Cushing notes:

A good scientific theory prohibits certain outcomes from occurring in nature and a stringent test of such a theory is an attempt to falsify or refute it by actually observing those prohibited results. Hence, for Popper, the hallmark of scientific theories is that they are (in principle) refutable or falsifiable. (This is not the same thing as saying that they are in fact constantly refuted. A successful scientific theory survives many serious attempts to refute it.)[128]

1.5.1.3   Scientific Revolution vs. Scientific Evolution

In the 1970s, the view that science proceeds through periods of relatively non-controversial "normal" science punctuated by episodes of "revolutionary" science was put forth by Thomas Kuhn in *The Structure of Scientific Revolutions*.[129] The periods of scientific revolution were described by Kuhn as, "non-cumulative developmental episodes in which an older paradigm is replaced in whole or in part by an incompatible new one."[130] The example put forth most frequently is the change in paradigm (view of the world) from the late 1800s of a universe everywhere filled by the luminiferous aether, in which light travels, to the special theory of relativity of Einstein in 1905, in which the concept of the aether would be considered "superfluous."[131] A second example would be Newton's unification of the heavens with the earth, in that the heavens no longer had a separate description (quintessence) from phenomena on the earth, but that the same laws apply in the heavens as apply on the earth. Between these episodes of revolutionary science were the periods of normal science where the accepted paradigm was applied with greater precision, with wider application, and with new predictions.

In contrast to the view of scientific revolutions in terms of paradigm shifts was the view of science proceeding in a more evolutionary manner. In the introduction to Pierre Duhem's *To Save the Phenomena*,[132] Stanley Jaki notes Duhem "knew that in intellectual history the beginnings are rarely abrupt.... Duhem's historical investigations on the origins of statics opened up for him the fascinating world of ancient Greek science, and with it the first phase in that great continuity which Duhem saw in the evolution of physical science."[133]

To the scientist, it matters little if the change is labeled evolutionary or if the change is labeled revolutionary. What matters is that the understanding of nature becomes deeper and more extensive. As noted above, Einstein's 1905 special theory of relativity was considered revolutionary by the scientific community, yet Einstein himself considered it more of an evolutionary step to modify Newton's work, making it correct near the speed of light. Abraham Pais reports that, in a lecture Einstein gave on relativity at King's College in 1921, Einstein himself "deprecated the idea that the new principle [relativity] was revolutionary. It was, he told his audience, the direct outcome and, in a sense, the natural completion of the work of Faraday, Maxwell, and Lorentz."[134]

1.5.2 Newton's Law of Gravitation from Kepler's Laws

Applying Newton's laws of motion to Kepler's three laws of planetary motion, from Kepler's law of equal areas it is determined the force between the sun and the planet is a central force. For a planet in an elliptical orbit due to a central force, Newton's mechanics derives the force to be an inverse square law, i.e., $F \propto 1/r^2$

1.5.2.1 Equal Area Rule => Conservation of Angular momentum => Central force

For planetary orbits Kepler's equal area rule is

$$\frac{dA}{dt} = \frac{1}{2}(R)(R\dot{\theta}) = \frac{1}{2}R^2\dot{\theta} = constant$$

The angular momentum of a planet in orbit is

$$\vec{L} = \vec{R} \times \vec{p} = \vec{R} \times m\vec{v} = (R)(mv_{perpendicular}) = (R)(mR\dot{\theta}) = mR^2\dot{\theta}$$

The angular momentum of a planet around the sun is constant since, by Kepler's equal area law, $R^2\dot{\theta} = constant$. Thus, any force from the sun on the planet must be a radial force, otherwise it would produce a torque on the planet and the angular momentum would not remain constant.

1.5.2.2   Elliptical Orbits+Central Force
=> $1/R^2$ Force

The equation of the elliptical orbit is $R = \frac{A}{1+\varepsilon\cos\theta}$.

Changing the variable from R to $u = 1/R$, the equation of the orbit is
$u(R) = \frac{1}{R} = \frac{1}{A}(1 + \varepsilon\cos\theta)$

Let P be the radial force (central force) of the discussion in Appendix 1.5.2.1:

$$-P = ma_R = m\left(\ddot{R} - R\dot{\theta}^2\right)$$

From angular momentum conservation and using the variable $u = 1/R$, the first term in the parentheses becomes

$$\dot{R} = \frac{dR}{dt} = \frac{d}{dt}\left(\frac{1}{u(\theta)}\right) = \frac{d}{d\theta}\left(\frac{1}{u}\right)\frac{d\theta}{dt} = -\frac{1}{u^2}\frac{du}{d\theta}\dot{\theta}$$

$$= -R^2\dot{\theta}\frac{du}{d\theta} = -\left(\frac{L}{m}\right)\frac{du}{d\theta}$$

$$\ddot{R} = \frac{d\dot{R}}{dt} = -\left(\frac{L}{m}\right)\frac{d}{dt}\left(\frac{du}{d\theta}\right) = -\left(\frac{L}{m}\right)\frac{d}{d\theta}\left(\frac{du}{d\theta}\right)\frac{d\theta}{dt}$$

$$= -\left(\frac{L}{m}\right)\left(\frac{d^2u}{d\theta^2}\right)\dot{\theta} = -\left(\frac{L}{m}\right)\left(\frac{d^2u}{d\theta^2}\right)\left(\frac{L}{mR^2}\right)$$

$$\Rightarrow \ddot{R} = \frac{-L^2u^2}{m^2}\frac{d^2u}{d\theta^2}$$

From the conservation of angular momentum the second term in the parentheses can be written as,

$$L = mR^2\dot{\theta}$$

$$\Rightarrow -R\dot{\theta}^2 = \frac{-L^2}{m^2R^3} = \frac{-L^2}{m^2}u^3$$

Substituting into the expression for P,

$$\Rightarrow P = \frac{L^2}{m}u^2\left(\frac{d^2u}{d\theta^2} + u\right)$$

But

$$\frac{d^2u}{d\theta^2} = \frac{d}{d\theta}\left[\frac{d}{d\theta}\left(\frac{1}{A}(1 + \varepsilon\cos\theta)\right)\right] = \frac{-1}{A}\varepsilon\cos\theta$$

Thus

$$P = \frac{L^2}{m}u^2\left[-\frac{1}{A}\varepsilon\cos\theta + \frac{1}{A}(1 + \varepsilon\cos\theta)\right] = \frac{L^2}{m}\frac{u^2}{A} \propto \frac{1}{R^2}$$

Thus we see that using Newton's laws of mechanics, from Kepler's laws of planetary motion it can be deduced that the force between the sun and the planet is a central force proportional to $\frac{1}{R^2}$.

1.6 Notes

1. Cushing, James T., *Philosophical Concepts in Physics*, Cambridge University Press, Cambridge, 1998, p. 360.

2. Sambursky, Shmuel, editor, *Physical Thought from the Presocratics to the Quantum Physicists*, Pica Press, New York, 1975, p. 12.

3. Dijksterhuis, E. J., *The Mechanization of the World Picture*, Princeton University Press, Princeton, NJ, 1986, p. 7.

4. Dijksterhuis, E. J., *Mechanization of the World Picture*, p. 13.

5. Dijksterhuis, E. J., *Mechanization of the World Picture*, p. 14.

6. Duhem, Pierre, *To Save the Phenomena*, The University of Chicago Press, Chicago, IL, 1969, p. xx.

7. Simplicius, *In Aristotelis quator libros de Coelo commemtaria* 2. 43, 46 (Karsten, ed., p. 219, col. A, and p. 221, col. A; Heiberg, ed., pp. 488, 493); Citation in Duhem, Pierre, *To Save the Phenomena*, p. 5.

8. Duhem, Pierre, *To Save the Phenomena*, pp. 5, 6.

9. Zeilik, Michael, *Astronomy, The Evolving Universe*, 9th edition, Cambridge University Press, New York, 2002, p. 36.

10. Dijksterhuis, E. J., *Mechanization of the World Picture*, p. 18.

11. Dijksterhuis, E. J., *Mechanization of the World Picture*, p. 28.

12. Dijksterhuis, E. J., *Mechanization of the World Picture*, p. 25.

13. Dijksterhuis, E. J., *Mechanization of the World Picture*, p. 32.

14. Dijksterhuis, E. J., *Mechanization of the World Picture*, p. 33.

15. Duhem, Pierre, *To Save the Phenomena*, p. xx.

16. Duhem, Pierre, *To Save the Phenomena*, p. xxi.

17. Dijksterhuis, E. J., *Mechanization of the World Picture*, p. 35.

18. Dijksterhuis, E. J., *Mechanization of the World Picture*, p. 56.

19. McGrew, Timothy, Alspector-Kelly, Marc, and Allhoff, Fritz, editors, *Philosophy of Science: An Historical Anthology*, Wiley-Blackwell, Chichester, England, 2009, p. 15.

20. Dijksterhuis, E. J., *Mechanization of the World Picture*, p. 65.

21. Dijksterhuis, E. J., *Mechanization of the World Picture*, p. 59; Zeilik, Michael, *Astronomy*, pp. 32, 33.

22. Sambursky, Shmuel, *Physical Thought*, p. 44.

23. Dijksterhuis, E. J., *Mechanization of the World Picture*, p. 181.

24. Dijksterhuis, E. J., *Mechanization of the World Picture*, p. 182.

25. Dijksterhuis, E. J., *Mechanization of the World Picture*, p. 294.

26. Layzer, David, *Constructing the Universe*, Scientific American Books, New York, 1984, p. 40.

27. Layzer, David, *Constructing the Universe*, p. 41.

28. Dijksterhuis, E. J., *Mechanization of the World Picture*, p. 289.

29. Spielberg, Nathan, and Anderson, Bryon D., *Seven Ideas That Shook the Universe*, 2nd edition, John Wiley and Sons, Inc., New York, 1995, p. 41. (Spielberg and Anderson, p. 41, state the accuracy to be four minutes of arc, while Dijksterhuis, E. J., *Mechanization of the World Picture*, p. 302, states it to be two minutes of arc. In either case it is significantly more accurate than the previous accuracy of ten minutes of arc.)

30. Dijksterhuis, E. J., *Mechanization of the World Picture*, p. 302.

31. Cushing, James T., *Philosophical Concepts*, p. 66.

32. Cushing, James T., *Philosophical Concepts*, p. 67.

33. Dijksterhuis, E. J., *Mechanization of the World Picture*, p. 320.

34. Dijksterhuis, E. J., *Mechanization of the World Picture*, p. 321.

35. Cushing, James T., *Philosophical Concepts*, pp. 67–69.
36. Zeilik, Michael, *Astronomy*, p. 55.
37. Zeilik, Michael, *Astronomy*, p. 56.
38. Spielberg, Nathan, and Anderson, Bryon D., *Seven Ideas*, p. 49; Zeilek, Michael, *Astronomy*, pp. 56, 57.
39. Zeilik, Michael, *Astronomy*, p. 56.
40. Zeilik, Michael, *Astronomy*, p. 57.
41. Dijksterhuis, E. J., *Mechanization of the World Picture*, pp. 334–337, 352. See also Duhem, Pierre, *To Save the Phenomena*, pp. 105, 111.
42. Dijksterhuis, E. J., *Mechanization of the World Picture*, p. 338.
43. Dijksterhuis, E. J., *Mechanization of the World Picture*, p. 345.
44. Cushing, James T., *Philosophical Concepts*, p. 77.
45. Dijksterhuis, E. J., *Mechanization of the World Picture*, p. 354.
46. Dijksterhuis, E. J., *Mechanization of the World Picture*, p. 348.
47. Cushing, James T., *Philosophical Concepts*, p. 80.
48. Dijksterhuis, E. J., *Mechanization of the World Picture*, pp. 366–367.
49. Zeilik, Michael, *Astronomy*, p. 65.
50. Newton, Isaac, *The Principia*, A new translation by I. Bernard Cohen and Anne Whitman, University of California Press, Berkeley, CA, 1999, paperback version.
51. Cushing, James T., *Philosophical Concepts*, p. 93.
52. Cushing, James T., *Philosophical Concepts*, p. 91.
53. Dijksterhuis, E. J., *Mechanization of the World Picture*, p. 464.
54. Dijksterhuis, E. J., *Mechanization of the World Picture*, pp. 469, 470, 479.
55. Newton, Isaac, *The Principia*, pp. 416–417.
56. Dijksterhuis, E. J., *Mechanization of the World Picture*, p. 468.
57. Newton, Isaac, *The Principia*, Corollary 5, p. 423.
58. Newton, Isaac, *The Principia*, Book 1, Section 2, Corollary 6, p. 451.
59. Newton, Isaac, *The Principia*, Proposition 11, pp. 462–463.
60. Newton, Isaac, *The Principia*, Proposition 15, p. 468.
61. Newton, Isaac, *The Principia*, Book 3, Proposition 2, p. 802.
62. Park, David, *The How and the Why*, Princeton University Press, Princeton, NJ, 1988, pp. 180–183; Dijksterhuis, E. J., *Mechanization of the World Picture*, pp. 477–478.
63. Park, David, *The How and the Why*, pp. 207–208.
64. Cushing, James T., *Philosophical Concepts*, pp. 107–109; Dijksterhuis, E. J., *Mechanization of the World Picture*, p. 478.
65. Casper, Barry M., and Noer, Richard J., *Revolutions in Physics*, W. W. Norton & Company, New York, 1972, pp. 238–240.
66. I thank Don Howard for bringing this to my attention.
67. For further reading on the history of the aether the reader is referred to Whittaker, Sir Edmund, *A History of the Theories of Aether and Electricity*, Volume I, Tomash Publishers, American Institute of Physics, New York, 1987, and to Cushing, James T., *Philosophical Concepts*, pp. 183–194.
68. Hoffmann, Banesh, *Relativity and Its Roots*, Scientific American Books, W. H. Freeman and Company, New York, 1983, p. 56.
69. Hoffmann, Banesh, *Relativity and Its Roots*, pp. 63–65.
70. Maxwell, James Clerk, A Dynamical Theory of the Electromagnetic Field, *Philosophical Transactions*, 155 (1865), 459–512.
71. Cushing, James T., *Philosophical Concepts*, p. 205.

72. Hoffmann, Banesh, *Relativity and Its Roots*, p. 74.

73. Lorentz, H. A., *The Theory of Electrons*, 2nd edition, Dover Publications, New York, 1952.

74. Lorentz, H. A., *The Theory of Electrons*, p. 230.

75. Lorentz, H. A., *The Theory of Electrons*, p. 11.

76. Boorse, Henry A., and Motz, Lloyd, Editors, *The World of the Atom, Volume 1*, Basic Books, Inc., New York, 1966.

77. Boorse, Henry A., and Motz, Lloyd. *The World of the Atom*, p. 516.

78. Cushing, James T., *Philosophical Concepts*, p. 205.

79. Miller, Arthur I., *Albert Einstein's Special Theory of Relativity*, Addison-Wesley Publishing Company, Reading, MA, 1981, p. 181, endnote 42–ii.

80. Winteler-Einstein, Maja, Albert Einstein: A Biographical Sketch; Reproduced in Stachel, John, ed., *The Collected Papers of Albert Einstein*, Volume 1 [CPAE1], Princeton University Press, Princeton, NJ, 1987, pp. xlviii–lii; English translation by Anna Beck [CPAE1 ET], pp. xv, xvi; Pais, Abraham, *Subtle is the Lord*, Oxford University Press, New York, 1982, pp. 35, 36; Clark, Ronald W., *Einstein:The Life and Times*, The World Publishing Company, New York, 1971, pp. 5, 8; Isaacson, Walter, *Einstein: His Life and Universe*, Simon & Schuster, New York, 2007; Fölsing, *Albert Einstein: A Biography*, Viking, New York, 1997.

81. Howard, Don, *Albert Einstein: Physicist, Philosopher, Humanitarian*, The Teaching Company, Chantilly, VA, 2008, pp. 10–12.

82. Seelig, Carl, *Albert Einstein: A Documentary Biography*, London, 1956, p. 71; citation from Clark, Ronald W., *Einstein: The Life and Times*, p. 10.

83. Winteler-Einstein, Maja, A Biographical Sketch; in Stachel, John, ed., [CPAE1 pp. lx–lxll; CPAE ET, p. xx].

84. Schilpp, Paul A., editor, *Albert Einstein: Philosopher–Scientist*, 3rd edition, Cambridge University Press, London, 1949 (1970). pp. 3, 5. Bracketing in the original.

85. Schilpp, Paul A., *Philosopher–Scientist*, p. 5

86. Schilpp, Paul A., *Philosopher–Scientist*, p. 9.

87. Schilpp, Paul A., *Philosopher–Scientist*, p. 53.

88. Winteler-Einstein, Maja, A Biographical Sketch; in Stachel, John, ed., [CPAE1 pp. lxiii–lxiv; CPAE ET, pp. xxi–xxii]; Pais, Abraham, *Subtle is the Lord*, p. 39; Clark, Ronald W., *Life and Times*, p. 20.

89. Howard, Don, and Stachel, John, *The Formative Years*, p. 46. After 1911 the "Zurich Polytechnic" was known as the ETH (Eidgenössische Technische Hochschule).

90. Stachel, John, ed., [CPAE1, pp. xxxvi, 11; CPAE1 ET, p. xxii]; Howard, Don, and Stachel, John, *The Formative Years*, p. 73; Pais, Abraham, *Subtle is the Lord*, p. 40; Clark, Ronald W., *Life and Times*, pp. 21, 22, 24, 25.

91. Stachel, John, ed., [CPAE1, pp. 11, 23–25].

92. Pais, Abraham, *Subtle is the Lord*, p. 41.

93. For further information on Einstein's time at the ETH, including his transcripts, the reader is referred to Stachel, John, ed., [CPAE1, pp. 43–50, 60–62] and to Howard, Don, and Stachel, John, *The Formative Years*, pp. 43–82.

94. Howard, Don, and Stachel, John, *The Formative Years*, pp. 43, 44, 64.

95. Pais, Abraham, *Subtle is the Lord*, p. 44.

96. Howard, Don, and Stachel, John, *The Formative Years*, pp. 45–46.

97. Howard, Don, and Stachel, John, *The Formative Years*, pp. 45–50.

98. Howard, Don, and Stachel, John, *The Formative Years*, pp. 63, 66, 67.

99. Howard, Don, and Stachel, John, *The Formative Years*, pp. 63, 64, 66.

100. Howard, Don, and Stachel, John, *The Formative Years*, pp. 72, 73.

101. Stachel, John, ed., [CPAE1, pp. xxxvi, xxxvii, 44]; Howard, Don, and Stachel, John, *The Formative Years*, pp. 73, 74; Clark, Ronald W., *Life and Times*, p. 40.

102. Pais, Abraham, *Subtle is the Lord*, p. 47.

103. Pais, Abraham, *Subtle is the Lord*, pp. 88, 89; Howard, Don, and Stachel, John, *The Formative Years*, pp. 74, 116, 120, 121; Stachel, John, CPAE2, p. 170.

104. Howard, Don, and Stachel, John, *The Formative Years*, pp. 2–4; Stachel, John, ed., [CPAE2, p. xxvii].

105. Howard, Don, and Stachel, John, *The Formative Years*, p. 126.

106. Stachel, John, editor, *The Collected Papers of Albert Einstein*, Volume 2 [CPAE2], Princeton University Press, Princeton, NJ, 1989, p. 41.

107. Einstein, Albert, Conclusions Drawn from the Phenomena of Capillarity, *Annalen der Physik* 309 (8), (1901), pp. 513–523; Stachel, John, ed., [CPAE2, doc. 1, pp. 9–21; CPAE2 ET, pp. 1–11].

108. Einstein, Albert, On the Thermodynamic Theory of the Difference in Potentials Between Metals and Fully Dissociated Solutions of Their Salts and On an Electrical Method for Investigating Molecular Forces, *Annalen der Physik* 313 (8), (1902), pp. 798–814; Stachel, John, ed., [CPAE2, doc. 2, pp. 22–40; CPAE2 ET, pp. 12–29].

109. Einstein, Albert, Kinetic Theory of Thermal Equilibrium and of the Second Law of Thermodynamics, *Annalen der Physik* 314 (10), (1902), pp. 417–433; Stachel, John, ed., [CPAE2, doc. 3, pp. 56–75; CPAE2 ET, pp. 30–47].

110. Einstein, Albert, A Theory of the Foundations of Thermodynamics, *Annalen der Physik* 316 (5), (1903), pp. 170–187; Stachel, John, ed., [CPAE2, doc. 4, pp. 76–97; CPAE2 ET, pp. 48–67].

111. Einstein, Albert, On the General Molecular Theory of Heat, *Annalen der Physik* 319 (7), (1904), pp. 354–362; Stachel, John, ed., [CPAE2, doc. 5, pp. 98–108; CPAE2 ET, pp. 68–77].

112. Stachel, John, ed., [CPAE2, p. 3].

113. Stachel, John, ed., [CPAE2, p. 41].

114. Howard, Don, and Stachel, John, *The Formative Years*, p. 109.

115. Stachel, John, ed., [CPAE2, p. 46].

116. Howard, Don, and Stachel, John, *The Formative Years*, pp. 4, 5, 9.

117. Pais, Abraham, *Subtle is the Lord*, p. 39.

118. Lorentz, H. A., *The Theory of Electrons*, 1952.

119. Lorentz, H. A., *The Theory of Electrons*, p. 230.

120. Lorentz, H. A., *The Theory of Electrons*, p. 11.

121. Lorentz, Hendrik, The Radiation of Light, *Nature*, 113 (26 April, 1924), 608–611.

122. Cushing, James T., *Philosophical Concepts*, pp. 30, 32.

123. Cushing, James T., *Philosophical Concepts*, p. 360.

124. Popper, Karl, *Conjectures and Refutations*, Routledge Classics, New York, 2002, pp. 43–44.

125. Popper, Karl, *Conjectures and Refutations*, p. 45.

126. Popper, Karl, *Conjectures and Refutations*, p. 47.

127. Popper, Karl, *Conjectures and Refutations*, p. 48.

128. Cushing, James T., *Philosophical Concepts*, p. 33.

129. Kuhn, Thomas, *The Structure of Scientific Revolutions*, The University of Chicago Press, Chicago, Il, 1970.
130. Kuhn, Thomas, *Scientific Revolutions*, p. 92.
131. Stachel, John, ed. [CPAE2, p. 277; CPAE2 ET, p. 141].
132. Duhem, Pierre, *To Save the Phenomena*.
133. Duhem, Pierre, *To Save the Phenomena*, p. xvii. In the Introduction by Stanley L. Jaki.
134. Einstein, Albert, Prof. Einstein's Lectures at King's College, London, and the University of Manchester, *Nature* 107 (16 June, 1921), p. 504; citation from Pais, Abraham, *Subtle is the Lord*, p. 30.

1.7   Bibliography

Boorse, Henry A., and Motz, Lloyd, editors, *The World of the Atom*, Volume 1, Basic Books, Inc., New York, 1966.

Casper, Barry M., and Noer, Richard J., *Revolutions in Physics*, W.W. Norton and Company, New York, 1972.

Clark, Ronald W., *Einstein: The Life and Times*, The World Publishing Company, New York, 1971.

Cushing, James T., *Philosophical Concepts in Physics*, Cambridge University Press, Cambridge, 1998.

Dijksterhuis, E. J., *The Mechanization of the World Picture*, Princeton University Press, Princeton, NJ, 1986.

Duhem, Pierre, *To Save the Phenomena*, The University of Chicago Press, Chicago, IL, 1969.

Einstein, Albert, Conclusions Drawn from the Phenomena of Capillarity, *Annalen der Physik* 309 (3), (1901), pp. 513–523 [CPAE2, doc. 1, pp. 9–21; CPAE2 ET, pp. 1–11].

Einstein, Albert, On the Thermodynamic Theory of the Difference in Potentials Between Metals and Fully Dissociated Solutions of Their Salts and On an Electrical Method for Investigating Molecular Forces, *Annalen der Physik* 313 (8), (1902), pp. 798–814 [CPAE2, pp. 22–40; CPAE2 ET, pp. 12–29].

Einstein, Albert, Kinetic Theory of Thermal Equilibrium and of the Second Law of Thermodynamics, *Annalen der Physik* 314 (10), (1902), pp. 417–433 [CPAE2, pp. 56–75; CPAE2 ET, pp. 30–47].

Einstein, Albert, A Theory of the Foundations of Thermodynamics, *Annalen der Physik* 316 (5), (1903), pp. 170–187 [CPAE2, pp. 76–97; CPAE2 ET, pp. 48–67].

Einstein, Albert, On the General Molecular Theory of Heat, *Annalen der Physik* 319 (7), (1904), pp. 354–362 [CPAE2, pp. 98–108; CPAE2 ET, pp. 68–77].

Einstein, Albert, Prof. Einstein's Lectures at King's College, London, and the University of Manchester, *Nature* 107, (1921), p. 504. (Citation in Pais, Abraham, *Subtle is the Lord*.)

Fölsing, Albrecht, *Albert Einstein: A Biography*, Viking, New York, 1997.

Hoffmann, Banesh, *Relativity and Its Roots*, Scientific American Books, W. H. Freeman and Company, New York, 1983.

Howard, Don, *Albert Einstein: Physicist, Philosopher, Humanitarian*, The Teaching Company, Chantilly, VA, 2008.

Howard, Don, and Stachel, John, editors, *Einstein: The Formative years*, 1879–1909, Birkhäuser, Boston, MA, 2000.

Isaacson, Walter, *Einstein: His Life and Universe*, Simon & Schuster, New York, 2007.

Karsten, editor, *In Aristotelis quator libros de Coelo commemtaria.* (Citation in Duhem, Pierre, *To Save the Phenomena*.)

Kuhn, Thomas S., *The Structure of Scientific Revolutions*, The University of Chicago Press, Chicago, IL, 1970.

Layzer, David, *Constructing the Universe*, Scientific American Books, New York, 1984.

Lorentz, H. A., *The Theory of Electrons*, 2nd edition, Dover Publications, New York, 1952.

Lorenz, Hendrik, The Radiation of Light, *Nature* 113 (1924).

Maxwell, James Clerk, A Dynamical Theory of the Electromagnetic Field, *Philos. Trans.* 155 (1865). (Citation in Boorse, Henry, and Motz, Lloyd, *The World of the Atom*.)

McGrew, Timothy, Alspector-Kelly, Marc, Allhoff, Fritz, editors, *Philosophy of Science: An Historical Anthology*, Wiley-Blackwell, Chichester, England, 2009.

Miller, Arthur, *Albert Einstein's Special Theory of Relativity*, Addison-Wesley Publishing Company, Reading, MA, 1981.

Newton, Isaac, *The Principia*, A new translation by I. Bernard Cohen and Anne Whitman, University of California Press, Berkeley, CA, 1999.

Pais, Abraham, *Subtle is the Lord*, Oxford University Press, London, 1982.

Park, David, *The How and the Why*, Princeton University Press, Princeton, NJ, 1988.

Popper, Karl, *Conjectures and Refutations*, Routledge Classics, New York, 2002.

Sambursky, Shmuel, *Physical Thought from the Presocratics to the Quantum Physicists*, Pica Press, New York, 1974.

Schilpp, Paul Arthur, *Albert Einstein: Philosopher–Scientist*, Cambridge University Press, Cambridge, 1970.

Seelig, Carl, *Albert Einstein: A Documentary Biography*, London, 1956, p. 71. (Citation in Clark, Ronald, *Einstein: The Life and Times*.)

Spielberg, Nathan, and Anderson, Bryon D., *Seven Ideas That Shook the Universe*, 2nd edition, John Wiley and Sons, Inc., New York, 1995.

Stachel, John, editor, *The Collected Papers of Albert Einstein*. Volume 1 [CPAE1], Princeton University Press, Princeton, NJ, 1987; English translation by Anna Beck [CPAE1 ET].

Stachel, John, editor, *The Collected Papers of Albert Einstein*, Volume 2 [CPAE2], Princeton University Press, Princeton, NJ, 1989; English translation by Anna Beck [CPAE2 ET].

Whittaker, Sir Edmund, *A History of the Theories of Aether and Electricity*, Volume 1, Tomash Publishers, American Institute of Physics, New York, 1987.

Winteler-Einstein, Maja, Albert Einstein: A Biographical Sketch [reproduced in CPAE1].

Zeilek, Michael, *Astronomy: The Evolving Universe*, 9th edition, Cambridge University Press, Cambridge, 2002.

Radiation and the Quanta

2

"On a Heuristic Point of View Concerning the Production and Transformation of Light"[1]

Received on 18 March, 1905, by *Annalen der Physik*

2.1 Historical Background   33

2.2 Albert Einstein's Paper, "On a Heuristic Point of View Concerning the Production and Transformation of Light"   39

2.3 Discussion and Comments   47

2.4 Appendices   49

2.5 Notes   52

2.6 Bibliography   55

2.1 Historical Background

Albert Einstein's paper, "On a Heuristic Point of View Concerning the Production and Transformation of Light," sometimes is referred to as the photoelectric effect paper, giving the impression its primary goal was to explain the photoelectric effect. To the contrary, the primary focus was to explain blackbody radiation and, at the end of the paper, the explanation of blackbody radiation was used to give a possible explanation of three experiments, one of them being the photoelectric effect. In this paper, he draws on currently accepted scientific ideas of electromagnetism and the electromagnetic aether (see Section 1.2.3.1); on thermodynamics and the newer ideas of statistical mechanics; on entropy; and on blackbody radiation, including the work of Max Planck of 1900.

2.1.1 Thermodynamics and Entropy

In nature, it was of interest why some processes are "irreversible." For example, why does heat transfer from a warmer object to a cooler object, but not from a cooler object to a warmer object? From the laws of classical mechanics there is nothing that prohibits this reverse process. To explain why a process such as this is not reversible, a new quantity called the entropy, S, was introduced, and a new law was formulated that "In any naturally occurring process the total entropy of the universe will increase or, at best, stay the same."[2] This law is an addition to the laws of mechanics, laws such as conservation of energy and conservation of momentum, and is not contained in mechanics. (See Appendix 2.4.1 for further discussion.)

In the mid 1800s, Ludwig Boltzmann, an ardent atomist, viewed the thermodynamic gas as composed of a very large number of independent molecules in random, chaotic motion. The motion of an individual molecule was determined by Newton's equations of mechanics through interactions (collisions) with the other molecules and with the walls of

the container. To each degree of freedom of the system there corresponded an equal portion of the energy of the system (each molecule
has three degrees of freedom associated with its translational motion,
one degree of freedom each for the x-, y-, and z-components of velocity,
plus possibly additional degrees of freedom if there is internal energy
associated with vibrational or rotational energy). This equal sharing of
the energy is known as the equipartition of energy.

Looking at the distribution of the molecules of the gas, not only in
terms of location within the container, but also in terms of distribution of
the velocities of the molecules, Boltzmann showed that for an arbitrary
initial distribution, the gas would tend to the equilibrium distribution,
i.e., to the Maxwell distribution function.[3] At any moment, the molecules
of the gas can be in any of a (nearly infinite) number of states, each
state with a calculable probability, W. For equilibrium states, Boltzmann
obtained the following relation between the entropy S of the system in
a given state and the probability W of that state[4]

$$S = k \log W$$

where k is a constant of proportionality, subsequently named the
Boltzmann constant. The relation $S = k \log W$ indicates that an increase
in the entropy of a system corresponds to the system moving from
a less probable state to a more probable state. Boltzmann calculated
the probability of a state occurring in terms of "complexions," the
number of ways a given state could occur.[5] In modern terminology,
Boltzmann developed the theory of the microcanonical ensemble, where
each possible configuration of the system corresponds to the same total
energy.

2.1.2 Blackbody Radiation

All matter emits electromagnetic radiation when it has a temperature
above absolute zero. The radiation a body emits because it is at some
temperature T is called thermal radiation. $E_\nu d\nu$ is the amount of
radiation energy emitted by the body per unit time per unit area in the
frequency interval $d\nu$. A_ν is the absorption coefficient at frequency ν. In
1860, Gustav Kirchhoff showed that the ratio E_ν/A_ν depended only on
the frequency ν and the temperature T of the body, and did not depend
on any other of its characteristics. The relation, $E_\nu/A_\nu = J(\nu, T)$, is
known as Kirchhoff's theorem. If the absorption coefficient is equal
to one, $A_\nu = 1$, the body absorbs all of the radiant energy incident
on it and is defined as a blackbody. For a blackbody, $E_\nu = J(\nu, T)$.
Kirchhoff stated, "It is a highly important task to find this function
$[J]$.... [T]here appear grounds for the hope that it has a simple form,
as do all functions which do not depend on the properties of individual
bodies ..."[6] Because of the experimental difficulty it was nearly forty

years before sufficient experimental data were available to determine the function $J(\nu, T)$.[7]

Consider a cavity, completely enclosed except for one very small opening. Radiation from the outside incident on the small opening will pass through the opening into the interior of the cavity, where it will be reflected internally and eventually be absorbed by the walls. Since the cross-sectional area of the opening is taken to be negligible in comparison to the area of the interior walls of the cavity, only a negligible amount of the incident radiation will be reflected back through the opening. Thus, the small opening, since virtually all of the incident radiation is "absorbed" and none is "reflected," acts like the surface of a blackbody. The small opening can therefore be treated the same as the surface of a blackbody.

Now consider the reverse process. If the cavity is heated, instead of absorbing radiation through the small opening, the cavity emits radiation through it. Because of the heating, the interior of the cavity will be filled with thermal radiation corresponding to some established equilibrium between the radiation in the cavity and the walls at the temperature T. The small opening in the wall of the container allows some of the radiation to escape the cavity and be measured. The opening is so small that the tiny bit of radiation escaping will have a negligible effect on the radiation within the cavity. Since (in absorption) the small opening was shown to behave as the surface of a blackbody, the spectrum of the escaping radiation must have the spectrum of a blackbody and, since the escaping radiation is but a sample of the radiation inside the cavity, the radiation inside the cavity must have the spectrum of a blackbody. The energy density, $\rho(\nu, T)$, is the amount of energy per unit volume per unit frequency of the radiation in the cavity. It was shown that $\rho(\nu, T)$ is related to $J(\nu, T)$ as $J(\nu, T) = (\frac{c}{8\pi})\rho(\nu, T)$.[8] Determining the function $\rho(\nu, T)$ would determine the function $J(\nu, T)$. Thus, to study blackbody radiation, scientists focused their attention on the radiation inside the cavity, called cavity radiation. See Figure 2.1.

The radiation in the container will consist of standing waves of certain wavelengths λ_n, where $\lambda_n = (2L)/n$, and n is an integer ($n = 1, 2, 3, \ldots$, but $n \neq 0$). See Figure 2.2.

Each wavelength corresponds to a degree of freedom of the radiation and, by the equipartition of energy theorem, shares equally in the energy and, "since there are infinitely many allowed waves of shorter and shorter wavelength, nearly all of the light should be at the short wavelength end of the spectrum."[9] It should be noted that the short wavelength region is the same as the high frequency region.[10]

In the years from 1890 to 1900, a number of experimentalists measured the cavity radiation density as a function of the frequency ν with increasing precision.[11] See Figure 2.3.

In 1893, Wilhelm Wien proved that $\rho(\nu, T)$ must be of the form $\rho(\nu, T) = \nu^3 f(\nu/T)$ and, in 1896, proposed that $\rho(\nu, T) = \alpha\nu^3 e^{-\beta\nu/T}$.[14] In 1897, this was verified to fit the known data. But, in 1900, new

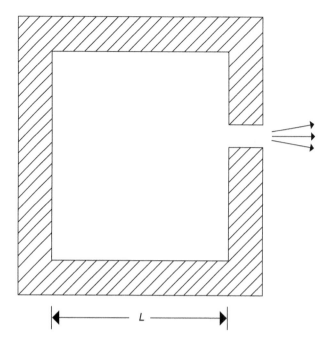

Fig. 2.1 Schematic of the cavity radiation experimental apparatus.

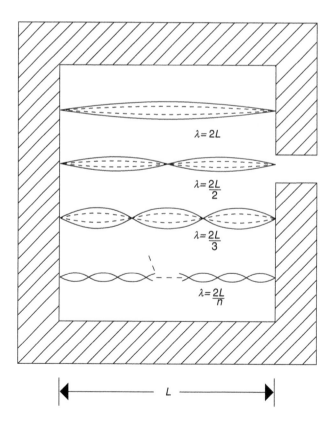

Fig. 2.2 Standing waves in a cavity of width L.

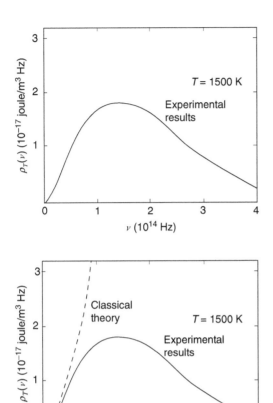

Fig. 2.3 Plot of energy density vs. frequency for cavity radiation.

Source: (Adapted from Eisberg, Robert, and Resnick, Robert, *Quantum Physics of Atoms, Molecules, Solids, Nuclei, and Particles*.[12])

Fig. 2.4 Rayleigh–Jeans theory and the ultraviolet catastrophe.

Source: (From Eisberg, Robert, and Resnick, Robert, *Quantum Physics of Atoms, Molecules, Solids, Nuclei, and Particles*.[13] Reproduced with permission.)

experimental data at lower frequencies showed discrepancies of 40% – 50% between the data and Wien's expression for the radiation energy density.[15]

In 1900, using the equipartition of energy, Lord Rayleigh obtained for $\rho(\nu, T)$ the expression, $\rho(\nu, T) = c\nu^2 T$. In 1905, Lord Rayleigh and James Jeans determined the value of the constant c. This became known as the Rayleigh–Jeans formula (see Section 2.1.1 for a derivation),[16]

$$\rho(\nu, T) = \frac{8\pi k T \nu^2}{c^3}$$

It should be noted that all constants in the Rayleigh–Jeans formula are determined – there are no parameters available to adjust in order to fit the experimental data. At low frequencies (long wavelengths) this formula agreed well with the experimental curves, but at high frequencies (short wavelengths) the formula became infinite and diverged dramatically from the experimental curves. This divergence became known as the ultraviolet catastrophe.[17] See Figure 2.4.

2.1.3   Max Planck's Derivation of the Radiation Density

Knowing of the success of the Wien law in the high frequency region and that $\rho(\nu, T)$ was proportional to T in the low frequency regime (although this was the same as the Rayleigh–Jeans result, Planck was responding to the work of Heinrich Rubens[18]), in 1900 Max Planck proposed that

$$\rho(\nu, T) = \frac{8\pi h \nu^3}{c^3} \frac{1}{e^{h\nu/kT} - 1}$$

This result was obtained by interpolating between the two expressions and requiring that the resulting expression reduce to the Wien law in the limit of large $h\nu/kT$, and be proportional to T in the limit of small $h\nu/kT$. As Cushing notes, "At this stage Planck had a formula that fit the data perfectly, but that had no theoretical justification."[19] Planck then set about to find a theoretical justification.[20] (See Appendix 2.4.2 for the details of this derivation.)

In 1900, the following was known about cavity radiation:[21]

1. The energy distribution of cavity radiation was independent of the material of which the walls were constructed.
2. Using general thermodynamic arguments, Wien had shown that any solution for the energy distribution for cavity radiation must be of the form $\rho(\nu, T) = \nu^3 f(\nu/T)$, where f is an arbitrary function to be determined. This relation became known as the Wien displacement law.

In his calculation for the energy density $\rho(\nu, T)$ of the cavity radiation, by the first statement Max Planck was free to choose the simplest model for the walls of the container that would allow the exchange of energy between the radiation and the walls. The walls of the container were an aggregate of simple resonators (oscillators), each with "one proper frequency" and capable of absorbing and/or emitting radiant energy near this frequency in discrete amounts, ε.[22] For the resonators in the walls in equilibrium with the radiation field, Planck calculated the average energy density of the radiation in the interval $d\nu$ to be

$$\rho(\nu, T)d\nu = \frac{8\pi\nu^2}{c^3} U d\nu$$

where U is the average energy of the resonator. It needs to be emphasized that $\rho(\nu, T)$ is the average energy density of the radiation field, while U is the average energy density of the resonators in the walls. Denoting the energy of an oscillator as ε, Planck obtained

$$U = \frac{\varepsilon}{e^{\varepsilon/kT} - 1}$$

To satisfy the Wien displacement law,

$$\rho(\nu, T) = \nu^3 f(\nu/T) = \frac{8\pi\nu^2}{c^3} U d\nu = \frac{8\pi\nu^2}{c^3} \left(\frac{\varepsilon}{e^{\varepsilon/kT} - 1} \right) d\nu$$

$\Rightarrow \frac{\varepsilon}{kT} = C\frac{\nu}{T}$, or $\varepsilon = Ck\nu = h\nu$, i.e., to satisfy the Wien displacement law the average energy of the resonator must be an integral multiple of $h\nu$. The resulting expression for the cavity radiation density is

$$\rho = \frac{A\nu^3}{e^{\frac{h\nu}{kT}} - 1}$$

Planck had imposed the condition that the resonators absorbed and emitted energy in discrete amounts ε, not that $\varepsilon = h\nu$. It was the constraint from the Wien displacement law that required the energy of an oscillator be a multiple of $h\nu$. The condition of $\varepsilon = h\nu$ is imposed on the resonators in the walls of the container, not on the radiation field. At high frequencies, this expression for ρ reduces to the Wien expression, while for low frequencies it reduces to the Rayleigh–Jeans expression.

Although the resulting expression for the energy density of the radiation matched well with experiment, questions about the derivation were raised by Wien.[23] In the derivation, the resonators needed to be independent of one another, yet (by the equipartition of energy) their total energy was to be distributed equally among them, indicating some dependence.

2.2 Albert Einstein's Paper, "On a Heuristic Point of View Concerning the Production and Transformation of Light"[24]

Einstein had come to the conclusion that "[light] energy is not distributed continuously over ever-increasing spaces, but consists of a finite number of energy quanta that are localized in points in space, move without dividing, and can be absorbed or generated only as a whole."[25] In this paper, he wished "to communicate my train of thought and present the facts that led me to this course."[26] His suggestion was that the failure of "Maxwell's theory to give an adequate account of radiation might be remedied by a theory in which radiant energy is distributed discontinuously in space."[27] Through an analysis of blackbody radiation, he shows how one can be led naturally to view the radiation as composed of a number of quanta of energy $h\nu$. The phrase "heuristic point of view" in the title indicates he does not consider the paper a "proof" of his assertion (of light quanta), but considers it as presenting a helpful procedure to enable a person to see the reasonableness of such a perspective.

Albert Einstein was concerned with the "profound formal differences" existing between the molecular theory of gases and Maxwell's theory of

electromagnetic radiation.[28] Gases were described in terms of discrete
particles (atoms and molecules), with the energy of the system a sum
over the energies of the finite number of individual particles. Maxwell's
theory described electromagnetism in terms of a continuous function,
the electromagnetic wave, and "a finite number of quantities cannot
be considered as sufficient for the complete determination of the electromagnetic state of a space."[29] The energy of an electromagnetic wave was
distributed continuously over the entire wave, and could not be reduced
to the sum of discrete parts, as was the case of the gas composed of
discrete particles. "The energy of a ponderable body cannot be broken
up into arbitrarily many, arbitrarily small parts, while according to
Maxwell's theory... the energy of a light ray emitted from a point source
of light spreads continuously over a steadily increasing volume."[30]

Einstein noted the wave theory of light worked "splendidly," but
pointed out "optical observations apply to time averages and not to
momentary values; and it is conceivable... [that the wave theory of
light]... may lead to contradictions with experience when it is applied
to the phenomena of production and transformation of light."[31]

Work on the blackbody radiation problem was well known to Einstein
in 1905, not only because it was a topic of much interest in the
physics community, but also because H. F. Weber, Einstein's professor at the ETH, was working in measuring emission radiation (from
carbon filament lamps). As Kuhn notes, "When Wien... published the
displacement law, his only reference to experiment was through Weber's
law."[32] In addition, in 1905, Einstein was aware of Planck's 1900 work
on blackbody radiation and his treatment of the entropy of the radiation
field through probability and the complexions. From his soon to be
published work on the special theory of relativity, Einstein knew the
concept of the aether to be superfluous, leaving no medium to trans-
port the electromagnetic waves. But his work on relativity also showed
electromagnetic radiation transfers inertial mass.[33] At heart, Einstein
was an atomist and, thus, predisposed to accepting the possibility of
elementary quanta.[34] Being an atomist, he was very well acquainted
with the work of Boltzmann in Statistical Mechanics and, in fact, had
clarified and expanded the field.[35]

As Einstein noted, in the low frequency regime (the Rayleigh–Jeans
region) the results were consistent with the classical treatment of Statistical Mechanics and Maxwell's theory. However in the high frequency
regime (Wien's distribution law regime) "these principles fail completely."[36] He then focused his efforts on the Wien region, avoiding use
of the potentially problematical Statistical Mechanics.

2.2.1   On a Difficulty Encountered in the Theory of "Blackbody Radiation"

Section 1 of Einstein's paper points out the limitations of our present
theoretical foundations for understanding radiation. Einstein showed

how the traditional treatment of radiation using Maxwell's theory leads to the Rayleigh–Jeans formula for the radiation energy density, ρ_ν.[37] He first considers a gas enclosed in a container that is in equilibrium with the walls of the container. After establishing certain conclusions for a gas in the container, he applies similar reasoning to radiation in a container.

Consider a gas composed of molecules interacting with electrons. There are two types of electrons: free electrons and bound (resonator) electrons. The bound electrons "are bound to points in space which are very far from each other by forces that are directed to these points and are proportional to the elongations from the points. These electrons, too, shall enter into conservative interactions with the free molecules and electrons when the latter come very close to them."[38] In equilibrium, the average kinetic energy of a molecule of the gas is equal to the average kinetic energy of a bound electron. For a linear (one-dimensional) electron, from kinetic theory/statistical mechanics, the average kinetic energy is $\frac{1}{2}kT$ (k = Boltzmann's constant = R/N used by Einstein). Boltzmann had shown that "the average kinetic energy equals the average potential energy for a system of particles each one of which oscillates under the influence of external harmonic forces"[39], giving a total average energy per electron of kT,

$$\langle E_{electron} \rangle = kT$$

Einstein then applies similar reasoning to the interaction between radiation in a cavity and the resonators in the walls of the cavity. Using the relation established by Max Planck between the energy density of the radiation and the energy density of the resonators in the walls of the cavity, $\rho(\nu, T) = \frac{8\pi\nu^2}{c^3} U = \frac{8\pi\nu^2}{c^3} \langle E_{electron} \rangle = \frac{8\pi\nu^2}{c^3} kT$,

$$\rho(\nu, T) = \frac{8\pi k T \nu^2}{c^3}$$

where k is the Boltzmann constant, ν is the frequency of the radiation, c is the speed of light, and T is the temperature of the system. He then pointed out that this expression not only "fails to agree with experience," but also leads to an infinite amount of energy in a given volume when all frequencies of radiation are included:[40]

$$E = \int_0^\infty \rho_\nu d\nu = k\frac{8\pi}{c^3}T \int_0^\infty \nu^2 d\nu = \infty$$

2.2.2 On Planck's Determination of the Elementary Quanta

By 1900 there existed several formulae that had been derived to explain the experimentally obtained curve for the radiation density of blackbody radiation. One of these formulae, the Rayleigh–Jeans law, matched well the experimental curve at low frequencies, while another formula, Wien's law, matched well the experimental curve for high frequencies. However, neither matched well the experimental curve over the entire frequency range.[41] (See Section 2.1.2.)

In 1900, Planck published a formula that fit the experimental curve over the entire frequency range.[42] To obtain his formula, Planck considered the radiation in the cavity interacting with resonators in the walls of the cavity (see Section 2.1.3). The resonators were bound electrons[43] that "emit and absorb electromagnetic waves of definite [frequencies]."[44] The radiation in the cavity, although composed of radiation of all frequencies, interacted with resonators only at the resonant frequencies of the resonators. Through this approach, Planck obtained for the radiation density in the cavity of a blackbody

$$\rho_\nu = \frac{\alpha \nu^3}{e^{\frac{\beta \nu}{T}} - 1}$$

For small values of $(\frac{\nu}{T})$, Planck's expression for ρ_ν becomes (expanding the exponential term in the denominator and keeping only the lowest order term)

$$\rho_\nu = \frac{\alpha}{\beta} \nu^2 T$$

in agreement with the expression obtained in Section 1, the Rayleigh–Jeans law. This was taken as justification that the lower the frequency, "the more useful the theoretical principles we have been using prove to be; however these principles fail completely in the case of [high frequencies]."[45]

2.2.3   On the Entropy of Radiation

In Section 3, Einstein reproduces some pertinent work from Wien, obtaining the expression for the entropy of radiation of frequency ν contained in a volume V.

The entropy of the radiation can be expressed as

$$S = V \int_0^\infty \phi(\rho_\nu, \nu) d\nu$$

where V is the volume, $\phi(\rho_\nu, \nu)$ is the entropy per unit volume per unit frequency range, and is a function of both the radiation energy density ρ_ν and the frequency ν.

Assume the system is in a state of dynamic equilibrium, that is, the system is in a state of maximum entropy. Maximizing the entropy of the system for a fixed energy, and using Lagrange's Method of Undetermined Multipliers, Wien showed the expression for ϕ satisfies (see Appendix 2.4.3 for the details of this derivation),[46]

$$\frac{\partial \phi}{\partial \rho_\nu} = \frac{1}{T}$$

This relation is known as Wien's principle.[47]

2.2.4 Limiting Law for the Entropy of Monochromatic Radiation at Low Radiation Density

In the limit of high frequencies, Wien had obtained for the radiation density ρ_ν the ad hoc expression[48]

$$\rho_\nu = \alpha \nu^3 e^{-\beta \frac{\nu}{T}}$$

which is in agreement with experiment for large ν. Inverting this expression,

$$\frac{1}{T} = -\frac{1}{\beta \nu} \ln \left(\frac{\rho_\nu}{\alpha \nu^3} \right)$$

Using the relation for $(\partial \phi / \partial \rho)$ obtained in Section 2.2.3, and integrating,

$$\frac{1}{T} = \frac{\partial \phi}{\partial \rho_\nu} = -\frac{1}{\beta \nu} \ln \left(\frac{\rho_\nu}{\alpha \nu^3} \right)$$

$$\phi(\rho_\nu, \nu) = -\frac{\rho_\nu}{\beta \nu} \left\{ \ln \left(\frac{\rho_\nu}{\alpha \nu^3} \right) - 1 \right\}$$

The energy in a volume V and frequency range $\Delta \nu$ is equal to $\rho_\nu V \Delta \nu$. The entropy S in the volume V and frequency range $\Delta \nu$ is equal to $\phi(\rho_\nu, \nu) V \Delta \nu$:

$$S = \phi(\rho_\nu, \nu) V \Delta \nu$$

$$= -\frac{\rho_\nu}{\beta \nu} \left\{ \ln \left(\frac{\rho_\nu}{\alpha \nu^3} \right) - 1 \right\} V \Delta \nu$$

$$= -\frac{1}{\beta \nu} (\rho_\nu V \Delta \nu) \left\{ \ln \left(\frac{\rho_\nu V \Delta \nu}{\alpha \nu^3 V \Delta \nu} \right) - 1 \right\}$$

$$= -\frac{1}{\beta \nu} (E) \left\{ \ln \left(\frac{E}{\alpha \nu^3 V \Delta \nu} \right) - 1 \right\}$$

Consider now only the dependence of the entropy on the volume V. For some V_0 designate the corresponding entropy as S_0, with S representing the entropy when the radiation occupies the volume V:[49]

$$S - S_0 = \frac{E}{\beta \nu} \ln \left(\frac{V}{V_0} \right)$$

2.2.5 Molecular-Theoretical Investigation of the Dependence of the Entropy of Gases and Dilute Solutions on the Volume

In Section 5, Einstein obtains the entropy of an ideal gas. (In Section 6 the expression for the entropy of radiation obtained in Section 4 will be compared with that of the ideal gas obtained in this section.)

A system moving to a state of higher entropy, S, corresponds to a transition to a state of higher probability, W. Ludwig Boltzmann showed the entropy of the system is related to the probability as

$$S = k \log W$$

where S is the entropy of the system, k is Boltzmann's constant, and W is the probability of the system being in the particular state.

Consider a container of volume V_0 containing a gas of N atoms. Consider also a subvolume V of the volume V_0. The probability W_1 that a specified atom is in the subvolume V at a given time is the ratio of the subvolume V to the total volume V_0:

$$W_1 = \left(\frac{V}{V_0} \right)$$

The probability W_2 that a specified set of two atoms is in the subvolume V at a given time is the probability that the first atom is in the subvolume multiplied by the probability that the second atom is in the subvolume V:

$$W_2 = \left(\frac{V}{V_0} \right) \times \left(\frac{V}{V_0} \right) = \left(\frac{V}{V_0} \right)^2$$

Generalizing, the probability all N atoms are in the subvolume V at a given time is

$$W_N = \left(\frac{V}{V_0} \right)^N$$

The entropy S corresponding to all N atoms being in the subvolume V, relative to the entropy S_0 of the atoms occupying the volume V_0 is[50]

$$S - S_0 = k \log W_N = k \log \left(\frac{V}{V_0} \right)^N$$

2.2.6   Interpretation of the Expression for the Dependence of the Entropy of Monochromatic Radiation on Volume According to Boltzmann's Principle

Section 6 is the key section, showing the entropy for radiation in the high frequency limit is of exactly the same form as the expression for the entropy of an ideal gas (if one makes certain straightforward identifications). Thus one can draw the conclusion, at least in the high frequency limit, that blackbody radiation behaves as if it were composed of a number of independent quanta, the same as an ideal gas is composed of a number of independent molecules.

For an ideal gas the expression for the entropy is (see Section 2.2.5),

$$S - S_0 = k\log W_N = k\log \left(\frac{V}{V_0}\right)^N$$

For monochromatic radiation, the entropy S is (see Section 2.2.4; replacing ln with log simply changes the value of β)

$$S - S_0 = \frac{E}{\beta\nu}\log\left(\frac{V}{V_0}\right)$$

$$S - S_0 = k\left(\frac{E}{k\beta\nu}\right)\log\left(\frac{V}{V_0}\right) = k\log\left(\frac{V}{V_0}\right)^{\frac{E}{k\beta\nu}}$$

Identifying this with Boltzmann's expression for the entropy, $S = k\log W$, the probability that the total radiation energy will be found in the subvolume V of the volume V_0 is[51]

$$W = \left(\frac{V}{V_0}\right)^{\frac{E}{k\beta\nu}}$$

Thus, at sufficiently high frequencies, the entropy of the radiation *behaves as if* it were an ideal gas composed of $n = (E/k\beta\nu)$ particles. Since E is the total energy of the system, the energy of each "particle" will be $k\beta\nu$ ($k\beta$ subsequently was shown to be Planck's constant h: $E = h\nu$). In the Wien (high frequency) region, the expression for the entropy of the radiation is what it would be if the radiation were composed of $n = (E/k\beta\nu) = (E/h\nu)$ independent particles (quanta), each with the energy $E = h\nu$.

2.2.7 On Stokes' Rule

In Sections 7, 8, and 9, Einstein shows how the concept of light quanta gives simple explanations of Stokes' rule for photoluminescence, of the photoelectric effect, and of the generation of cathode rays by photo-ionization.

In Section 7, Einstein discusses photoluminescence. In photoluminescence, i.e., phosphorescence and fluorescence, the frequency of the emitted light is equal to, or less than, the frequency of the absorbed light. This is known as Stokes' rule.

Einstein's light quanta could explain Stokes' rule through the conservation of energy. A light quantum of energy $h\nu_1$ is absorbed, transmitting an amount of energy $E_1 = h\nu_1$ to the absorbing body. Absorption of the incoming light quantum triggers the emission of another light quantum of frequency ν_2, possibly giving rise also to other emitted light quanta of frequencies ν_3, ν_4, etc., as well as heat. From the conservation of energy, the energy of the initial emitted light quantum cannot be greater than the initial incoming light quantum that produced it. As Einstein noted, "If the photoluminescent substance is not to be regarded as a permanent source of energy, then, according to the energy principle, the energy of a produced energy quantum cannot be greater than that of a producing light quantum; hence we must have...."[52]

$$E_{emitted} \leq E_{incident}$$

$$h\nu_{emitted} \leq h\nu_{incident}$$

$$\nu_{emitted} \leq \nu_{incident}$$

Thus is Stokes' rule obtained in a direct and straightforward manner using the light quanta of energy $h\nu$ introduced by Albert Einstein.

2.2.8 On the Generation of Cathode Rays by Illumination of Solid Bodies

Section 8 applies the concept of light quanta to what is known as the photoelectric effect, providing not only a straightforward explanation of the effect, but also answering some questions that were problematical with the wave theory of light.

When certain metals are illuminated by light they emit electrons. This emission of electrons is known as the photoelectric effect. To escape from a metal, the electrons must possess an extra amount of energy to pass through the surface of the metal, a quantity known as the work function.

In 1905, known features of the photoelectric effect included:[53]

a. When light is shone on a metal, electrons sometimes are emitted, sometimes not.
b. Bright light shining on a metal ejects a greater number of electrons than dim light.
c. The brightness of the light does not affect the kinetic energy of the ejected electrons.
d. For a given metal illuminated by a particular light source, the kinetic energy of the ejected electrons never exceeded a particular maximum.

These aspects of the photoelectric effect were contrary to the predictions of the wave theory of light. From the wave theory, no matter the frequency, if the intensity were strong enough the electrons would receive the energy needed to be emitted from the metal. As the electron intercepted the light waves, each wave front passing the electron would contribute a certain amount of energy to the electron. After a sufficient number of wave fronts had each given some small amount of energy to the electron, the electron would have acquired sufficient energy to escape through the surface of the metal. The energy of the illuminating light would be proportional to the intensity of the light, with more intense light providing more energy to the electrons. Yet experiment showed that more intense light of the same frequency resulted in more emitted electrons, but each electron had the same maximum energy.

Einstein's light quanta picture addressed these concerns with a simple explanation. For the electron to be emitted from the metal it needed to acquire a certain amount of energy. If it received an amount equal to the work function, or more, the electron would be emitted. For a quantum to have a certain amount of energy, or more, it would correspond to a certain frequency, or higher, because for the quantum, $E = h\nu$. Once light hit the metal, if one quantum of light hit an electron, the electron would have enough energy to be emitted through the surface. And, lastly, no matter how intense the light, if it is of the same frequency, each quantum will have the same energy to give to an electron, $E = h\nu$, and all electrons will have the same maximum energy as long as the frequency of the absorbed light remains the same.

For a single light quantum transferring its entire energy to a single electron, the said electron loses a portion of its energy in its travel to the surface of the metal and another portion to escape through the surface of the metal ($P = $ Work Function). The maximum energy of an electron leaving the metal is[54]

$$E_{max} = h\nu - P$$

The physics community did not immediately accept this explanation, partly because there was scant experimental evidence to support it (or to refute it). Lenard's data of 1902 did, though, "provide qualitative evidence for an increase of E_{max} with frequency."[55]

2.2.9   On the Ionization of Gases by Ultraviolet Light

As a third application of the light quantum hypothesis, Einstein considered the ionization of gases by light. Assuming "...one quantum of light energy is used for the ionization of one molecule of gas...it follows that the work of ionization (i.e., the work theoretically required for ionization) of one molecule cannot be greater than the energy of one effective quantum of light absorbed."[56] He then shows the energy of the smallest frequency (largest wavelength) for ionization of air corresponds to the smallest measured ionization potentials for air.[57]

2.3   Discussion and Comments

In this paper, Einstein focused on the entropy of blackbody radiation, calculating an expression for the entropy of the radiation in the Wien limit. From the expression for the entropy of the radiation, he obtained the expression for the probability W, and showed it is of the same form as the probability for an ideal gas of independent particles. This suggested the quanta of the radiation were independent of one another, just as were the atoms of the ideal gas.

In June 1906, Max Laue wrote to Einstein, denying Einstein's assertion that the radiation field was composed of quanta. Laue claimed the radiation was a continuous wave, and the quantum was a characteristic of the emission or absorption of radiation by matter, not a characteristic of the wave itself.[58] This was similar to Planck's 1900 application of the quantum effect only to the "resonators," i.e., to the electrons in the walls of the cavity.[59] In 1907, Planck wrote to Einstein,

> For I do not seek the meaning of the quantum of action (light quantum) in the vacuum, but at the sites of absorption and emission, and assume that the processes in the vacuum are described *exactly* by Maxwell's equations.[60]

In 1913, in his nomination of Einstein for membership in the Prussian Academy of Sciences, Planck reiterates his view regarding Einstein's quanta for the radiation field:

> In sum, one can say that there is hardly one among the great problems in which modern physics is so rich to which Einstein has not made a remarkable contribution. That he may sometimes have missed the target in his specula- tions, as, for instance, in his hypothesis of light-quanta, cannot really be held too much against him, for it is not possible to introduce really new ideas even in the most exact sciences without sometimes taking a risk.[61]

The explanation given by Einstein described not only what was known of the photoelectric effect in 1905, but anticipated what was to come.[62]

a. Using the wave model, a straightforward calculation shows it would require at least five minutes for the electron to acquire sufficient energy to escape the metal. With the light energy concentrated in the light quanta, that energy can be given up in a single collision with an electron, with the electron being emitted immediately. Experimentation later showed the electrons were ejected immedi- ately from the metal.[63]

b. Since the energy of the light quantum is $E = h\nu$, the energy available to transfer to the electron is proportional to ν, so the kinetic energy of the emitted electrons would be proportional to the frequency of the incoming light. This was verified experimentally.[64]

c. If the frequency of the incoming light is too low, there will not be enough energy to allow the electrons to escape from the metal. The frequency at which this occurs is called the "cutoff frequency," and was verified by experiment.[65]

Since 1905, Robert Millikan at the University of Chicago had been working on the photoelectric effect, unaware of Einstein's photoelectric equation. When, in 1912, he became aware of the equation, and its basis, he set about to disprove it. But, by 1915, Millikan had ended up verifying the predictions of the equation: a cutoff frequency and the kinetic energy of the emitted electrons proportional to the frequency of the incident light. Millikan asserted "... its unambiguous experimental verification in spite of its unreasonableness since it seemed to violate everything that we knew about the interference of light."[66] In a 1916 publication, Millikan writes similarly, "...the Einstein equation accurately represents the energy of electron emission under irradiation with light," but continues, saying he considers "...the physical theory upon which the equation is based to be totally untenable."[67]

Even with this evidence, as John Stachel notes, "for almost two decades [after 1905] the argument failed to persuade most physicists of the validity of the light quantum hypothesis."[68] The concept of the quantum was accepted, but as due to the interaction of the radiation field with matter, not as a property of the radiation itself.

Planck's derivation of the expression for ρ_ν rested upon "all the complexions corresponding to a given energy being equally probable."[69] However, as was later shown by Einstein, the validity of Planck's law implies that all complexions cannot be equally probable.[70]

The independence of the quanta was valid only in the Wien (high ν) region. Einstein wanted to extend his results to the full range of frequencies, but the quanta were not independent in the low ν region.[71] Einstein pursued this concern of the non-independence in the low ν region for the next (nearly) 20 years. In 1924, he received a paper from Bose that solved the problem of the counting of complexions, and led to a new method of counting, now called Bose–Einstein statistics (see Section 6.2.6).[72]

It was not until 1923 that the Compton effect gave undeniable evidence of the quanta of radiation (see Section 6.2.5). In 1926, Gilbert Lewis introduced the term "photon" for the light quantum, the term used today.[73]

2.4 Appendices

2.4.1 Entropy and Irreversibility

Consider two bodies identical in every respect, with the exception that one is at a temperature of 60°C and the other is at a temperature of 80°C. See Figure 2.5.

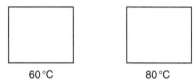

60 °C   80 °C

Fig. 2.5 Two identical bodies with one at 60°C, the other at 80°C.

Our everyday experience tells us that when the two bodies are brought into thermal contact, heat will flow from the warmer body at 80°C to the cooler body at 60°C. Heat will continue flowing between the two bodies until they have reached a common temperature of 70°C.

For a process in which heat flows into a body, the entropy change is defined as $\Delta S = \Delta Q/T$, where ΔS is the change in entropy, ΔQ is the heat added to the body, and T is the temperature of the body on the absolute temperature scale (degrees C plus 273). To analyze the above process in terms of entropy, we consider the case in which all of the heat that leaves the body at 80°C is taken up by the body at 60°C. In the first instant of time a small amount of heat, ΔQ, leaves the 80°C body and enters the 60°C body. For this "short duration of time," the temperature of each body will be assumed to change so little that it can be considered constant. During this short duration the change in entropy of the 80°C body is negative since heat left the body, $\Delta S_{80} = -\Delta Q/(80+273)$, while the 60°C body has a positive change in entropy since heat entered the body, $\Delta S_{60} = +\Delta Q/(60+273)$. In magnitude, $\Delta Q/(80+273)$ is less than $\Delta Q/(60+273)$, i.e., the decrease in entropy of the warmer body is less than the increase in entropy of the cooler body. Thus the entropy of the system increases. After the first short duration of time, the temperatures of the systems will have changed slightly, and over the next short duration of time the entropy change will be calculated using these slightly changed temperatures, continuing the process until the temperatures of the two systems have become equal. In each short duration of time, the entropy of the system increases. The total change in entropy is the sum of these short duration processes. Thus the change in entropy for the entire process is an increase in entropy.

For the reverse process of both bodies starting at 70°C and exchanging heat until one has risen to 80°C and the other has fallen to 60°C, heat will be going into the warmer body and leaving the cooler body. Each step in the calculation of the change in entropy of the system will be the reverse of the previously completed calculation. In this case, the change in the entropy would be negative, and would violate the Second Law of Thermodynamics.

2.4.2   Planck's derivation of $\rho(\nu, T)$

For the resonators in the walls in equilibrium with the radiation field, Planck calculated the average energy density of the radiation in the interval $d\nu$ to be

$$\rho(\nu.T) = \frac{8\pi\nu^2}{c^3} U d\nu$$

where U is the average energy of the resonator.

Planck considered a system consisting of N_1 resonators of frequency ν_1, N_2 resonators of frequency ν_2, etc. A portion E_1 of the total energy of the system is in the N_1 resonators of frequency ν_1, a portion E_2 in the N_2 resonators of frequency ν_2, etc. Assuming E_1 is a sum of P_1 equal discrete energy elements ε_1 (not necessarily one element ε_1 to each resonator; some resonators could have none while others had more than one), E_2 is a sum of P_2 equal discrete energy elements ε_2, ..., Planck calculated the number of complexions, i.e., the number of ways of distributing the P_1 *indistinguishable* equal discrete energy elements ε_1 into the *distinguishable* N_1 resonators of frequency ν_1. Equating the probability of the state of the N_1 resonators to the number of complexions corresponding to that state, and using the Boltzmann relation $S = k\log W$, Planck arrived at the formula,

$$U = \frac{\varepsilon}{e^{\varepsilon/kT} - 1}$$

To satisfy the Wien relation, $\frac{\varepsilon}{kT} = C\frac{\nu}{T}$, or $\varepsilon = Ck\nu = h\nu$, i.e., to satisfy the Wien displacement law, the average energy of the resonator must be an integral multiple of $h\nu$. The resulting expression for the blackbody radiation density is

$$\rho = \frac{A\nu^3}{e^{\frac{h\nu}{kT}} - 1}$$

2.4.3 Wien's Expression for Entropy

The entropy of the radiation in a volume V can be expressed as

$$S = V \int_0^\infty \varphi(\rho, \nu) d\nu$$

where $\varphi(\rho, \nu)$ is the entropy per unit volume per unit frequency range $d\nu$, and ρ is the radiation density (the radiation energy per unit volume per unit frequency range $d\nu$). The total energy of the radiation is

$$E = V \int_0^\infty \rho \, d\nu$$

"In the case of 'blackbody radiation' ρ is such a function of ν that the entropy is a maximum for a given energy, i.e.,"[74]

$$S = Max \Rightarrow \delta S = 0$$

$$\delta \int_0^\infty \varphi(\rho, \nu) d\nu = 0 \qquad (2.1)$$

Since the system has a "given energy," i.e., the energy is constant, $\delta E = 0$.

$$\delta \int_0^\infty \rho \, d\nu = 0 \qquad (2.2)$$

Using Lagrange's method of undetermined multipliers to solve Eq. (2.1) subject to the constraint Eq. (2.2), varying the distribution ρ,

(Eq. 2.1) $- \lambda$(Eq. 2.2)$=0$

The Lagrange multiplier λ can be moved inside the integral as long as it is independent of ν:

$$\int \left(\frac{\partial \varphi}{\partial \rho} - \lambda \right) \delta \rho \, d\nu = 0$$

This can be equal to zero only if the expression in the parenthesis is equal to zero:

$$\frac{\partial \varphi}{\partial \rho} = \lambda$$

Since λ is independent of ν, the expression $\partial \varphi / \partial \rho$ also is independent of ν:

$$dS = d \left[S = V \int_0^\infty \varphi(\rho, \nu) d\nu \right] = V \int_0^\infty (\frac{\partial \varphi}{\partial \rho} d\rho) d\nu = V \frac{\partial \varphi}{\partial \rho} \int_o^\infty d\rho \, d\nu$$

But from $E = V \int_0^\infty \rho \, d\nu$, we have that $dE = V \int_0^\infty d\rho \, d\nu$, and

$$dS = \frac{\partial \varphi}{\partial \rho} dE$$

But from Maxwell's relations in thermodynamics, $\partial E/\partial S = T \Rightarrow dS = dE/T$. Equating the two expressions for dS,

$$\frac{\partial \varphi}{\partial \rho} = \frac{1}{T}$$

2.5   Notes

1. Einstein, Albert, On a Heuristic Point of View Concerning the Production and Transformation of Light, *Annalen der Physik* 322 (6), (1905), pp. 132–148; Stachel, John, ed., *The Collected Papers of Albert Einstein*, Volume 2 [CPAE2], Princeton University Press, Princeton, NJ, 1989, pp. 150–166; English translation by Anna Beck [CPAE2 ET], pp. 86–103. The original text contains a number of editorial comments and introductory comments (pp. 134–148) that are quite informative.
2. Atkins, P. W., *The Second Law*, Scientific American Books, New York, 1984, p. 32. This phrasing is slightly modified from that of Atkins.
3. Boltzmann, Ludwig, *Vorlesungen über Gastheorie. I. Teil: Theorie der Gase mit einatomigen Molekülen, deren Dimensionen gegen die mittlere Weglänge verschwinden* (Barth, Leipzig, 1896), p. 32. Citation from Kuhn, Thomas S., *Black-Body Theory and the Quantum Discontinuity, 1894–1912*, Oxford University Press, Oxford, 1978, (paperback edition), p. 41.
4. Boltzmann, Ludwig, *Gastheorie I*, p. 32. Citation from Kuhn, Thomas S., *The Quantum Discontinuity*, pp. 38–47, 58–61.
5. Kuhn, Thomas S., *The Quantum Discontinuity*, pp. 48–49.
6. Kirchhoff, Gustav, *Ann. Phys. Chem.* 109, 275 (1860); Citation from Pais, Abraham, *Subtle is the Lord*, Oxford University Press, Oxford, 1982, pp. 365, 387.
7. Pais, Abraham, *Subtle is the Lord*, p. 365.
8. Pais, Abraham, *Subtle is the Lord*, p. 365.
9. Cushing, James T., *Philosophical Concepts in Physics*, Cambridge University Press, Cambridge, 1998, p. 274.
10. The wavelength, λ, of a wave and its frequency, ν, are related as $\lambda\nu = c$ = speed of light for radiation.
11. Cushing, James T., *Philosophical Concepts*, p. 273.
12. Eisberg, Robert, and Resnick, Robert, *Quantum Physics of Atoms, Molecules, Solids, Nuclei, and Particles*, 2nd edition, John Wiley and Sons, New York, 1985, p. 13.
13. Eisberg, Robert, and Resnick, Robert, *Quantum Physics*, p. 13.
14. Wien, Wilhelm, PAW, 1893, p. 55, and Wien, Wilhelm, *AdP.*, 58, 662 (1896), Citations from Pais, Abraham, *Subtle is the Lord*, pp. 365, 366.
15. Kuhn, Thomas S., *The Quantum Discontinuity*, p. 96, endnote 10 on page 281.
16. Einstein, Albert, Heuristic View of Light; [CPAE2, p. 154; CPAE2 ET, p. 90].
17. Cushing, James T., *Philosophical Concepts*, pp. 274, 275. Eisberg, Robert, and Resnick, Robert, *Quantum Physics of Atoms, Molecules, Solids, Nuclei, and Particles*, 2nd edition, John Wiley and Sons, New York, 1985, p. 13.
18. Pais, Abraham, *Subtle is the Lord*, p. 368.
19. Cushing, James T., *Philosophical Concepts*, p. 277.

20. Whittaker, Edmund, *A History of the Theories of Aether and Electricity II: The Modern Theories,* Volume 7, Tomash Publishers, American Institute of Physics, 1954, p. 80; Kuhn, Thomas S., *The Quantum Discontinuity,* pp. 78–87.

21. Whittaker, Edmund, *A History of the Theories of Aether and Electricity II,* p. 80; Kuhn, Thomas S., *The Quantum Discontinuity,* pp. 6–7.

22. Whittaker, Edmund, *A History of the Theories of Aether and Electricity II,* p. 79. See also Cushing, James T., *Philosophical Concepts,* p. 277.

23. Kuhn, Thomas S., *The Quantum Discontinuity,* p. 99.

24. Einstein, Albert, Heuristic View of Light; [CPAE2, pp. 150–166; CPAE2 ET, pp. 86–103].

25. Einstein, Albert, Heuristic View of Light; [CPAE2, p. 151; CPAE2 ET, p. 87].

26. Einstein, Albert, Heuristic View of Light; [CPAE2, p. 151; CPAE2 ET, p. 87].

27. Stachel, John, editor, *Einstein's Miraculous Year,* Princeton University Press, Princeton, NJ, 1998, p. 173.

28. Einstein, Albert, Heuristic View of Light; [CPAE2, p. 150; CPAE2 ET, p. 86].

29. Einstein, Albert, Heuristic View of Light; CPAE2, p. 150; CPAE2 ET, p. 86.

30. Einstein, Albert, Heuristic View of Light; [CPAE2, p. 150; CPAE2 ET, p. 86].

31. Einstein, Albert, Heuristic View of Light; [CPAE2, pp. 150–151; CPAE2 ET, p. 86].

32. Kuhn, Thomas S., *The Quantum Discontinuity,* p. 8.

33. Stachel, John, ed., [CPAE2, p. 139].

34. Stachel, John, ed., [CPAE2, p. 46].

35. Howard, Don, and Stachel, John, editors, *Einstein: The Formative Years, 1879–1909,* Birkhäuser, Boston, MA, 2000, pp. 107–109.

36. Stachel, John, ed., [CPAE2, pp. 139, 155; CPAE2 ET, p. 91].

37. Einstein, Albert, Heuristic View of Light; [CPAE2, p. 154; CPAE2 ET, pp. 88, 89].

38. Einstein, Albert, Heuristic View of Light; [CPAE2, p. 152; CPAE2 ET, p. 87].

39. Boltzmann, L., *Wiener Ber.* 63, 679 (1871); Reprinted in *Wissenschaftliche Abhandlugen von L. Boltzmann* (F. Hasenohrl, ed.) Volume 1, p. 259; Reprinted by Chelsea, New York, 1968; Citation from Pais, Abraham, *Subtle is the Lord,* pp. 392, 400.

40. Einstein, Albert, Heuristic View of Light; [CPAE2, p. 154; CPAE2 ET, p. 90].

41. Pais, Abraham, *Subtle is the Lord,* p. 367.

42. Einstein, Albert, Heuristic View of Light; [CPAE2, p. 154; CPAE2 ET, p. 90]. Citation to Planck, *Annalen der Physik,* 4 (1901), p. 561.

43. I thank Don Howard for bringing to my attention that, in the period around 1900, the term "electron" referred to a hypothetical unit of electricity, which today could be an electron or an ion, or another unit of electricity. It was not necessarily associated with a particle with mass. It could be a massless entity, with a "mass" arising from its motion through the aether.

44. Einstein, Albert, Heuristic View of Light; [CPAE2, p. 152; CPAE2 ET, p. 87].

45. Einstein, Albert, Heuristic View of Light; [CPAE2, p. 155; CPAE2 ET, p. 91].
46. Einstein, Albert, Heuristic View of Light; [CPAE2, p. 156; CPAE2 ET, pp. 92–93].
47. I thank Don Howard for bringing this to my attention.
48. Cushing, James T., *Philosophical Concepts*, p. 277.
49. Einstein, Albert, Heuristic View of Light; [CPAE2, p. 157; CPAE2 ET, p. 94].
50. Einstein, Albert, Heuristic View of Light; [CPAE2, p. 160; CPAE2 ET, p. 96].
51. Einstein, Albert, Heuristic View of Light; [CPAE2, p. 161; CPAE2 ET, p. 97].
52. Einstein, Albert, Heuristic View of Light; [CPAE2, p. 162; CPAE2 ET, p. 98].
53. Rigden, John S., *Einstein 1905: The Standard of Greatness*, Harvard University Press, Cambridge, MA, 2005, p. 32.
54. Einstein, Albert, Heuristic View of Light; [CPAE2, p. 164; CPAE2 ET, p. 100].
55. Stachel, John, ed., [CPAE2, p. 142].
56. Einstein, Albert, Heuristic View of Light; [CPAE2, pp. 165–166; CPAE2 ET, p. 102].
57. Einstein, Albert, Heuristic View of Light; [CPAE2, p. 166; CPAE2 ET, p. 102].
58. Rigden, John S., *Einstein 1905*, p.20.
59. Rigden, John S., *Einstein 1905*, p.25.
60. Klein, Martin J., Kox, A. J., and Schulmann, Robert, editors., *The Collected Papers of Albert Einstein*, Volume 5 [CPAE5], Princeton University Press, Princeton, NJ, 1989, pp. 150–166; English translation by Anna Beck [CPAE5 ET], p. 31. Letter from Max Planck to Albert Einstein, 6 July, 1907.
61. Kirsten, G., and Körber, H., *Physiker über Physiker*, p. 201, Akademie Verlag, Berlin, 1975. Citation from Pais, Abraham, *Subtle is the Lord*, pp. 382, 387.
62. Rigden, John S., *Einstein 1905*, pp. 33–34.
63. Rigden, John S., *Einstein 1905*, p. 33.
64. Rigden, John S., *Einstein 1905*, p. 36.
65. Rigden, John S., *Einstein 1905*, p. 35.
66. Millikan, R. A., *Reviews of Modern Physics*, Volume 21, 1949, p. 343. Citation from Fölsing, Albrecht, *Albert Einstein*, Viking (Penguin Group), New York, 1997, p. 148.
67. Millikan, R. A., Quantenbeziehungen beim photoelektrischen Effekt, in *PhZ.*, 17 (1916), pp. 217–221. Citation from Fölsing, Albrecht, *Albert Einstein*, p. 149.
68. Stachel, John, ed., [CPAE2, p. 142].
69. Stachel, John, *Einstein's Miraculous Year*, p. 171.
70. Stachel, John, *Einstein's Miraculous Year*, p. 171.
71. Howard, Don, and Stachel, John, *The Formative Years*, p. 244.
72. Howard, Don, and Stachel, John, *The Formative Years*, p. 244.
73. Rigden, John S., *Einstein 1905*, p. 38.
74. Einstein, Albert, Heuristic View of Light; [CPAE2, p. 156; CPAE2 ET, p. 92].

2.6 Bibliography

Atkins, P. W., *The Second Law*, Scientific American Books, New York, 1984.

Boltzmann, Ludwig, *Vorlesungen über Gastheorie. I. Teil: Theorie der Gase mit einatomigen Molekülen, deren Dimensionen gegen die mittlere Weglänge verschwinden*, Barth, Leipzig, 1896. (Citation from Kuhn, Thomas S., *Black-Body Theory and the Quantum Discontinuity.*)

Boltzmann, L., *Wiener Ber.* 63, 679, (1871); reprinted in *Wissenschaftliche Abhandlugen von L. Boltzmann* (F. Hasenohrl, ed.), Volume 1, p. 259; Reprinted by Chelsea, New York, 1968. (Citation in Pais, Abraham, *Subtle is the Lord.*)

Cushing, James T., *Philosophical Concepts in Physics*, Cambridge University Press, Cambridge, 1998.

Einstein, Albert, On a Heuristic Point of View Concerning the Production and Transformation of Light, *Annalen der Physik* 17 (1905), pp. 132–148 [CPAE2, pp. 150–166; CPAE2 ET, pp. 86–103].

Eisberg, Robert, and Resnick, Robert, *Quantum Physics*, Second edition, John Wiley and Sons, New York, 1985.

Fölsing, Albrecht, *Albert Einstein*, Viking, New York, 1997.

Howard, Don, and Stachel, John, editors, *Einstein: The Formative Years, 1879-1909*, Birkhäuser, Boston, MA, 2000.

Kirchhoff, Gustav, *Ann. Phys. Chem.* 109 (1860). (Citation from Pais, Abraham, *Subtle is the Lord*, pp. 365, 387.)

Kirsten, G, and Körber, H., *Physiker über Physiker*, Akademie Verlag, Berlin, 1975. (Citation from Pais, Abraham, *Subtle is the Lord.*)

Klein, Martin J., Kox, A. J., and Schulmann, Robert, editors, *The Collected Papers of Albert Einstein*, Volume 5 [CPAE5], Princeton University Press, Princeton, NJ, 1989; English translation by Anna Beck [CPAE5 ET].

Kuhn, Thomas S., *Black-Body Theory and the Quantum Discontinuity, 1894-1912*, Oxford University Press, Oxford, 1978.

Millikan, R. A., *Reviews of Modern Physics*, Volume 21, 1949, p. 343. (Citation from Fölsing, Albrecht, *Albert Einstein.*)

Millikan, R. A., Quantenbeziehungen beim photoelektrischen Effekt, *PhZ.* 17 (1916), pp. 217–221. (Citation from Fölsing, Albrecht, *Albert Einstein.*)

Pais, Abraham, *Subtle is the Lord*, Oxford University Press, Oxford, 1982.

Rigden, John S., *Einstein 1905*, Harvard University Press, Cambridge, MA, 2005.

Rigden, John S., *Einstein 1905: The Standard of Greatness*, Harvard University Press, Cambridge, MA, 2005.

Stachel, John, editor, *The Collected Papers of Albert Einstein*, Volume 2 [CPAE2], Princeton University Press, Princeton, NJ, 1989; English translation by Anna Beck [CPAE2 ET].

Stachel, John, editor, *Einstein's Miraculous Year*, Princeton University Press, Princeton, NJ, 1998, p. 173.

Whittaker, Edmund, *A History of the Theories of Aether and Electricity II: The Modern Theories*, Volume 7, Tomash Publishers, American Institute of Physics, 1954.

Wien, Wilhelm, *PAW* (1893), p. 55. (Citation from Pais, Abraham, *Subtle is the Lord*, pp. 365, 366.)

Wien, Wilhelm, *AdP.* 58, 662 (1896). (Citation from Pais, Abraham, *Subtle is the Lord*, pp. 365, 366.)

The Atom and Brownian Motion

<table>
<tr><td>**3**</td><td></td></tr>
</table>

3.1 Historical Background   56

3.2 Albert Einstein's
Paper, "A New
Determination of
Molecular Dimensions"   62

3.3 Albert Einstein's
Paper, "On the
Movement of Small
Particles Suspended in
Stationary Liquids
Required by the
Molecular-Kinetic
Theory of Heat"   70

3.4 Discussion and Comments   76

3.5 Appendices   78

3.6 Notes   98

3.7 Bibliography   103

"A New Determination of Molecular Dimensions"[1]

30 April, 1905, dissertation submitted to the University of Zurich

"On the Movement of Small Particles Suspended in Stationary Liquids Required by the Molecular-Kinetic Theory of Heat"[2]

Received on 11 May, 1905, by *Annalen der Physik*

3.1 Historical Background

By 1905 a number of experiments were producing results for the dimension of an atom or molecule that were "in more or less satisfactory agreement with each other."[3] Many of the experiments giving good results were based on the kinetic theory of gases. Others, based on the theory of fluids, did not give as consistent results. In his thesis, Albert Einstein devised a method based on fluids that gave results consistent with, and as precise as, the results based on the kinetic theory of gases.[4] Beyond the obvious goal of determining the size of molecules, he was looking for evidence for the atomic hypothesis beyond that of the kinetic theory of gases.[5] Of the three worldviews of 1905, the mechanical, the electromagnetic, and the energetics, the energetics worldview was adamantly anti-atomistic. Evidence for the reality of atoms not only would verify the mechanical worldview, it also would refute (falsify) the energetics worldview.

The second of these papers, "On the Movement of Small Particles Suspended in Stationary Liquids Required by the Molecular-Kinetic Theory of Heat," sometimes is referred to as the Brownian motion paper, giving the impression the primary goal was to explain this motion. Speaking specifically of Brownian motion, in the second paper Einstein writes, "It is possible that the motions to be discussed here are identical with the so-called 'Brownian molecular motion'; however, the data available to me on the latter are so imprecise that I could not form a definite opinion on this matter."[6]

In these papers, Albert Einstein addresses the scientific idea of the atom, including the work of Boltzmann and the philosophy of Ernst

Mach; Robert Brown's work that became known as Brownian motion; and the three worldviews known as the mechanical, the electromagnetic, and the energetic worldviews.

3.1.1 The Atom

"There was one question of eminently epistemological (or rather metaphysical) character that occupied the minds of [the] early philosophers. 'What is the immutable principle that underlies changing phenomena – what is the single root of the multiplicity of things in the physical world?'"[7] Heraclitus (c. 480 BC) believed that "*to be* was to change." Parmenides, a contemporary of Heraclitus, held the opposite view, that "permanency is real and change is only an illusion."[8]

Democritus, some 70 years after the time of Heraclitus and Parmenides, was strongly inclined to the view of Parmenides, yet saw that the science of nature was a process of change. Democritus asserted that "being is not *one*, but is divided into many 'beings,' each permanent and indivisible. These 'beings' he called atoms. Democritus also asserted that *nothing* does exist – *nothing* implying what we would call a void. Thus, *to be*, for him, meant to be *atom* or to be *void*. According to this view, change consists only in the rearrangement of eternal and unchanging atoms. This part of his argument agreed with Parmenides – the permanence of atoms is real and the change we see is an illusion – for the permanent atoms always [sic] remain."[9] Democritus believed the atoms to be "so small as to be imperceptible to us, and to take all kinds of shapes and all kinds of forms and differences in size. Out of them, like out of elements, he now lets combine and originate the visible and perceptible bodies."[10] Epicurus described the atoms as "indivisible and unchangeable, ... [they] vary indefinitely in their shapes; ... [they] are in continual motion through all eternity.... Of all this there is no beginning, since both atoms and void exist from everlasting."[11]

To Aristotle, the concept of the void was inconceivable: nonexistence cannot be a part of existence.[12] To Aristotle all of space was permeated by the qualities of hot, cold, wet, and dry. These qualities were grouped into the basic elements of matter: air (hot and wet), earth (cold and dry), fire (hot and dry), and water (cold and wet). See Figure 3.1.

Earth	is	dry and cold		
Water	is		cold and moist	
Air	is			moist and hot
Fire	is			hot and dry

To Aristotle, "the elements are not permanent but can change into one another. This is especially simple if the two share a quality: Water cannot easily change to Fire but it can easily, by evaporation, become Air."[13] The heavens were composed of a fifth element, the quintessence.

For the next 2000 years, the ideas of Aristotle predominated, with a revival of an atomic theory in the seventeenth century. The resurgence of atomic theory owed much of its revival to the work of Pierre Gassendi, a French priest "thoroughly orthodox in his religious beliefs."[14] There had

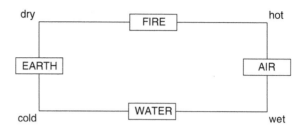

Fig. 3.1 The four elements of Aristotle.

been theological objections to the atom being uncreated and eternal, attributing to it properties belonging only to God. In *Observations on the Tenth Book of Diogenes Laertius* (1649), Gassendi "removed the...philosophy that affronted theologians and explained, clearly and with no use of scholastic vocabulary, the physical theory of Democritus, Epicurus, and Lucretius. He explained that there is no reason to insist that atoms are uncreated, eternal, or infinitely numerous...[In England] in the tolerant atmosphere of the Stuart Monarchy, atoms took root and flourished."[15]

In 1662, Robert Boyle published the results of his experiments on the compression and expansion of a given quantity of air as the pressure is changed.[16] From these experiments came Boyle's law, which states that, at a constant temperature, the product of the volume of the enclosed air and the (absolute) pressure of the air remains constant ($pV = $ constant). Though an atomist, Boyle did not explain this law in terms of atoms. In 1687, Newton explained Boyle's law in terms of "gas atoms [that] were static but mutually repulsive with a force that varied as the distance."[17] In 1738, Daniel Bernoulli explained Boyle's law in terms of the kinetic theory of atoms.[18]

By the mid 1700s the atomic nature of matter was still highly speculative. As Boorse and Motz comment in *The World of the Atom*:

It is a sobering thought that two thousand years of atomic speculation produced no mind able to formulate atomic theory in questions simple enough for direct experimental answers. Even Newton's derivation of Boyle's law on the assumption of...[static particles]...and, some half century later, Daniel Bernoulli's brilliant deduction of the same law on a kinetic particle hypothesis opened no doors to the world of the atom. It would be logical to expect that if the greatest minds of two millennia had found no way to question nature in atomic terms, only a demigod might be expected to see where mortals were blind.[19]

John Dalton showed Newton's particle concept led naturally to the law of partial pressures, and explained the law of definite proportions as "the combination of an atom of one substance with that of another. Thus was the atomic theory at last placed on an experimental basis."[20] Dalton's papers on the atomic hypothesis were published in 1805. "Gay-Lussac's experiments, published in 1809, demonstrated that gases combined in equal volumes, or in small volume ratios. These experiments were explained in 1811 in terms of Dalton's theory by...Amedeo

Avogadro."[21] Avogadro added to Dalton's hypothesis the distinction between what today we call atoms and molecules. From these ideas Avogadro also deduced that equal volumes of different gases at the same temperature and pressure contain the same number of molecules. "Even though many chemists...did not accept the reality of atoms, nevertheless the atomic theory became widely accepted as a potent tool in understanding chemical reactions."[22]

In 1860, James Clerk Maxwell published the paper "Illustrations of the Dynamical Theory of Gases"[23] on the kinetic theory of gases (the mechanical model that a gas is composed of small hard particles in constant chaotic motion), obtaining an expression for the number of particles in the gas with velocities that lie between the values v_x and $v_x + dv_x$. Today, this expression is known as the Maxwell velocity distribution function.[24]

Ludwig Boltzmann showed that the Maxwell velocity distribution is the only possible equilibrium distribution function.[25] In the 1860s, Boltzmann introduced an area of physics known as statistical mechanics, forging "a link between the properties of matter in bulk...and the behavior of matter's individual particles, its atoms.... Boltzmann perceived that identifying the cooperation between atoms which showed itself to the observer as the properties of bulk matter would take him to the innermost workings of Nature.... [H]e saw further into the workings of the world than most of his contemporaries, and he began to discover the deep structure of change; furthermore he did all this before the existence of atoms was generally accepted."[26]

Boltzmann developed a statistical theory of entropy.[27] He saw the Second Law of Thermodynamics as macroscopic, and agreed with Maxwell's statement that, "The truth of the second law is therefore a statistical, not a mathematical, truth, for it depends upon the fact that the bodies we deal with consist of millions of molecules."[28]

In the late 1800s, Ernst Mach represented a widely held view that science is based on observation, and that it was necessary to reject anything that transcended observation or sense experience.[29] And atoms had never been observed. For all of the explanatory success that the atomic theory provided, it remained a topic of much controversy. Reasons for the continuing controversy included the following:

1. Beginning with the atomic assumption, the specific heat of a system can be calculated. Experiment had shown this value to be true in general. However, measurements were beginning to show anomalous behavior at low temperatures (See Section 6.2.1), casting some doubt about the underlying assumption of an atomic structure.

2. The prevailing view was that, philosophically, science was grounded in the "directly observable." Atoms were not directly observable and must, on philosophical grounds, be excluded.

3. With the success of Newton's mechanics providing a deterministic understanding of nature, scientists were not inclined to leave determinism for probability and statistics.

4. The Second Law of Thermodynamics states that the entropy of an isolated system will increase or remain constant. The entropy will not decrease. The statistical interpretation of entropy says a system will move to a state of higher, or equal, probability, but the statistical interpretation also allows the possibility of moving to a state of lower probability. This would be a violation of the Second Law and, therefore, cannot occur. It needs to be noted that statistical arguments were a new approach and were not familiar to most physicists at this time.

5. The atomic theory assumed the deterministic Newton's laws were operational at the atomic level. Since Newton's laws are time-reversible, all interactions at the atomic level will be time-reversible. In the words of Wilhelm Ostwald, a strong opponent of the reality of atoms, from his address at the 1895 meeting of the Deutsche Gesellschaft für Naturforscher und Ärzte:

The proposition that all natural phenomena can ultimately be reduced to mechanical ones cannot even be taken as a useful working hypothesis: it is simply a mistake. This mistake is most clearly revealed by the following fact. All the equations of mechanics have the property that they admit of sign inversion in the temporal quantities. That is to say, theoretically perfectly mechanical processes can develop equally well forward and backward [in time]. Thus, in a purely mechanical world there would not be a before and an after as we have in our world; the tree would become a shoot and a seed again, ... The actual irreversibility of natural phenomena thus proves the existence of processes that cannot be described by mechanical equations; and with this the verdict on scientific materialism [atoms] is settled.[30]

3.1.2   Brownian Motion

In 1827, Robert Brown, a botanist, discovered what he termed "active molecules" while examining the behavior of pollen grains suspended in water. Some of the pollen grains were in "active, chaotic motion." Initially Brown attributed the motion to "the vitality of the pollen ... [but] soon found that all small particles under the same conditions behaved the same way."[31] Brown extended his observations to other plants with the same result. Eventually, he "inferred that [this motion] was not limited to organic bodies ...."[32] Brown repeated the observations with "minute fragments of window-glass, ... [with] rocks of all ages, ... [even with] a fragment of the Sphinx being one of the specimens examined."[33] After repeated experiments, Brown was unable to determine the cause of the active, chaotic motion of small particles suspended in water. This active, chaotic motion of the small suspended particles became known as Brownian motion.

3.1.3   The Worldview in 1900

By the late 1800s, the mechanics of Newton was well entrenched, and scientists were attempting to reduce their understanding of nature to a mechanical foundation. In particular were the attempts to provide a mechanical basis for electromagnetic phenomena, i.e., to provide a mechanical underpinning for Maxwell's equations. By 1900, the electromagnetism of Maxwell was gaining strength, and Lorentz was developing his theory of electrons. Because of the stunning success of the electromagnetic theory (i.e., Maxwell's equations), things were beginning to be reversed – scientists were now contemplating an electromagnetic foundation for mechanics, in particular, an electromagnetic explanation of the mass of the electron.[34] Based on the success of thermodynamics in describing the world in terms of energy, a third worldview, the energetics worldview, looked at the underlying structure of the world to be forms of energy and energy transformations.

3.1.3.1   The Mechanical Worldview

From the late 1600s through the late 1800s, the mechanics of Newton had shown itself ever more powerful and ever more inclusive for understanding the physical world in which we live, from unifying the heavens and earth under the one system (Newton's three laws and the law of gravitation), to predicting the existence of new planets, to providing an understanding of thermodynamics in terms of atoms that follow Newton's laws of motion. Regarding this mechanical worldview, Cushing writes:

Boyle certainly applied this line of argument to support an atomistic and mechanical natural philosophy.... [H]is influence...became evident in the natural philosophy of Newton that dominated science for nearly two hundred years. A key element of reality that emerged from this Newtonian tradition was a mechanistic view with atoms possessing simple quantitative properties (mass, location, etc.) and interacting causally in accord with definite mathematical laws and equations of motion. This mechanistic philosophy...replaced the organismic philosophy of the Middle Ages...Newtonian physics was so successful in the practical sphere that it engendered a philosophical outlook that claimed that the laws of classical physics were necessarily or a priori true, as in some sense held by the philosopher Immanuel Kant (1724–1804).[35]

Regarding the electrons of Lorentz, Boorse and Motz note that even though electrons were an entirely new microscopic entity to explain matter and its interaction with electromagnetic fields, the electrons were assumed to follow the same laws of mechanics as ordinary macroscopic matter:

Aside from their being charged spheres, the electrons of Lorentz were to be treated like any other particles, and the same mechanical laws were to apply to them as apply to ordinary bits of matter. In particular, they were supposed to move, when acted upon by a force, according to Newton's laws of motion: if a force were applied to an electron, it was supposed to experience an acceleration proportional to the force and inversely proportional to its mass.[36]

3.1.3.2   The Electromagnetic Worldview

By 1900, because of the success of the electromagnetic theory, scientists were contemplating an electromagnetic foundation for mechanics.

Commenting on the success of Lorentz's theory, Miller says:

By 1900 attempts to deduce the laws of electromagnetism from increasingly complex mechanical models of the ether (i.e. a mechanical world-picture) paled before the successes of Lorentz's electromagnetic theory which took the equations of electromagnetism as axiomatic.[37]

In the same vein, Cushing writes:

Abraham's basic idea was to provide an electromagnetic foundation for all mechanics. This was essentially a complete reversal from an earlier tendency on the part of theorists, such as Maxwell, to provide a mechanical basis for electromagnetic phenomena via mechanical models of the aether.[38]

And later, again in *Philosophical Concepts in Physics*, Cushing writes:

By the end of the nineteenth century, classical physics was encountering difficulties of its own (for example, the aether . . .), so that some began to question the absolute necessity of the attendant mechanist philosophy.[39]

3.1.3.3   The Energetics Worldview

One of the foremost critics of the mechanical worldview was Wilhelm Ostwald (the winner of the 1909 Nobel Prize for chemistry). His outlook, typical of others, was that what was needed were general laws, similar to those of thermodynamics in which detailed knowledge of substances is unnecessary (for example, heat engines and the Second Law of Thermodynamics). It should not be necessary to introduce hypothetical quantities such as atoms and molecules.[40]

Based on the success of thermodynamics in describing the world in terms of energy, the energetics worldview looked at the underlying structure of the world to be forms of energy and energy transformations. Thus assumptions on the constitution of matter were unnecessary. Like the positivists, the energeticists were adamant in their demand that physical theories contain no metaphysical quantities, and so both groups were anti-atomistic.[41]

3.2   Albert Einstein's Paper, "A New Determination of Molecular Dimensions"[42]

Likening the molecules of a substance dissolved in a dilute solution to a "solid body suspended in a solvent,"[43] Einstein showed the size of the molecules could be determined from the viscosity of the solution and the diffusion rate of the solute in the solvent. Assuming the solute molecule is large compared to the solvent molecules, Einstein proceeded to treat the solvent as a continuous, homogeneous fluid, subject to the equations of hydrodynamics, while treating the solute molecule as a discrete particle. The "solid body" representing the solute molecule is taken to be a sphere.

To estimate the size of a molecule, Einstein derived two equations, one for the viscosity of a fluid with molecules distributed randomly in it (see Sections 3.2.1 and 3.2.2), the other for the coefficient of diffusion (see Section 3.2.4). Each of these equations contained the molecular radius P and Avogadro's number N (Einstein denoted Avogadro's number as N, not as N_A. We will use Einstein's notation.). Using data from standard tables for the viscosity and for the coefficient of diffusion, Einstein solved these two equations for the molecular radius P and Avogadro's number N (see Section 3.2.5).

To obtain the expression for the viscosity of a fluid, Einstein calculated first the influence of a single sphere suspended in a fluid (Section 3.2.1), then used this result to obtain an expression for the viscosity of a liquid containing many such spheres (Section 3.2.2). To obtain the expression for the coefficient of diffusion, he introduced a force K that will give the molecule a velocity ω when in balance with the viscous force, then set K equal to the pressure gradient in the fluid. Eliminating the force K from his equations Einstein obtained the expression for the coefficient of diffusion. The force K was introduced as a mathematical ploy, as it was introduced only for the purposes of calculation and could be eliminated before the final result was obtained (Section 3.2.4).

3.2.1 On the Influence on the Motion of a Liquid Exercised by a Very Small Sphere Suspended in It

Consider a large container of volume V, filled with an incompressible homogeneous liquid with coefficient of viscosity k. The velocity components of the fluid (u, v, w) are a given function of the location in the container, (x, y, z), and of the time t. Consider an arbitrary point x_0, y_0, z_0 within the container. Around this point, the velocity components (u, v, w) can be expanded in a Taylor series in terms of $x - x_0$, $y - y_0$, and $z - z_0$. Further, consider a region G around the point x_0, y_0, z_0, so small that only the linear terms in the Taylor series expansion need be considered within G. See Figure 3.2.

Einstein describes the motion of the fluid within G as a superposition of three components:

1. Translation of all of the particles in the liquid without a change in their relative positions.
2. Rotation of all of the particles in the liquid without a change in their relative positions.
3. A dilatational motion in three mutually perpendicular directions.

Within G insert a rigid sphere centered on the point (x_0, y_0, z_0). The dimension of the sphere is much less than the dimension of G which is, in turn, much less than the dimension of the volume V. It is assumed "the

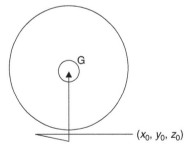

Fig. 3.2 The region G in the volume V. The region G is centered on the point x_0, y_0, z_0.

velocity components of a surface element of the sphere coincide with the corresponding velocity components of the adjacent liquid particles."[44] Since, in components 1 and 2 of the velocity, the "liquid moves like a rigid body," the presence of the sphere will not alter the motion of the neighboring liquid particles. However, the presence of the sphere does alter the third component, the dilatational motion. It is on the dilatational motion that Einstein sets his focus.

Introduce the coordinates (ξ, η, ζ) (the Greek letters xi, eta, zeta) as

$$\xi = x - x_0$$

$$\eta = y - y_0$$

$$\zeta = z - z_0$$

When the sphere is not present, within the region G the velocity components are:

$$u_0 = A\xi$$
$$v_0 = B\eta \tag{3.1}$$
$$w_0 = C\zeta$$

with, from hydrodynamics, because the liquid is incompressible,[45]

$$A + B + C = 0 \tag{3.2}$$

Introduce now a rigid sphere, centered on the point $(x_0,\ y_0,\ z_0)$. Denote the radius of the sphere as P (the Greek letter rho). "Due to the symmetry of the motion of the liquid, it is clear that the sphere can perform neither a translation nor a rotation during the motion considered, and we obtain the boundary conditions."[46]

$u = v = w = 0$ when $\rho = P$ ($u = v = w = 0$ relative to the surface of the sphere) where

$$\rho = \sqrt{\xi^2 + \eta^2 + \zeta^2}$$

This is *boundary condition 1* that any solution for the velocity components u, v, and w must satisfy.

With the sphere present, the velocity components of the liquid can be written as:

$$u = A\xi + u_1$$
$$v = B\eta + v_1 \tag{3.3}$$
$$w = C\zeta + w_1$$

where u_1, v_1, and w_1 are the "alterations" to the liquid velocity due to the presence of the sphere. At "infinity," these alterations must vanish as Eq. (3.3) must reduce to Eq. (3.1). This is *boundary condition 2* that any solution for the velocity components u, v, and w must satisfy.

The velocity components must satisfy the hydrodynamic equations:

$$\frac{\partial p}{\partial \xi} = k \left[\frac{\partial^2 u}{\partial \xi^2} + \frac{\partial^2 u}{\partial \eta^2} + \frac{\partial^2 u}{\partial \zeta^2} \right]$$

$$\frac{\partial p}{\partial \eta} = k \left[\frac{\partial^2 v}{\partial \xi^2} + \frac{\partial^2 v}{\partial \eta^2} + \frac{\partial^2 v}{\partial \zeta^2} \right]$$

$$\frac{\partial p}{\partial \zeta} = k \left[\frac{\partial^2 w}{\partial \xi^2} + \frac{\partial^2 w}{\partial \eta^2} + \frac{\partial^2 w}{\partial \zeta^2} \right]$$

$$\frac{\partial u}{\partial \xi} + \frac{\partial v}{\partial \eta} + \frac{\partial w}{\partial \zeta} = 0 \qquad (3.4)$$

where k is the viscosity of the liquid and p is the hydrostatic pressure within the liquid. The first three equations are the components of the Navier–Stokes equation that describe the conservation of momentum for a viscous incompressible fluid, with no external forces, and a low velocity. The fourth equation is the equation of continuity for the liquid.[47]

Since $u_0 = A\xi, v_0 = B\eta, w_0 = C\zeta$ are solutions to Eqs.(3.4), since Eqs. (3.4) are linear in u, v, w, and since $u = u_0 + u_1, v = v_0 + v_1, w = w_0 + w_1$, it necessarily follows that u_1, v_1, w_1 also must satisfy Eqs. (3.4). Solving the hydrodynamic equations for the velocity components, Einstein obtained for u the expression (see Appendix 3.5.1 for the derivation)[48]

$$u = A\xi - \frac{5}{2}\frac{P^3}{\rho^5}\xi\left(A\xi^2 + B\eta^2 + C\zeta^2\right) + \frac{5}{2}\frac{P^5}{\rho^7}\xi\left(A\xi^2 + B\eta^2 + C\zeta^2\right) - \frac{P^5}{\rho^5}A\xi$$

$$(3.5)$$

By symmetry, similar solutions exist for v and w.

For $\rho = P$ it can be seen this equation reduces to zero, satisfying boundary condition 1. For infinitely large values of ρ, it can be seen that Eq. (3.5) reduces to Eq. (3.1), satisfying boundary condition 2. This demonstrates that at least one solution exists that satisfies the hydrodynamical equations and satisfies the boundary conditions. Einstein then showed that the solution to the hydrodynamical equations satisfying the boundary conditions was unique, i.e. this is the only solution.[49]

Construct a sphere of radius R around the point (x_0, y_0, z_0), where $R \gg P$ (P is the radius of the spherical body). Denote by W the energy per unit time converted into heat within the sphere. W is equal to the work done on the liquid in the sphere per unit time. "If X_n, Y_n, Z_n denote the components of the pressure exerted on the surface of the sphere of radius R, we have"[50] (ds is the element of surface area)

$$W = \int (X_n u + Y_n v + Z_n w) ds$$

After a series of approximations neglecting higher order terms in (P/R), Einstein obtains for W. (See Appendix 3.5.2 for details of this calculation. N.B. there were some numerical mistakes in Einstein's paper. Einstein's original expressions will be given, with the corrected expression noted in the Endnotes. Calculations of these corrected expressions are given in Appendices 3.5.2, 3.5.3, and 3.5.4. The corrected expression for W is $W = 2\delta^2 k \left(V + \frac{\Phi}{2}\right).$[51])[52]

$$W = 2\delta^2 k (V - \Phi) \tag{3.6}$$

where V is the volume of the constructed sphere of radius R ($V = \frac{4}{3}\pi R^3$), Φ is the volume of the spherical body ($\Phi = \frac{4}{3}\pi P^3$), and $\delta^2 = A^2 + B^2 + C^2$. If the spherical body were not present ($\Phi = 0$) the energy dissipated would be[53]

$$W = 2\delta^2 k V \tag{3.7}$$

The presence of the sphere of volume Φ results in a change of heat production by $-2\delta^2 k \Phi$ per unit time.[54]

3.2.2 Calculation of the Coefficient of Viscosity of a Liquid in Which Very Many Irregularly Distributed Small Spheres are Suspended

Einstein calculates the viscosity k in an indirect manner. He considers a volume V filled with a fluid in which there are suspended very many irregularly distributed small spheres. He first derives an expression for the heat production in the volume V with no spheres present, then an expression for the heat production with the spheres present. The two expressions for the heat production are then solved for the viscosity.

In Section 3.2.1, we considered a region G with a sphere suspended in it that was very small compared to the dimension of G. Consider again the same region G, but this time containing "infinitely many randomly distributed spheres of equal radius, and that this radius is so small that the combined volume of all the spheres is very small compared with the region G."[55] The number of spheres per unit volume is denoted as n, with n being a constant. The distance between the spheres is taken to be large compared to their radius. At a given point in the liquid, the "alteration" of the velocity components of the liquid compared to the case with no spheres present will be small but non-negligible. These alterations are the sum of the contributions noted in Eq. (3.5) for a single sphere. The net result of these alterations is situated in a modification of the constants A, B and C, with (see Appendix 3.5.3 for details of these calculations)

$$A \to A^* = A(1 - \varphi)$$
$$B \to B^* = B(1 - \varphi)$$
$$C \to C^* = C(1 - \varphi)$$

where φ is the fraction of the total volume occupied by all of the spheres.[56]

$$\varphi = \left(\frac{number\ of\ spheres}{volume}\right) \times (volume\ of\ one\ sphere) = n\left(\frac{4}{3}\pi P^3\right)$$

In the previous section it was shown that "the presence of each sphere results in a decrease of heat production by $2\delta^2 k\Phi$ per unit time."[57] From Eq. (3.6), the energy converted to heat per unit volume (dividing by V) per unit time by *all of the spheres* is[58]

$$W = 2\delta^2 k(1 - \varphi) \qquad (3.8)$$

Analogous to $\delta^2 = A^2 + B^2 + C^2$ we define $\delta^{*2} = A^{*2} + B^{*2} + C^{*2}$. Neglecting higher order terms in φ,

$$\delta^{*2} = \delta^2(1 - 2\varphi)$$

With no spheres present, the heat generated per unit time and per unit volume is, from Eq. (3.7), $W = 2\delta^2 k$. With "very many spheres present," the heat generated per unit time and per unit volume can be written in the same form, $W^* = 2\delta^{*2}k^*$, with k^* the viscosity when the spheres are present. But, from Eq. (3.8), we know the form of W^* is $W^* = 2\delta^2 k(1 - \varphi)$.[59] Equating these two expressions for W^*, and neglecting higher order terms in φ,[60]

$$k^* = k(1 + \varphi) \qquad (3.9)$$

3.2.3 On the Volume of a Dissolved Substance Whose Molecular Volume is Large Compared to that of the Solvent

As John Stachel notes, "An outstanding current problem of the theory of solutions was whether molecules of the solvent are attached to the molecules or ions of the solute."[61] Einstein addresses this question in section 3 of his paper. From Section 3.2.2 (see Appendix 3.5.4 for details of this calculation),

$$\frac{k^*}{k} = (1 + \varphi)$$

Using data for a 1% aqueous solution of sugar (1 gm of sugar in 100 cm^3 of water), and using Einstein's relation, $\frac{k^*}{k} = 1 + \varphi$, Einstein found $k^*/k = 1.0245$, giving $\varphi = 0.0245$. "Thus, one gram of sugar dissolved in water has the same effect on the coefficient of viscosity as do small suspended rigid spheres of a total volume 2.45 cm^3."[62]

For comparison, Einstein notes one gram of solid sugar has a volume of 0.61 cm^3. Also, from the density of a 1% aqueous sugar solution, he calculates the specific volume s of the sugar in solution to be $s = 0.61$. He concludes, "Thus, while the sugar solution behaves as a mixture of water and solid sugar with respect to its density [0.61 cm^3/gm], the effect on internal friction [viscosity] is four times larger [2.45/0.61 = 4.02] than that which would result from the suspension of the same amount of sugar. It seems to me that from the point of view of the molecular theory, this result can hardly be interpreted otherwise than by assuming that the sugar molecule in the solution impedes the mobility of the water in its immediate vicinity, so that an amount of water whose volume is about three times larger than the volume of the sugar molecule is attached to the sugar molecule."[63] [The corrected calculations show the volume effect on viscosity to be $0.98\,\mathrm{cm}^3$, and a volume of water attached to each sugar molecule of about 60% the volume of each water molecule. Details in Appendix 3.5.4.]

3.2.4   On the Diffusion of an Undissociated Substance in a Liquid Solution

In Section 3.2.5, the expression $k^* = k(1 + \varphi)$ will be solved to obtain an expression for NP^3. In this section (3.2.4), Einstein obtains an expression for the coefficient of diffusion which, in turn, is solved to obtain an expression for NP. Using data from standard tables for the viscosity and for the coefficient of diffusion, Einstein solves these two equations for the molecular radius P and Avogadro's Number N (see Section 3.2.5).

The coefficient of diffusion D is defined by the equation $\omega\rho = -D\frac{\partial p}{\partial x}$, with $\omega =$ the migration velocity, $\rho =$ the mass density, $D =$ the coefficient of diffusion, $p =$ the pressure, and $x =$ the distance. If there is a pressure gradient $\frac{\partial p}{\partial x}$, the velocity of the diffusing particles reaches its terminal value when the pressure gradient is balanced by a retarding viscous force, given by Stokes' law $F = 6\pi kP\omega$, with P the radius of the spherical particles. Einstein begins this section of his paper with an expression for ω, the migration velocity, for a molecule acted on by a constant force K. He than identifies this force as due to a pressure gradient. Substituting in an expression for the pressure gradient, he then obtains the expression for the coefficient of diffusion.

Consider a solution similar to the sugar solution of Section 3.2.3. "If a force K acts upon the molecule, which we consider as a sphere of radius P, the molecule will move with a velocity ω which is determined by P and the coefficient of viscosity k of the solvent, since we have the equation"[64]

$$\omega = \frac{K}{6\pi kP} \tag{3.10}$$

The viscosity does not produce motion, it only retards it. The only force capable of producing motion is a pressure gradient due to the dissolved substance. This is the osmotic pressure, the pressure exerted by the solute above the pressure exerted by the solvent alone.[65] Aligning the x-axis with the direction of the pressure gradient, the motion-producing force acting on the dissolved substance is [66]

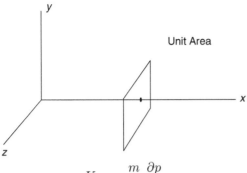

Fig. 3.3 Unit cross-sectional area perpendicular to the x-axis.

$$K = -\frac{m}{\rho N}\frac{\partial p}{\partial x} \qquad (3.11)$$

Van't Hoff's law says that the osmotic pressure of a dissolved substance is equal to the gas pressure it would exert if it were an ideal gas that occupied the same volume.[67] Assuming the validity of van't Hoff's law, Einstein wrote for the osmotic pressure[68]

$$p = \frac{R}{m}\rho T \quad \Rightarrow \quad \frac{\partial p}{\partial x} = \frac{RT}{m}\frac{\partial \rho}{\partial x} \qquad (3.12)$$

with R the universal gas constant and T the absolute temperature. Substituting Eqs. (3.11) and (3.12) into Eq. (3.10), the force K introduced can be eliminated from the expression for ω, the migration velocity:

$$\omega = -\frac{RT}{6\pi k}\frac{1}{NP}\frac{1}{\rho}\frac{\partial \rho}{\partial x} \qquad (3.13)$$

The amount of material passing through a unit cross section per unit time is (see Figure 3.3)

$$\omega\rho = -\frac{RT}{6\pi k}\frac{1}{NP}\frac{\partial p}{\partial x} = -D\frac{\partial p}{\partial x}$$

where D is the coefficient of diffusion. Thus the coefficient of diffusion is

$$D = \frac{RT}{6\pi k}\frac{1}{NP} \qquad (3.14)$$

In Section 3.2.5, this expression for the coefficient of diffusion and the expression for the viscosity from Section 3.2.2 will be used together to determine both the size of the molecules, P, and the number of molecules in one gram molecule, N.

3.2.5 Determination of the Molecular Dimensions with the Help of the Relations Obtained

From Section 3.2.2, Eq. (3.9), the coefficients of viscosity with the spheres present, and the spheres not present, are related as

$$\frac{k^*}{k} = (1+\varphi) = 1 + \frac{\left(N\frac{4}{3}\pi\mathrm{P}^3\right)}{V} = 1 + n\left(\frac{4}{3}\pi\mathrm{P}^3\right)$$

Using the relation $(n/N) = (\rho/m)$ to substitute for n:

$$NP^3 = \frac{3}{4\pi}\frac{m}{\rho}\left(\frac{k^*}{k} - 1\right) \tag{3.15}$$

From section 3.2.4, Eq. (3.14), the coefficient of diffusion is

$$D = \frac{RT}{6\pi k}\frac{1}{NP}$$

Solving this for NP,

$$NP = \frac{RT}{6\pi k}\frac{1}{D} \tag{3.16}$$

The two equations, Eq. (3.15) for NP^3 and Eq. (3.16) for NP, are solved to obtain the values of N and of P.

Returning to the aqueous solution of sugar, $NP^3 = 200$ for the 1% solution.[69] The coefficient of diffusion is $0.384\frac{cm^2}{day} = 0.384\frac{cm^2}{24 \times 3600s} = 4.44 \times 10^{-6}$ in cgs units. The expression for NP is calculated to be

$$NP = \frac{RT}{6\pi kD} = \frac{(8.31 \times 10^7)(9.5 + 273)}{6\pi(0.0135)(4.44 \times 10^{-6})} = 2.08 \times 10^{16}$$

Einstein noted this value was from a 10% solution, and the "strict validity of our formula cannot be expected at such high concentrations."[70]

Solving these equations for NP^3 and for NP, N, and P are calculated to be[71]

$$P = 9.9 \times 10^{-8} \text{ cm}$$

$$N = 2.1 \times 10^{23}$$

3.3   Albert Einstein's Paper, "On the Movement of Small Particles Suspended in Stationary Liquids Required by the Molecular-Kinetic Theory of Heat"[72]

Albert Einstein continued his quest for verification of the atomic theory of matter, from "an exact determination of the real size of atoms ... [to verification of] ... the molecular-kinetic conception of heat."[73] He noted that if his predictions proved to be wrong it would be an argument against the molecular conception of heat. "It is possible that the motions to be discussed here are identical with the so-called 'Brownian molecular motion'; however, the data available to me on the latter are so imprecise that I could not form a judgment on this matter."[74]

3.3.1 On the Osmotic Pressure Attributable to Suspended Particles

A container of volume V is divided into two parts separated by a semipermeable membrane. The volume of one part is V^*, while the volume of the second part is $V - V^*$. See Figure 3.4.

The membrane allows passage of the solvent from one part to the other, but not the solute. Solvent is put into both parts of the container, while solute is put into only the volume V^*. The wall is subject to an osmotic pressure p. Using van't Hoff's law, for low densities ($= z/V^*$, where z is one gram weight of the solute, i.e., one mole) the osmotic pressure is given by the ideal gas equation of state:[75]

$$pV^* = RTz$$

Einstein points out that the osmotic pressure is due to the solute molecules in the solvent in V^*. However, according to thermodynamics, if the solute is replaced by small suspended bodies that also cannot pass through the semi-permeable membranes separating the two portions of the volume V, "we should not expect...that a force [pressure] be exerted on the wall....."[76] But, he continues, "from the standpoint of the molecular-kinetic theory of heat we are led to a different conception. According to this theory, a dissolved molecule differs from a suspended body in size *alone*, and it is difficult to see why suspended bodies should not produce the same osmotic pressure as an equal number of dissolved molecules...they will exert forces [pressure] upon the wall exactly as dissolved molecules do...there will correspond to them an osmotic pressure p of magnitude"[77]

$$p = \frac{RT}{V^*}\frac{n}{N} = \frac{RT}{N}\nu \tag{3.17}$$

where n is the number of suspended bodies in V^*, $n/V^* = \nu$ is the number of suspended bodies per unit volume, and N is Avogadro's number. In the next section, Einstein shows how one can obtain this expression from the molecular-kinetic theory of heat, i.e., van't Hoff's law can be obtained from the molecular-kinetic theory of heat.

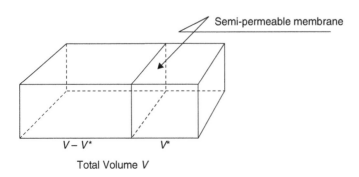

Total Volume V

Fig. 3.4 The container of volume V divided into two sections, one of volume V^*, the other of volume $V–V^*$, separated by a semi-permeable membrane.

3.3.2   Osmotic Pressure from the Standpoint of the Molecular-Kinetic Theory of Heat

In Einstein's paper, section 2 will show "that the existence of osmotic pressure is a consequence of the molecular-kinetic theory of heat, and that, according to this theory, at great dilutions numerically equal quantities of dissolved molecules and suspended particles behave completely identically with regard to osmotic pressure."[78] Section 2 also provides some thermodynamic background and relations for use in later sections of the paper.[79]

Let $p_1, p_2, \ldots p_{3n}, q_1, q_2, \ldots q_{3n}$ be the variables that completely determine the state of a system at any given time.[80] Typically, these would be the three position and three momentum coordinates of each of the n atoms in the volume V^*. The changes in each of the p and q variables are determined by Hamilton's equations. The entropy of the system is given by (see Appendix 3.5.5 for the details of this derivation)[81]

$$S = \frac{\bar{E}}{T} + k \ln \left\{ \int e^{-\frac{E}{kT}} dp_1 \, dp_2 \, \ldots \, dp_{3n} \, dq_1 \, dq_2 \, \ldots \, dq_{3n} \right\}$$

where $\bar{E}$ denotes the energy of the system, E denotes the energy as a function of the variables of the system $p_1, p_2, \ldots p_{3n}, q_1, q_2, \ldots q_{3n}$, and k is Boltzmann's constant. The Helmholtz Free Energy, F, is defined as $F = E - TS$:

$$F = -kT \ln \left\{ \int e^{-\frac{E}{kT}} dp_1 \, dp_2 \ldots dp_{3n} \, dq_1 \, dq_2 \, \ldots \, dq_{3n} \right\} = -kT \ln B$$

Einstein points out "the calculation of the integral B would be so difficult as to make an exact calculation of F all but inconceivable. However, here we only have to know how F depends on the size of the volume $V^* \ldots$"[82] B is then shown to be of the form $B = JV^{*n}$, where J does not depend on the volume V^* (see Appendix 3.5.6 for the details of this derivation):[83]

$$F = -kT \ln B = -kT \{ \ln J + n \ln V^* \}$$

$$p = -\frac{\partial F}{\partial V^*} = \frac{kT}{V^*} n$$

Replacing $k = R/N$,

$$p = \frac{R}{N} \frac{T}{V^*} n = \frac{RT}{V^*} \frac{n}{N} = \frac{RT}{N} \nu \tag{3.18}$$

This agrees with the expression for the osmotic pressure in Section 3.3.1, Eq. (3.17). This "demonstrates that osmotic pressure is a consequence of the molecular-kinetic theory of heat, and that ... numerically equal quantities of dissolved molecules and suspended particles behave completely identically with regard to osmotic pressure."[84] This completes the demonstration that van't Hoff's law is a consequence of the molecular-kinetic theory of heat.

3.3.3   Theory of Diffusion of Small Suspended Spheres

Consider the volume V^* filled with a liquid. In the liquid are suspended randomly distributed spheres. Einstein introduces a force K strictly for the convenience of calculation (it will subsequently drop out). The force K acts on the individual particles, is a function of position but not of time, and is parallel to the x-axis.

The variation of the free energy, F, is zero since the system is in thermodynamic equilibrium:

$$\delta F = \delta(E - TS) = \delta E - T\delta S = 0$$

Einstein assumes the liquid has a unit cross section perpendicular to the x-axis, and that it is bounded by the planes $x = 0$ and $x = l.$[85] See Figure 3.5.

We then have[86]

$$\delta E_{particle} = -K\delta x$$

$$\delta E_{system} = -\sum_{all\ particles} \delta E_{particle} \left(\frac{\#\ particles}{volume} = \nu \right) (dV^* = unit\ area \bullet dx)$$

$$= -\int_0^l K\nu\delta x dx$$

Writing

$$\delta S = \frac{\delta E}{T}$$

and considering a narrow cross section of the volume V^* of width dx, a virtual displacement of δx would produce a virtual energy change of $p_L\delta x \bullet unit\ area = p_L\delta x$ due to the forces on the face at $x = 0$, and a virtual energy change of $-p_R\delta x \bullet unit\ area = -p_R\delta x$ due to the forces on the face at dx. See Figure 3.6.

For this slice, the virtual energy change is $-(p_R - p_L)\delta x = -\frac{\partial p}{\partial x} dx\delta x$. Summing (integrating) over all slices of width dx from $x = 0$ to $x = l$:

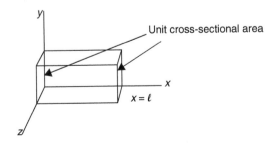

Fig. 3.5 Volume of liquid showing the unit cross-sectional areas, bounded by planes at $x = 0$ and at $x = l$.

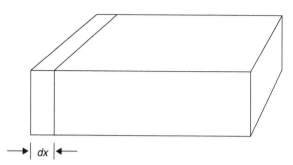

Fig. 3.6 The narrow cross section of the volume V^*.

$$\delta E = -\sum_{x=0}^{x=l} \frac{\partial p}{\partial x} dx \delta x \rightarrow -\int_{x=0}^{x=l} \frac{\partial p}{\partial x} dx \delta x$$

$$\delta S = \frac{\delta E}{T} = -\int_{x=0}^{x=l} \frac{\partial}{\partial x} \left(\frac{p}{T} = \nu \frac{R}{N} \right) dx \delta x = -\frac{R}{N} \int_{x=0}^{x=l} \frac{\partial}{\partial x} (\nu) \, dx \delta x$$

Substituting these expressions, the condition of equilibrium is

$$0 = \delta E - T \delta S = -\int_{0}^{l} K \nu \delta x dx + \frac{RT}{N} \int_{x=0}^{x=l} \frac{\partial}{\partial x} (\nu) \, dx \delta x$$

$$0 = \int_{0}^{l} \left(-K\nu + \frac{RT}{N} \frac{\partial v}{\partial x} \right) \delta x dx \Rightarrow -K\nu + \frac{RT}{N} \frac{\partial v}{\partial x} = 0 \qquad (3.19)$$

From Section 3.3.2, Eq. (3.18), $p = \frac{RT}{N} \nu \Rightarrow \frac{\partial p}{\partial x} = \frac{RT}{N} \frac{\partial \nu}{\partial x}$

$$K\nu - \frac{\partial p}{\partial x} = 0$$

This equation "states that the force K is balanced by the forces of osmotic pressure."[87]

Equation (3.19) is used to determine the diffusion coefficient of the suspended substance. Consider a unit cross-sectional area perpendicular to the x-axis (see Figure 3.3). Since dynamic equilibrium exists, the number of particles passing through this unit area because of the force K will be balanced by an equal number passing through the unit area in the opposite direction because of diffusion.

"If the suspended particles are of spherical shape (where P is the radius of the sphere) and the coefficient of friction [viscosity] of the liquid is k, then the force K imparts to an individual particle the velocity..."[88]

$$\omega = \frac{K}{6\pi k P}$$

The number of particles of velocity ω that pass through the unit area per unit time is

$$\nu\omega = \frac{\nu K}{6\pi kP}$$

Balancing this is the number of particles passing through the unit area per unit time because of diffusion. From the definition of the coefficient of diffusion, this expression is $-D\frac{\partial \nu}{\partial x}$. From the condition of dynamic equilibrium:

$$\frac{\nu K}{6\pi kP} - D\frac{\partial \nu}{\partial x} = 0 \qquad (3.20)$$

Eliminating $\partial \nu / \partial x$ from Eqs. (3.19) and (3.20), the diffusion coefficient is

$$D = \frac{RT}{N}\frac{1}{6\pi kP} \qquad (3.21)$$

At this stage, the force K has been eliminated from the expression for the diffusion constant D.

3.3.4 On the Random Motion of Particles Suspended in a Liquid and Their Relation to Diffusion

Section 4 of Einstein's paper examines more closely the thermal molecular motion that gives rise to diffusion. In this calculation, it is assumed that

1. the motion of each particle is independent of the motions of the other particles, and
2. the motions of the same particle in different time periods are independent of each other, provided the time intervals are not too small.

A time interval τ is introduced, where

1. τ is much less than observable time intervals, and
2. τ is so large that "the motions performed by a particle during two consecutive time intervals τ can be considered as mutually independent events."[89]

Consider a system with n suspended particles in a liquid. Focus on the x-coordinate of an individual particle. In a time interval τ the x-coordinate will change by some amount Δ where Δ could be positive or negative. This is true for each of the n particles, with Δ (in general) having a different value for each of the n particles. "A certain frequency law [probability distribution law] will hold for Δ: the number dn of particles experiencing a displacement lying between Δ and $\Delta + d\Delta$ in the time interval τ will be expressed by an equation of the form..."[90]

$$dn = n\varphi\left(\Delta\right)d\Delta$$

Since $n = \int dn = \int n\varphi(\Delta)d\Delta \Rightarrow \int \varphi(\Delta)d\Delta = 1$

φ differs from zero only for very small values of Δ and is symmetric in Δ, i.e., $\varphi(\Delta) = \varphi(-\Delta)$.

The number of particles per unit volume, ν, is a function of position x and time t. Setting $\nu = f(x,t)$, Einstein determined $f(x,t)$ to be of the form (see Appendix 3.5.7 for the details of this derivation),

$$f(x,t) = \frac{n}{\sqrt{4\pi D}} \frac{e^{-\frac{x^2}{4Dt}}}{\sqrt{t}}$$

"With the help of this equation we now calculate the displacement λ_x in the direction of the x-axis that a particle experiences on the average, or, to be more precise, the square root of the arithmetic mean of the squares of displacements [the root-mean-square displacement] in the direction of the x-axis; we get ..."[91] (see Appendix 3.5.8)

$$\lambda_x = \sqrt{\langle x^2 \rangle} = \sqrt{2Dt} \qquad (3.22)$$

3.3.5   Formula for the Mean Displacement of Suspended Particles. A New Method of Determining the True Size of Atoms

From Eq. (3.21), $D = \frac{RT}{N} \frac{1}{6\pi kP}$

From Eq. (3.22) , $\lambda_x = \sqrt{2Dt}$

Eliminating D, $\lambda_x = \sqrt{t}\sqrt{\frac{RT}{N} \frac{1}{3\pi kP}}$

This equation for the root-mean-square displacement is the main result of this paper.

Using water at $17°\,\mathrm{C}$ as the liquid with suspended particles of diameter $0.001\,\mathrm{mm}$, for a one-second time interval Einstein calculates $\lambda_x \approx 8 \times 10^{-5}\,\mathrm{cm} = 0.8\,\mathrm{microns}$. Using cgs units, $\lambda_x = \sqrt{1}\sqrt{\frac{(8.31\times 10^7)(17+273)}{6\times 10^{23}} \frac{1}{3\pi(1.35\times 10^{-2})(5\times 19^{-5})}}$

$$= \sqrt{6.15 \times 10^{-9}} = 7.8 \times 10^{-5}\mathrm{cm} \approx 0.8\,\mathrm{micron}$$

The equation can be "inverted" to determine Avogadro's number by experimentally measuring λ_x:

$$N = \frac{t}{\lambda_x^2} \frac{RT}{3\pi kP}$$

Einstein concludes the paper with the words, "Let us hope that a researcher will soon succeed in solving the problem posed here, which is of such importance in the theory of heat!"[92]

3.4   Discussion and Comments

The goal of the second paper was to find some evidence showing the effects of the atoms. Einstein believed that Brownian motion might be the evidence needed to establish the reality of atoms but, he noted, "It is possible that the motions to be discussed here are identical with so-called 'Brownian molecular motion'; however, the data available to me on the latter are so imprecise that I could not form a judgment on this matter."[93] Einstein's hope that "a researcher will soon succeed in solving the problem posed here"[94] was answered by Jean Perrin and his co-workers. In 1908, Jean Perrin verified Einstein's Brownian motion formula and, for this work, was awarded the 1926 Nobel Prize in physics. In his Nobel Prize address Jean Perrin commented,[95]

These theories (of Einstein) can be judged by experiment if we know how to *prepare spherules of a measurable radius.* I was, therefore, in a position to attempt this check as soon as I knew ... of the work of Einstein.
. . .
   Having such grains, I was able to check Einstein's formulae by seeing whether they always led to the same value for Avogadro's Number and whether it was appreciably equal to the value already found [the accepted value for Avogadro's Number was 60.2×10^{22}].
. . .
   In several series of measurements I varied, with the aid of several collaborators, the size of the grains (in the ratio of 1 to 70,000) as well as the nature of the liquid (water, solutions of sugar or urea, glycerol) and its viscosity (in the ratio of 1 to 125). They gave values between 55×10^{22} and 72×10^{22}, with differences which could be explained by experimental errors. The agreement is such that it is impossible to doubt the correctness of the kinetic theory of the translational Brownian movement.

   The experimental work of Perrin was the final piece of evidence for belief in the reality of atoms. Although most of the physics community now accepted the reality of atoms, Mach remained unconvinced.[96]
   Albert Einstein was an atomist. The kinetic theory of gases was enjoying great success. Boltzmann had developed the area of statistical mechanics to forge a link between the properties of matter in bulk and the behavior of individual atoms.[97] Einstein expanded and developed the ideas of statistical mechanics. However, partly in response to the philosophical position espoused by Mach and others, the reality of atoms and molecules remained speculative. Einstein's dissertation "marked the first major success in Einstein's effort to find further evidence for the atomic hypothesis, an effort that culminated in his explanation of Brownian motion."[98] That these papers tipped the scales solidly in favor of the reality of atoms is seen in the comments of Henri Poincaré and Wilhelm Ostwald (the words of Ostwald are the more compelling as he had been an outspoken opponent of the reality of atoms):

[Around 1908, Henri Poincaré wrote] the atomic hypothesis has recently acquired enough credence to cease being a mere hypothesis. Atoms are no longer just useful fiction; we can rightfully claim to see them, since we can actually count them.[99]

[In 1909, Ostwald wrote] I have convinced myself that we have recently come into possession of experimental proof of the discrete or grainy nature of matter, for which the atomic hypothesis had vainly sought for centuries, even millennia.[100]

Einstein recalled that " the agreement of these considerations [theory of Brownian motion] with experience...convinced the skeptics, who were quite numerous at that time (Ostwald, Mach) of the reality of atoms. The antipathy of these scholars towards atomic theory can indubitably be traced back to their positivistic philosophical attitude. This is an interesting example of the fact that even scholars of audacious spirit and fine instinct can be obstructed in the interpretation of facts by philosophical prejudices."[101] Einstein's satisfaction is apparent in attaining his goal of having refuted the energetics worldview.

3.5 Appendices

3.5.1 Derivation of the Expressions for u, v, and w

From Section 3.2.1, Eqs. (3.1) to (3.4), we have the following:
  The Navier–Stokes equations for an incompressible fluid are

$$\frac{\partial p}{\partial \xi} = k\nabla^2 u; \quad \frac{\partial p}{\partial \eta} = k\nabla^2 v; \quad \frac{\partial p}{\partial \zeta} = k\nabla^2 w \tag{3.4}$$

with $\nabla^2 = \frac{\partial^2}{\partial \xi^2} + \frac{\partial^2}{\partial \eta^2} + \frac{\partial^2}{\partial \zeta^2}$
  The equation of continuity is

$$\frac{\partial u}{\partial \xi} + \frac{\partial v}{\partial \eta} + \frac{\partial w}{\partial \zeta} = 0 \tag{3.4}$$

With no spheres present, $u = A\xi; v = B\eta; w = C\zeta \Rightarrow A + B + C = 0.$ (3.1); (3.2)
  With spheres present, $u = A\xi + u_1; v = B\eta + v_1; w = C\zeta + w_1.$ (3.3)
  Taking the derivative of the Navier–Stokes equations, Eqs. (3.4),

$$\frac{\partial}{\partial \xi}\left(\frac{\partial p}{\partial \xi} = k\nabla^2 u\right) \quad \Rightarrow \frac{\partial^2 p}{\partial \xi^2} = k\nabla^2\left(\frac{\partial u}{\partial \xi}\right)$$

$$\frac{\partial}{\partial \eta}\left(\frac{\partial p}{\partial \eta} = k\nabla^2 v\right) \quad \Rightarrow \frac{\partial^2 p}{\partial \eta^2} = k\nabla^2\left(\frac{\partial v}{\partial \eta}\right)$$

$$\frac{\partial}{\partial \zeta}\left(\frac{\partial p}{\partial \zeta} = k\nabla^2 w\right) \quad \Rightarrow \frac{\partial^2 p}{\partial \zeta^2} = k\nabla^2\left(\frac{\partial w}{\partial \zeta}\right)$$

Adding, and using the equation of continuity

$$\frac{\partial^2 p}{\partial \xi^2} + \frac{\partial^2 p}{\partial \eta^2} + \frac{\partial^2 p}{\partial \zeta^2} = k\nabla^2\left(\frac{\partial u}{\partial \xi} + \frac{\partial v}{\partial \eta} + \frac{\partial w}{\partial \zeta}\right) = k\nabla^2\,(0)$$

$$\nabla^2 p = 0$$

The following adopts a method developed by Gustav Kirchhoff[102] to determine the function p that satisfies the equation $\nabla^2 p = 0$. Einstein defines a function $V(\xi, \eta, \zeta)$ such that $p = k\nabla^2 V$. Taking the derivative of p and using the Navier–Stokes equations,

$$\frac{\partial p}{\partial \xi} = \frac{\partial}{\partial \xi}(\nabla^2 V) = k\nabla^2\left(\frac{\partial V}{\partial \xi}\right) = k\nabla^2 u \quad \Rightarrow u = \frac{\partial V}{\partial \xi} + u'$$

with $\nabla^2 u' = 0$. In general, $u' \neq u_1$.

Similarly, one obtains $v = \frac{\partial V}{\partial \eta} + v'$ with $\nabla^2 v' = 0$ and $w = \frac{\partial V}{\partial \zeta} + w'$ with $\nabla^2 w' = 0$.

Using the equation of continuity and the definition of V,

$$\frac{\partial u}{\partial \xi} + \frac{\partial v}{\partial \eta} + \frac{\partial w}{\partial \zeta} = 0 \Rightarrow \frac{\partial}{\partial \xi}\left(\frac{\partial V}{\partial \xi} + u'\right) + \frac{\partial}{\partial \eta}\left(\frac{\partial V}{\partial \eta} + v'\right) + \frac{\partial}{\partial \zeta}\left(\frac{\partial V}{\partial \zeta} + w'\right) = 0$$

$$\nabla^2 V + \left(\frac{\partial u'}{\partial \xi} + \frac{\partial v'}{\partial \eta} + \frac{\partial w'}{\partial \zeta}\right) = 0 \Rightarrow \frac{p}{k} + \left(\frac{\partial u'}{\partial \xi} + \frac{\partial v'}{\partial \eta} + \frac{\partial w'}{\partial \zeta}\right) = 0$$

$$\left(\frac{\partial u'}{\partial \xi} + \frac{\partial v'}{\partial \eta} + \frac{\partial w'}{\partial \zeta}\right) = -\frac{p}{k}$$

Combining the previous results one obtains

$$\frac{p}{k} = \nabla^2 V = \frac{\partial^2 V}{\partial \xi^2} + \frac{\partial^2 V}{\partial \eta^2} + \frac{\partial^2 V}{\partial \zeta^2} = -\left(\frac{\partial u'}{\partial \xi} + \frac{\partial v'}{\partial \eta} + \frac{\partial w'}{\partial \zeta}\right)$$

This implies $u' = -\frac{\partial V}{\partial \xi} + f$, where $\frac{\partial f}{\partial \xi} = 0$. Einstein sets $f = 0$. Similar results obtain for v' and for w'.

From Kirchhoff, the radially symmetric solution to $\nabla^2(\frac{p}{k}) = 0$ is $\frac{p_0}{k} = a + \frac{2c}{\rho}$.

It is to be noted that, if $\frac{p_0(\rho)}{k}$ is a valid solution to $\nabla^2(\frac{p}{k}) = 0$, so also are the derivatives of $\frac{p_0(\rho)}{k}$. Consider the function $f(\rho) = \frac{\partial}{\partial \xi}(\frac{p_0(\rho)}{k})$.

$$\nabla^2 f(\rho) = \nabla^2 \frac{\partial}{\partial \xi}\left(\frac{p_0(\rho)}{k}\right) = \frac{\partial}{\partial \xi}\left(\nabla^2 \frac{p_0(\rho)}{k}\right) = \frac{\partial}{\partial \xi}(0) = 0$$

Similar results are obtained for higher order derivatives of $\frac{p_0(\rho)}{k}$.

Using the identities $\nabla^2 \rho = \frac{2}{\rho}$ and $\nabla^2(\frac{1}{\rho}) = 0$, the expression for the potential $V_0(\xi, \eta, \zeta)$ is obtained as $\frac{p_0}{k} = a + \frac{2c}{\rho} = a + c\nabla^2 \rho = \nabla^2(a\frac{\xi^2}{2} + c\rho + \frac{b}{\rho}) = \nabla^2 V_0$:

$$V_0 = a\frac{\xi^2}{2} + c\rho + b\left(\frac{1}{\rho}\right)$$

Einstein selects the second derivative of the solution for $\frac{p_0(r)}{k}$ which, as was pointed out, also is a valid solution to the equation, $\nabla^2(\frac{p}{k}) = 0$.

$$\frac{p_E}{k} = \frac{\partial^2}{\partial \xi^2}\left(\frac{p_0}{k}\right) = \frac{\partial^2}{\partial \xi^2}\left(a + \frac{2c}{\rho}\right) = 2c\frac{\partial^2}{\partial \xi^2}\left(\frac{1}{\rho}\right)$$

Einstein looks first for a solution of the velocity components such that $u' \neq 0, v' = 0, w' = 0$, and then for similar solutions for $v' \neq 0$ and

$w' \neq 0$. These three separate solutions for $u = \frac{\partial V}{\partial x} + u'$, $v = \frac{\partial V}{\partial y} + v'$, $w = \frac{\partial V}{\partial z} + w'$ then will be superimposed to obtain the sought-for solution.

For $u' \neq 0, v' = 0, w' = 0$, $v' = 0 \Rightarrow \frac{\partial v'}{\partial x} = 0 \Rightarrow \frac{\partial^2 V}{\partial x^2} = 0$. Similarly $w' = 0 \Rightarrow \frac{\partial^2 V}{\partial y^2} = 0$. But, since the solution for $\frac{p_E}{k}$ is $\frac{p_E}{k} = 2c\frac{\partial^2}{\partial \xi^2}(\frac{1}{\rho})$, and $\frac{p_E}{k} = -\frac{\partial u'}{\partial \xi}$, Einstein sets $u' = -\frac{\partial}{\partial \xi}(\frac{2c}{\rho})$.

Since the solution to $\nabla^2(\frac{p}{k}) = 0$ is $\frac{p_0}{k} = a + \frac{2c}{\rho}$, with $V_0 = a\frac{\xi^2}{2} + c\rho + b(\frac{1}{\rho})$, taking the second derivative of the equation,

$$\frac{\partial^2}{\partial \xi^2}\left[\nabla^2\left(\frac{p_0}{k}\right)\right] = \nabla^2\left[\frac{\partial^2}{\partial \xi^2}\left(\frac{p_0}{k}\right)\right] = \nabla^2\left(\frac{p_E}{k}\right) = \frac{\partial^2}{\partial \xi^2}\left[\nabla^2 V_0\right]$$

$$= \nabla^2\left[\frac{\partial^2 V_0}{\partial \xi^2}\right] = \nabla^2 V_E$$

$$V_{Einstein} = V_E = \frac{\partial^2 V_0}{\partial \xi^2} = c\frac{\partial^2 \rho}{\partial \xi^2} + b\frac{\partial^2}{\partial \xi^2}\left(\frac{1}{\rho}\right) + \frac{a}{2}\left[\xi^2 - \frac{\eta^2}{2} - \frac{\zeta^2}{2}\right]$$

The first two terms are from those of Kirchhoff. The term $\frac{a}{2}[\xi^2 - \frac{\eta^2}{2} - \frac{\zeta^2}{2}]$ gives, for $u = \frac{\partial V_E}{\partial \xi} + u'$, the contribution $A\xi$ to u, and for $\nabla^2 V_E$ a contribution of zero.

Following the method of Kirchhoff, Einstein solves first the equations for the case of

$$u' = -\frac{\partial}{\partial \xi}\left(\frac{2c}{\rho}\right), \quad v' = 0, \quad w' = 0$$

For $V_E = c\frac{\partial^2 \rho}{\partial \xi^2} + b\frac{\partial^2}{\partial \xi^2}\left(\frac{1}{\rho}\right) + \frac{a}{2}\left[\xi^2 - \frac{\eta^2}{2} - \frac{\zeta^2}{2}\right]$

$$u = \frac{\partial V_E}{\partial \xi} + u' = \frac{\partial V_E}{\partial \xi} - \frac{\partial}{\partial \xi}\left(\frac{2c}{\rho}\right)$$

$$= \frac{\partial}{\partial \xi}\left\{c\frac{\partial^2 \rho}{\partial \xi^2} + b\frac{\partial^2}{\partial \rho^2}\left(\frac{1}{\rho}\right) + \frac{a}{2}\left[\xi^2 - \frac{\eta^2}{2} - \frac{\zeta^2}{2}\right]\right\} - 2c\frac{\partial}{\partial \xi}\left(\frac{1}{\rho}\right)$$

$$u = A\xi + 2c\frac{\xi}{\rho^3} + \frac{\partial}{\partial \xi}\left\{c\frac{\partial^2 \rho}{\partial \xi^2} + b\frac{\partial^2}{\partial \xi^2}\left(\frac{1}{\rho}\right)\right\}$$

$$= A\left[\xi + 2c'\frac{\xi}{\rho^3} + \frac{\partial}{\partial \xi}\left\{c'\frac{\partial^2 \rho}{\partial \xi^2} + b'\frac{\partial^2}{\partial \xi^2}\left(\frac{1}{\rho}\right)\right\}\right]$$

where the constant a has been replaced by A, and where $c = Ac', b = Ab'$. In a similar manner, for v and w, one obtains

$$v = B\left[\eta + 2c'\frac{\eta}{\rho^3} + \frac{\partial}{\partial \eta}\left\{c'\frac{\partial^2 \rho}{\partial \eta^2} + b'\frac{\partial^2}{\partial \eta^2}\left(\frac{1}{\rho}\right)\right\}\right]$$

$$w = C\left[\zeta + 2c'\frac{\zeta}{\rho^3} + \frac{\partial}{\partial \zeta}\left\{c'\frac{\partial^2 \rho}{\partial \zeta^2} + b'\frac{\partial^2}{\partial \zeta^2}\left(\frac{1}{\rho}\right)\right\}\right]$$

It is important to note that, because of the cyclic nature of the variables, each expression has the same constants b and c, but the coefficient of the first term is distinct for each expression. The sum, $A + B + C = 0$.

The expressions for the velocity components u, v, w were calculated for the particular situation of the other two components being zero. But we need to find one expression for the general case. The term $\frac{a}{2}[\xi^2 + \frac{\eta^2}{2} + \frac{\zeta^2}{2}]$ has been accounted for in the expression for u (the term $a\xi$), with analogous reasoning for the expressions for v and w. The remaining terms are summed together as

$$F = A\left\{c'\frac{\partial^2 \rho}{\partial \xi^2} + b'\frac{\partial^2}{\partial \xi^2}\left(\frac{1}{\rho}\right)\right\} + B\left\{c'\frac{\partial^2 \rho}{\partial \eta^2} + b'\frac{\partial^2}{\partial \eta^2}\left(\frac{1}{\rho}\right)\right\}$$
$$+ C\left\{c'\frac{\partial^2 \rho}{\partial \zeta^2} + b'\frac{\partial^2}{\partial \zeta^2}\left(\frac{1}{\rho}\right)\right\}$$

With the general solution for u now given as

$$u = A\xi + 2c\frac{\xi}{\rho^3} + \frac{\partial F}{\partial \xi}$$

And similar expressions for v and w.

The constants b and c are determined by writing out explicitly the expression for u and requiring it to be zero on the surface of the sphere, $\rho = P$. Looking first at the expression $\frac{\partial F}{\partial \xi}$, we have

$$\frac{\partial F}{\partial \xi} = A\left\{c'\frac{\partial}{\partial \xi}\left(\frac{\partial^2 \rho}{\partial \xi^2}\right) + b'\frac{\partial}{\partial \xi}\left(\frac{\partial^2}{\partial \xi^2}\left(\frac{1}{\rho}\right)\right)\right\}$$
$$+ B\left\{c'\frac{\partial}{\partial \xi}\left(\frac{\partial^2 \rho}{\partial \eta^2}\right) + b'\frac{\partial}{\partial \xi}\left(\frac{\partial^2}{\partial \eta^2}\left(\frac{1}{\rho}\right)\right)\right\}$$
$$+ C\left\{c'\frac{\partial}{\partial \xi}\left(\frac{\partial^2 \rho}{\partial \zeta^2}\right)\right\} + b'\frac{\partial}{\partial \xi}\left(\frac{\partial^2}{\partial \zeta^2}\left(\frac{1}{\rho}\right)\right)$$
$$= Ac'\left(-\frac{\xi}{\rho^3} - \frac{2\xi}{\rho^3} + \frac{3\xi^3}{\rho^5}\right) + Ab'\left(\frac{3\xi}{\rho^5} + \frac{6\xi}{\rho^5} - \frac{15\xi^3}{\rho^7}\right)$$
$$+ Bc'\left(-\frac{\xi}{\rho^3} + \frac{3\eta^2\xi}{\rho^5}\right) + Bb'\left(\frac{3\xi}{\rho^5} - \frac{15\eta^2\xi}{\rho^7}\right)$$
$$+ Cc'\left(-\frac{\xi}{\rho^3} + \frac{3\zeta^2\xi}{\rho^5}\right) + Cb'\left(\frac{3\xi}{\rho^5} - \frac{15\zeta^2\xi}{\rho^7}\right)$$
$$= -\frac{c'\xi}{\rho^3}(A + B + C) + \frac{3c'\xi}{\rho^5}\left(A\xi^2 + B\eta^2 + C\zeta^2\right) - \frac{2Ac'\xi}{\rho^3}$$
$$+ \frac{3b'\xi}{\rho^5}(A + B + C) - \frac{15b'\xi}{\rho^7}\left(A\xi^2 + B\eta^2 + C\zeta^2\right) + \frac{6Ab'\xi}{\rho^5}$$

Using $A + B + C = 0$, the expressions for $\frac{\partial F}{\partial \xi}$ and for u become

$$\frac{\partial F}{\partial \xi} = \frac{3c'\xi}{\rho^5}\left(A\xi^2 + B\eta^2 + C\zeta^2\right) - \frac{2Ac'\xi}{\rho^3} - \frac{15b'\xi}{\rho^7}\left(A\xi^2 + B\eta^2 + C\zeta^2\right)$$
$$+ \frac{6Ab'\xi}{\rho^5}$$

$$u = A\xi + 2c\frac{\xi}{\rho^3} + \left(A\xi^2 + B\eta^2 + C\zeta^2\right)\left(\frac{3\xi}{\rho^5}\right)\left(c' - \frac{5b'}{\rho^2}\right) - \frac{2c'A\xi}{\rho^3} + \frac{6b'A\xi}{\rho^5}$$

$$= A\xi\left(1 + \frac{2c'}{\rho^3} - \frac{2c'}{\rho^3} + \frac{6b'}{\rho^5}\right) + \left(A\xi^2 + B\eta^2 + C\zeta^2\right)\left(\frac{3\xi}{\rho^5}\right)\left(c' - \frac{5b'}{\rho^2}\right)$$
$$+ \frac{6Ab'\xi}{\rho^5}$$

On the surface of the sphere $u = 0$. Setting $\rho = P$, the expression for u becomes (with $A = a$)

$$u = 0 = A\xi\left(1 + \frac{6b'}{P^5}\right) + \left(A\xi^2 + B\eta^2 + C\zeta^2\right)\left(\frac{3\xi}{P^5}\right)\left(c' - \frac{5b'}{P^2}\right)$$

For this to be valid for all values of ξ, $1 + \frac{6b'}{P^5} = 0$ and $c' - \frac{5b'}{P^2} = 0$. Solving, $b' = -\frac{P^5}{6}, c' = -\frac{5P^3}{6}$, and $b = Ab' = -\frac{AP^5}{6}, c = Ac' = -\frac{5AP^3}{6}$. (It needs to be noted these values differ by a factor of two from those in Einstein's paper: $b = -\frac{1}{12}P^5 a$ and $c = -\frac{5}{12}P^3 a$.)[103]

With these values for b and c, the expression for u becomes

$$u = A\xi + 2\left(-\frac{5AP^3}{6}\right)\frac{\xi}{\rho^3} + \left(A\xi^2 + B\eta^2 + C\zeta^2\right)\left(\frac{3\xi}{\rho^5}\right)\left(\frac{-5P^3}{6}\right)$$

$$- 5\left(A\xi^2 + B\eta^2 + C\zeta^2\right)\left(\frac{3\xi}{\rho^7}\right)\left(\frac{-P^5}{6}\right) - \frac{2A\xi}{\rho^3}\left(-\frac{5P^3}{6}\right) + \frac{6A\xi}{\rho^5}\left(-\frac{P^5}{6}\right)$$

$$u = A\xi - \frac{5}{2}\frac{P^3}{\rho^5}\xi\left(A\xi^2 + B\eta^2 + C\zeta^2\right) + \frac{5}{2}\frac{P^5}{\rho^7}\xi\left(A\xi^2 + B\eta^2 + C\zeta^2\right) - \frac{P^5}{\rho^5}A\xi$$

Analogous expressions are obtained for v and w.

$$u = A\xi + 2c\frac{\xi}{\rho^3} + \frac{\partial F}{\partial \xi}$$

With

$$F = A\left\{c'\frac{\partial^2 \rho}{\partial \xi^2} + b'\frac{\partial^2}{\partial \xi^2}\left(\frac{1}{\rho}\right)\right\} + B\left\{c'\frac{\partial^2 \rho}{\partial \eta^2} + b'\frac{\partial^2}{\partial \eta^2}\left(\frac{1}{\rho}\right)\right\}$$
$$+ C\left\{c'\frac{\partial^2 \rho}{\partial \zeta^2} + b'\frac{\partial^2}{\partial \zeta^2}\left(\frac{1}{\rho}\right)\right\}$$

$$F = A \left\{ \left(-\frac{5P^3}{6} \right) \frac{\partial^2 \rho}{\partial \xi^2} + \left(-\frac{P^5}{6} \right) \frac{\partial^2}{\partial \xi^2} \left(\frac{1}{\rho} \right) \right\}$$

$$+ B \left\{ \left(-\frac{5P^3}{6} \right) \frac{\partial^2 \rho}{\partial \eta^2} + \left(-\frac{P^5}{6} \right) \frac{\partial^2}{\partial \eta^2} \left(\frac{1}{\rho} \right) \right\}$$

$$+ C \left\{ \left(-\frac{5P^3}{6} \right) \frac{\partial^2 \rho}{\partial \zeta^2} + \left(-\frac{P^5}{6} \right) \frac{\partial^2}{\partial \zeta^2} \left(\frac{1}{\rho} \right) \right\}$$

To be rid of the negative signs in the definition of F, define a quantity D as the negative of F, i.e., $D = -F$.

$$u = A\xi - \frac{5P^3 A}{3} \frac{\xi}{\rho^3} - \frac{\partial D}{\partial \xi} \qquad (3.23)$$

The expression for the pressure p is obtained by substituting the expressions for b and for c, and remembering $a = A$,

$$p = 2ck \frac{\partial^2}{\partial \xi^2} \left(\frac{1}{\rho} \right) = 2 \left(-\frac{5}{6} P^3 A \right) k \frac{\partial^2}{\partial \xi^2} \left(\frac{1}{\rho} \right) = -\frac{5}{3} k P^3 A \frac{\partial^2}{\partial \xi^2} \left(\frac{1}{\rho} \right)$$

Using symmetry, one obtains similar results for the situations of $u' = 0, v' = -\frac{\partial}{\partial \eta} \left(\frac{2c}{\rho} \right), w' = 0$, and of $u' = 0, v' = 0, w' = -\frac{\partial}{\partial \zeta} \left(\frac{2c}{\rho} \right)$. The superposition of these three solutions gives for the pressure

$$p = -\frac{5}{3} k P^3 \left[A \frac{\partial^2}{\partial \xi^2} \left(\frac{1}{\rho} \right) + B \frac{\partial^2}{\partial \eta^2} \left(\frac{1}{\rho} \right) + C \frac{\partial^2}{\partial \zeta^2} \left(\frac{1}{\rho} \right) \right] \qquad (3.24)$$

To verify that these solutions satisfy Eq. (3.4), we look first at $\frac{\partial p}{\partial \xi} = k\nabla^2 u$. From Eq. (3.23),

$$k\nabla^2 u = k\nabla^2 \left[A\xi - \frac{5}{3} P^3 A \frac{\xi}{\rho^3} - \frac{\partial D}{\partial \xi} \right] = k\nabla^2 \left[0 + 0 - \frac{\partial D}{\partial \xi} \right] = -k \frac{\partial}{\partial \xi} \left(\nabla^2 D \right)$$

$$= -k \frac{\partial}{\partial \xi} \left[\nabla^2 \left\{ A \left(\frac{5}{6} P^3 \frac{\partial^2 \rho}{\partial \xi^2} + \frac{1}{6} P^5 \frac{\partial^2}{\partial \xi^2} \left(\frac{1}{\rho} \right) \right) + B \left(\ldots \right) + C \left(\ldots \right) \right\} \right]$$

$$= -k \frac{\partial}{\partial \xi} \left\{ A \left[\frac{5}{6} P^3 \frac{\partial^2}{\partial \xi^2} \left(\nabla^2 \rho \right) + \frac{1}{6} P^5 \frac{\partial^2}{\partial \xi^2} \left(\nabla^2 \left(\frac{1}{\rho} \right) \right) \right] + \nabla^2 B \left(\ldots \right) \right.$$

$$\left. + \nabla^2 C \left(\ldots \right) \right\}$$

$$= -k \frac{\partial}{\partial \xi} \left\{ A \left[\frac{5}{6} P^3 \frac{\partial^2}{\partial \xi^2} \left(\frac{2}{\rho} \right) + 0 \right] + \nabla^2 B \left(\ldots \right) + \nabla^2 C \left(\ldots \right) \right\}$$

$$= -k \frac{\partial}{\partial \xi} \left\{ \frac{5}{3} P^3 A \frac{\partial^2}{\partial \xi^2} \left(\frac{1}{\rho} \right) + \frac{5}{3} P^3 B \frac{\partial^2}{\partial \eta^2} \left(\frac{1}{\rho} \right) + \frac{5}{3} P^3 C \frac{\partial^2}{\partial \zeta^2} \left(\frac{1}{\rho} \right) \right\}$$

From Eq. (3.24),

$$\frac{\partial p}{\partial \xi} = \frac{\partial}{\partial \xi} \left\{ -\frac{5}{3}k\mathrm{P}^3 \left[A\frac{\partial^2}{\partial \xi^2}\left(\frac{1}{\rho}\right) + B\frac{\partial^2}{\partial \eta^2}\left(\frac{1}{\rho}\right) + C\frac{\partial^2}{\partial \zeta^2}\left(\frac{1}{\rho}\right) \right] \right\}$$

$$= -k\frac{\partial}{\partial \xi} \left\{ \frac{5}{3}\mathrm{P}^3 A\frac{\partial^2}{\partial \xi^2}\left(\frac{1}{\rho}\right) + \frac{5}{3}\mathrm{P}^3 B\frac{\partial^2}{\partial \eta^2}\left(\frac{1}{\rho}\right) + \frac{5}{3}\mathrm{P}^3 C\frac{\partial^2}{\partial \zeta^2}\left(\frac{1}{\rho}\right) \right\}$$

It is seen this is the same as the expression for $\nabla^2 u$. In a similar manner, one verifies that $\frac{\partial p}{\partial \eta} = k\nabla^2 v$ and that $\frac{\partial p}{\partial \zeta} = k\nabla^2 w$.

The remaining equation of Eq. (3.4) is $\frac{\partial u}{\partial \xi} + \frac{\partial v}{\partial \eta} + \frac{\partial w}{\partial \zeta} = 0$. From Eq. (3.23),

$$\frac{\partial u}{\partial \xi} = \frac{\partial}{\partial \xi}\left\{ A\xi - \frac{5}{3}\mathrm{P}^3 A\left(\frac{\xi}{\rho^3}\right) - \frac{\partial D}{\partial \xi} \right\} = A - \frac{5}{3}\mathrm{P}^3 A\frac{\partial}{\partial \xi}\left(\frac{\xi}{\rho^3}\right) - \frac{\partial^2 D}{\partial \xi^2}$$

$$= A - \frac{5}{3}\mathrm{P}^3 A\left(\frac{1}{\rho^3} - \frac{3\xi^2}{\rho^5}\right) - \frac{\partial D}{\partial \xi^2} = A - \frac{5}{3}\mathrm{P}^3\frac{A}{\rho^3} + 5\mathrm{P}^3 A\frac{\xi^2}{\rho^5} - \frac{\partial^2 D}{\partial \xi^2}$$

In a similar manner,

$$\frac{\partial v}{\partial \eta} = B - \frac{5}{3}\mathrm{P}^3 B\frac{1}{\rho^3} + 5\mathrm{P}^3 B\frac{\eta^2}{\rho^5} - \frac{\partial^2 D}{\partial \eta^2}$$

$$\frac{\partial w}{\partial \zeta} = C - \frac{5}{3}\mathrm{P}^3 C\frac{1}{\rho^3} + 5\mathrm{P}^3 C\frac{\zeta^2}{\rho^5} - \frac{\partial^2 D}{\partial \zeta^2}$$

Adding these expressions,

$$\frac{\partial u}{\partial \xi} + \frac{\partial v}{\partial \eta} + \frac{\partial w}{\partial \zeta} = (A + B + C+) - \frac{5}{3}\frac{\mathrm{P}^3}{\rho^3}(A + B + C)$$

$$+ 5\frac{\mathrm{P}^3}{\rho^5}\left(A\xi^2 + B\eta^2 + C\zeta^2\right) - \nabla^2 D$$

But,

$$\nabla^2 D = \nabla^2 A\left[\frac{5}{6}\mathrm{P}^3\frac{\partial^2 \rho}{\partial \xi^2} + \frac{1}{6}\mathrm{P}^5\frac{\partial^2}{\partial \xi^2}\left(\frac{1}{\rho}\right)\right] + \nabla^2 B\left[\ldots\right] + \nabla^2 C\left[\ldots\right]$$

$$= A\left[\frac{5}{6}\mathrm{P}^3\frac{\partial^2}{\partial \xi^2}\left(\nabla^2\rho\right) + \frac{1}{6}\mathrm{P}^5\frac{\partial^2}{\partial \xi^2}\left(\nabla^2\frac{1}{\rho}\right)\right] + \nabla^2 B\left[\ldots\right] + \nabla^2 C\left[\ldots\right]$$

$$= A\left[\frac{5}{6}\mathrm{P}^3\frac{\partial^2}{\partial \xi^2}\left(\frac{2}{\rho}\right) + 0\right] + \nabla^2 B\left[\ldots\right] + \nabla^2 C\left[\ldots\right]$$

$$= \frac{5}{3}\mathrm{P}^3\left[A\frac{\partial^2}{\partial \xi^2}\left(\frac{1}{\rho}\right) + B\frac{\partial^2}{\partial \eta^2}\left(\frac{1}{\rho}\right) + C\frac{\partial^2}{\partial \zeta^2}\left(\frac{1}{\rho}\right)\right]$$

$$= \frac{5}{3}\mathrm{P}^3\left[A\left(-\frac{1}{\rho^3} + \frac{3\xi^2}{\rho^5}\right) + B\left(-\frac{1}{\rho^3} + \frac{3\eta^2}{\rho^5}\right) + C\left(-\frac{1}{\rho^3} + \frac{3\zeta^2}{\rho^5}\right)\right]$$

$$= -\frac{5}{3}\frac{\mathrm{P}^3}{\rho^3}(A + B + C) + 5\frac{\mathrm{P}^3}{\rho^5}\left(A\xi^2 + B\eta^2 + C\zeta^2\right)$$

Thus, $\frac{\partial u}{\partial \xi} + \frac{\partial v}{\partial \eta} + \frac{\partial w}{\partial \zeta} = A + B + C = 0$.

This shows that the solution, Eq. (3.23), satisfies Eq. (3.4).
The expression for u is

$$u = A\xi - \frac{5}{2}\frac{P^3}{\rho^5}\xi\left(A\xi^2 + B\eta^2 + C\zeta^2\right) + \frac{5}{2}\frac{P^5}{\rho^7}\xi\left(A\xi^2 + B\eta^2 + C\zeta^2\right) - \frac{P^5}{\rho^5}A\xi$$
$$(3.25)$$

For infinitely large values of ρ the expression for u reduces to Eq. (3.1), $u = A\xi$ and, for $\rho = P$, the expression for u reduces to $u = 0$. Similar results hold for v and w. This proves that the solution, Eq. (3.5), satisfies Eq. (3.4) as well as the boundary conditions.

3.5.2 Derivation of the Expression for $W =$ Energy per Unit Time Converted into Heat

Construct a sphere of radius R around the point (x_0, y_0, z_0). Let X, Y, and Z be the components of the pressure at some point (ξ, η, ζ) on the surface of the sphere in the liquid, and let $(\hat{e}_\xi, \hat{e}_\eta, \hat{e}_\zeta)$ be the unit vectors corresponding to the variables (ξ, η, ζ). The energy supplied mechanically to the liquid is $\int_1^2 \vec{F} \cdot d\vec{r} = \int p\,dV$ along a path from point 1 to point 2. For the sphere, the dV is a contraction/expansion of the sphere. This energy is converted into heat at the rate

$$W = \frac{d}{dt}\left(\int_1^2 \vec{F} \cdot d\vec{r}\right) = \int_1^2 \vec{F} \cdot \frac{d\vec{r}}{dt}$$

$$= \int\int_{surface} (Xds\hat{e}_\xi + Yds\hat{e}_\eta + Zds\hat{e}_\zeta) \cdot (u\hat{e}_\xi + v\hat{e}_\eta + w\hat{e}_\zeta)$$

For contraction/expansion, the path is normal to the surface element ds. The unit vector normal to the surface of the sphere is $\hat{e}_n = e_{n\xi}\hat{e}_\xi + e_{n\eta}\hat{e}_\eta + e_{n\zeta}\hat{e}_{N\zeta} = \frac{\xi}{\rho}\hat{e}_\xi + \frac{\eta}{\rho}\hat{e}_\eta + \frac{\zeta}{\rho}\hat{e}_\zeta$.

$$W = \int\int_{surface} (Xu\cos\alpha + Yv\cos\beta + Zw\cos\beta)\,ds$$

where $\cos\alpha, \cos\beta, \cos\gamma$ are the direction cosines between the velocity components and the unit normal vector $\hat{e}_n$. However, $X_n = X\cos\alpha$ is the component of X along the normal. Similarly for Y_n and for Z_n. Thus W can be written as

$$W = \int\int_{surface} (X_n u + Y_n v + Z_n w)\,ds$$

The x-component of the pressure can be written, $X = X_\xi\hat{e}_\xi + X_\eta\hat{e}_\eta + X_\zeta\hat{e}_\zeta,$[104] with corresponding expressions for Y and for Z.

$$X_n = \vec{X} \cdot \hat{e}_n = (X_\xi \hat{e}_\xi + X_\eta \hat{e}_\eta + X_\zeta \hat{e}_\zeta) \cdot \left(\frac{\xi}{\rho} \hat{e}_\xi + \frac{\eta}{\rho} \hat{e}_\eta + \frac{\zeta}{\rho} \hat{e}_\zeta \right)$$

$$X_n = X_\xi \frac{\xi}{\rho} + X_\eta \frac{\eta}{\rho} + X_\zeta \frac{\zeta}{\rho}$$

$$Y_n = Y_\xi \frac{\xi}{\rho} + Y_\eta \frac{\eta}{\rho} + Y_\zeta \frac{\zeta}{\rho}$$

$$Z_n = Z_\xi \frac{\xi}{\rho} + Z_\eta \frac{\mu}{\rho} + Z_\zeta \frac{\zeta}{\rho}$$

For a viscous liquid, the expressions for $X_\xi, X_\eta, \ldots$ are[105]

$$X_\xi = p - 2k\frac{\partial u}{\partial \xi} \qquad\qquad Y_\zeta = Z_\eta = -k \left[\frac{\partial v}{\partial \zeta} + \frac{\partial w}{\partial \eta} \right]$$

$$Y_\eta = p - 2k\frac{\partial v}{\partial \eta} \qquad\qquad Z_\xi = X_\zeta = -k \left[\frac{\partial w}{\partial \xi} + \frac{\partial u}{\partial \zeta} \right]$$

$$Z_\zeta = p - 2k\frac{\partial w}{\partial \zeta} \qquad\qquad X_\eta = Y_\xi = -k \left[\frac{\partial u}{\partial \eta} + \frac{\partial v}{\partial \xi} \right]$$

Recalling that $R \gg P$, where P is the radius of the spherical body, and that on the surface of the sphere of radius R, $\rho = R$, Einstein approximated the expressions for u, v, and w (Eq. 3.5), by dropping terms proportional to powers of P/ρ higher than three.

$$u \approx A\xi - \frac{5}{2}\frac{P^3}{\rho^5}\xi \left(A\xi^2 + B\eta^2 + C\zeta^2 \right)$$

$$v \approx B\eta - \frac{5}{2}\frac{P^3}{\rho^5}\eta \left(A\xi^2 + B\eta^2 + C\zeta^2 \right) \qquad (3.26)$$

$$w \approx C\zeta - \frac{5}{2}\frac{P^3}{\rho^5}\zeta \left(A\xi^2 + B\eta^2 + C\zeta^2 \right)$$

Remembering $\rho = \sqrt{\xi^2 + \eta^2 + \zeta^2}$, the expression for the pressure, Eq. (3.24), becomes

$$p = -\frac{5}{3}k P^3 \left\{ A\frac{\partial^2}{\partial \xi^2}\left(\frac{1}{\rho}\right) + B\frac{\partial^2}{\partial \eta^2}\left(\frac{1}{\rho}\right) + C\frac{\partial^2}{\partial \zeta^2}\left(\frac{1}{\rho}\right) \right\}$$

$$\frac{\partial^2}{\partial \xi^2}\left(\frac{1}{\rho}\right) = \frac{\partial}{\partial \xi}\left[\frac{\partial}{\partial \xi}\left(\frac{1}{\rho}\right) \right] = \frac{\partial}{\partial \xi}\left[-\frac{1}{\rho^2}\frac{\partial \rho}{\partial \xi} \right] = \frac{\partial}{\partial \xi}\left[-\frac{1}{\rho^2}\frac{\xi}{\rho} \right] = \frac{\partial}{\partial \xi}\left[-\frac{\xi}{\rho^3} \right]$$

$$= -\frac{1}{\rho^3} - \left(-\frac{3\xi}{\rho^4} \right)\left(\frac{\xi}{\rho}\right) = -\frac{1}{\rho^3} + \frac{3\xi^2}{\rho^5}$$

$$p = -\frac{5}{3}kP^3 \left\{ A\left(-\frac{1}{\rho^3} + \frac{3\xi^2}{\rho^5}\right) + B\left(-\frac{1}{\rho^3} + \frac{3\eta^2}{\rho^5}\right) \right.$$

$$\left. +C\left(-\frac{1}{\rho^3} + \frac{3\zeta^2}{\rho^5}\right) \right\} + constant$$

$$= -5k\frac{P^3}{\rho^5}\left(A\xi^2 + B\eta^2 + C\zeta^2\right) + \frac{5}{3}k\frac{P^3}{\rho^3}\left(A + B + C\right) + constant$$

But, from Eq. (3.2), $A + B + C = 0$, giving

$$p = -5k\frac{P^3}{\rho^5}\left(A\xi^2 + B\eta^2 + C\zeta^2\right) + constant$$

The expressions for X_ξ, X_η, X_ζ are calculated to be[106]

$$X_\xi = p - 2k\frac{\partial u}{\partial \xi}$$

$$= \left[-5k\frac{P^3}{\rho^5}\left(A\xi^2 + B\eta^2 + C\zeta^2\right)\right]$$

$$- 2k\left[A - \frac{5}{2}P^3\left(\frac{-5}{\rho^6}\frac{\xi}{\rho}\right)\xi\left(A\xi^2 + B\eta^2 + C\zeta^2\right)\right.$$

$$\left. -\frac{5}{2}\frac{P^3}{\rho^5}\left(A\xi^2 + B\eta^2 + C\zeta^2\right) - \frac{5}{2}\frac{P^3}{\rho^5}\xi\left(2A\xi\right)\right]$$

$$= -2kA + 10k\frac{P^3}{\rho^5}A\xi^2 - 25k\frac{P^3}{\rho^7}\xi^2\left(A\xi^2 + B\eta^2 + C\zeta^2\right)$$

$$X_\eta = -k\left[\frac{\partial u}{\partial \eta} + \frac{\partial v}{\partial \xi}\right]$$

$$= -k\left[-\frac{5}{2}P^3\left\{\left(-\frac{5}{\rho^6}\frac{\eta}{\rho}\right)\xi\left(A\xi^2 + B\eta^2 + C\zeta^2\right) + \frac{\xi}{\rho^5}2B\eta\right\}\right]$$

$$- k\left[-\frac{5}{2}P^3\left\{\left(-\frac{5}{\rho^6}\frac{\xi}{\rho}\right)\eta\left(A\xi^2 + B\eta^2 + C\zeta^2\right)\right\} + \frac{\eta}{\rho^5}2A\xi\right]$$

$$= 5k\frac{P^3}{\rho^5}\xi\eta\left(A + B\right) - 25k\frac{P^3}{\rho^7}\eta\xi\left(A\xi^2 + B\eta^2 + C\zeta^2\right)$$

Similarly,

$$X_\zeta = 5k\frac{P^3}{\rho^5}\xi\zeta\left(A + C\right) - 25k\frac{P^3}{\rho^7}\xi\zeta\left(A\xi^2 + B\eta^2 + C\zeta^2\right)$$

The expression for X_n is calculated to be[107]

$$X_n = -\left[X_\xi \frac{\xi}{\rho} + X_\eta \frac{\eta}{\rho} + X_\zeta \frac{\zeta}{\rho} \right]$$

$$= -\left[-2kA + 10k\frac{P^3}{\rho^5} A\xi^2 - 25k\frac{P^3}{\rho^7}\xi^2 \left(A\xi^2 + B\eta^2 + C\zeta^2\right) \right] \times \frac{\xi}{\rho}$$

$$- \left[-5k\frac{P^3}{\rho^5}\xi\eta\left(A + B\right) - 25k\frac{P^3}{\rho^7}\eta\xi\left(a\xi^2 + B\eta^2 + C\zeta^2\right) \right] \times \frac{\eta}{\rho}$$

$$- \left[-5k\frac{P^3}{\rho^5}\xi\zeta\left(A + C\right) - 25k\frac{P^3}{\rho^7}\xi\zeta\left(a\xi^2 + B\eta^2 + C\zeta^2\right) \right] \times \frac{\zeta}{\rho}$$

$$= 2Ak\frac{\xi}{\rho} - k\xi\frac{P^3}{\rho^6}\left[10A\xi^2 + 5\eta^2\left(A + B\right) + 5\zeta^2\left(A + C\right)\right]$$

$$+ 25\frac{P^3}{\rho^8}\left(A\xi^2 + B\eta^2 + C\zeta^2\right)\left(\xi^2 + \eta^2 + \zeta^2\right)$$

$$= 2Ak\frac{\xi}{\rho} - k\xi\frac{P^3}{\rho^6}\left[5A\left(\xi^2 + \eta^2 + \zeta^2\right) + 5\left(A\xi^2 + B\eta^2 + C\zeta^2\right)\right]$$

$$+ 25\frac{P^3}{\rho^8}\left(A\xi^2 + B\eta^2 + C\zeta^2\right)\left(\xi^2 + \eta^2 + \zeta^2\right)$$

Since $\xi^2 + \eta^2 + \zeta^2 = \rho^2$,

$$X_n = 2Ak\frac{\xi}{\rho} - 5Ak\frac{P^3}{\rho^4}\xi + 20k\frac{P^3}{\rho^6}\xi\left(A\xi^2 + B\eta^2 + C\zeta^2\right)$$

In a similar manner,

$$Y_n = 2Bk\frac{\eta}{\rho} - 5Bk\frac{P^3}{\rho^4}\eta + 20k\frac{P^3}{\rho^6}\eta\left(A\xi^2 + B\eta^2 + C\zeta^2\right)$$

$$Z_n = 2Ck\frac{\zeta}{\rho} - 5Ck\frac{P^3}{\rho^4}\zeta + 20k\frac{P^3}{\rho^6}\zeta\left(A\xi^2 + B\eta^2 + C\zeta^2\right)$$

Putting together these pieces for the function in the integral for W, and neglecting all terms of higher order than three in $P\rho$,[108]

$$X_n u + Y_n v + Z_n w$$

$$= \left(2Ak\frac{\xi}{\rho} - 5Ak\frac{P^3}{\rho^4}\xi + 20k\frac{P^3}{\rho^6}\xi\left(A\xi^2 + B\eta^2 + C\zeta^2\right)\right)$$

$$\times \left(A\xi - \frac{5}{2}\frac{P^3}{\rho^5}\xi\left(A\xi^2 + B\eta^2 + C\zeta^2\right)\right)$$

$$+ \left(2Bk\frac{\eta}{\rho} - 4Bk\frac{P^3}{\rho^4}\eta + 20k\frac{P^3}{\rho^6}\eta\left(A\xi^2 + B\eta^2 + C\zeta^2\right)\right)$$

$$\times \left(B\eta - \frac{5}{2}\frac{P^3}{\rho^5}\eta\left(A\xi^2 + B\eta^2 + C\zeta^2\right)\right)$$

$$+ \left(2Ck\frac{\zeta}{\rho} - 5Ck\frac{P^3}{\rho^4}\zeta + 20k\frac{P^3}{\rho^6}\zeta\left(A\xi^2 + B\eta^2 + C\zeta^2\right)\right)$$

$$\times \left(C\zeta - \frac{5}{2}\frac{P^3}{\rho^5}\zeta\left(A\xi^2 + B\eta^2 + C\zeta^2\right)\right)$$

$$= \frac{2k}{\rho}\left(A^2\xi^2 + B^2\eta^2 + C^2\zeta^2\right) - 5k\frac{P^3}{\rho^4}\left(A^2\xi^2 + B^2\eta^2 + C^2\zeta^2\right)$$

$$- 15k\frac{P^3}{\rho^6}\left(A\xi^2 + B\eta^2 + C\zeta^2\right)^2 + \cdots$$

Using the following integrals over the surface of the sphere (with $R^2 = \xi^2 + \eta^2 + \zeta^2$, $ds = R^2\sin\theta d\theta d\phi$, $\xi = R\sin\theta\cos\phi$, $\eta = R\sin\theta\sin\phi$, $\zeta = R\cos\theta$, with $0 \le \theta \le \pi$, and $0 \le \phi \le 2\pi$),[109]

$$\int ds = 4\pi R^2$$

$$\int \xi^2 ds = \int \eta^2 ds = \int \zeta^2 ds = \frac{4}{3}\pi R^4$$

$$\int \xi^4 ds = \int \eta^4 ds = \int \zeta^4 ds = \frac{4}{5}\pi R^6$$

$$\int \eta^2\zeta^2 ds = \int \eta^2\xi^2 ds = \int \xi^2\eta^2 ds = \frac{4}{15}\pi R^6$$

$$\int \left(A\xi^2 + B\eta^2 + C\zeta^2\right)^2 ds$$

$$= \int \left(A^2\xi^4 + B^2\eta^4 + C\zeta^4 + 2AB\xi^2\eta^2 + 2AC\xi^2\zeta^2 + 2BC\eta^2\zeta^2\right) ds$$

$$= \left(A^2 + B^2 + C^2\right)\frac{4}{5}\pi R^6 + 2\left(AB + AC + BC\right)\frac{4}{15}\pi R^6$$

Using $A + B + C = 0 \Rightarrow (A + B + C)^2 = A^2 + B^2 + C^2 + 2(AB + AC + BC) = 0$,

$$\Rightarrow 2(AB + AC + BC) = -(A^2 + B^2 + C^2)$$

$$\int \left(A\xi^2 + B\eta^2 + C\zeta^2\right)^2 ds = \left(A^2 + B^2 + C^2\right) \frac{4}{5}\pi R^6 - \left(A^2 + B^2 + C^2\right)$$

$$\times \frac{4}{15}\pi R^6 = \frac{8}{15}\pi R^6 \left(A^2 + B^2 + C^2\right)$$

On the surface of the sphere, $\rho = R$. The expression for W, is calculated to be[110]

$$W = \int (X_n u + Y_n v + Z_n w)\, ds = \int \left[\frac{2k}{\rho} \left(A^2\xi^2 + B^2\eta^2 + C^2\zeta^2\right) \right.$$

$$\left. -5k\frac{P^3}{\rho^4} \left(A^2\xi^2 + B^2\eta^2 + C^2\zeta^2\right) + 15k\frac{P^3}{\rho^6} \left(A^2\xi^2 + B^2\eta^2 + C^2\zeta^2\right) \right] ds$$

$$= 2k\delta^2 \left(\frac{4}{3}\pi R^3\right) + k\delta^2 \left(\frac{4}{3}\pi P^3\right)$$

$$= 2\delta^2 k \left(V + \frac{\Phi}{2}\right) \tag{3.27}$$

with $\delta^2 = A^2 + B^2 + C^2$, $V = \frac{4}{3}\pi R^3$, and $\Phi = \frac{4}{3}\pi P^3$, where V is the volume of the space enclosed and Φ is the volume of the suspended sphere. This is the corrected version of Eq. (3.6).

3.5.3 Derivation of the Coefficient of Viscosity of a Liquid in Which Very Many Irregularly Distributed Spheres are Suspended

Consider a spherical region G containing a large number of randomly distributed spheres. See Figure 3.7.

For one sphere present, the expressions for the components of the velocity were given by Eq. (3.3) and Eq. (3.6). A sphere at location

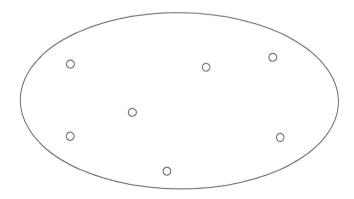

Fig. 3.7 The region G containing a large number of randomly distributed spheres.

x_ν, y_ν, z_ν will affect the velocity according to these equations. For several spheres, assuming the average distance between spheres to be large compared with their radius, the expression for the components of velocity at location x, y, z from all of the spheres is just the sum of the individual contributions:

$$u = Ax + \sum_\nu u_\nu$$

$$= \sum_\nu \left[\frac{5}{2} \frac{P^3}{\rho_\nu^2} \frac{\xi_\nu \left(A\xi_\nu^2 + B\eta_\nu^2 + C\zeta_\nu^2 \right)}{\rho_\nu^3} - \frac{5}{2} \frac{P^5}{\rho_\nu^4} \frac{\xi_\nu \left(A\xi_\nu^2 + B\eta_\nu^2 + C\zeta_\nu^2 \right)}{\rho_\nu^3} + \frac{P^5}{\rho_\nu^4} \frac{A\xi_\nu}{\rho_\nu} \right]$$

$$v = By + \sum_\nu v_\nu$$

$$= \sum_\nu \left[\frac{5}{2} \frac{P^3}{\rho_\nu^2} \frac{\eta_\nu \left(A\xi_\nu^2 + B\eta_\nu^2 + C\zeta_\nu^2 \right)}{\rho_\nu^3} - \frac{5}{2} \frac{P^5}{\rho_\nu^4} \frac{\eta_\nu \left(A\xi_\nu^2 + B\eta_\nu^2 + C\zeta_\nu^2 \right)}{\rho_\nu^3} + \frac{P^5}{\rho_\nu^4} \frac{B\eta_\nu}{\rho_\nu} \right]$$

$$w = Cz + \sum_\nu w_\nu$$

$$= \sum_\nu \left[\frac{5}{2} \frac{P^3}{\rho_\nu^2} \frac{\zeta_\nu \left(A\xi_\nu^2 + B\eta_\nu^2 + C\zeta_\nu^2 \right)}{\rho_\nu^3} - \frac{5}{2} \frac{P^5}{\rho_\nu^4} \frac{\zeta_\nu \left(A\xi_\nu^2 + B\eta_\nu^2 + C\zeta_\nu^2 \right)}{\rho_\nu^3} + \frac{P^5}{\rho_\nu^4} \frac{C\zeta_\nu}{\rho_\nu} \right]$$

$$\text{(3.8)}$$

with $\xi_\nu = x - x_\nu, \eta_\nu = y - y_\nu, \zeta = z - z_\nu, \rho_\nu = \sqrt{\xi_\nu^2 + \eta_\nu^2 + \zeta_\nu^2}$. From Eq. (3.27), for one sphere present, the heat production per unit time is $W = 2\delta^2 k (V + \frac{\Phi}{2})$. For N spheres present, the heat production per unit time becomes $W = 2\delta^2 k (V + \frac{N\Phi}{2})$ for the volume V. Dividing by V, the heat production per unit time per unit volume is

$$W^* = 2\delta^2 k \left(1 + \frac{1}{2} \frac{N\Phi}{V} \right) = 2\delta^2 k \left(1 + \frac{\varphi}{2} \right) \qquad \text{(3.28)}$$

where $\varphi = \frac{N\Phi}{V}$ is the fraction of the volume that is occupied by the spheres. This is the corrected version of Eq. (3.8).

The principal dilatations of the liquid motion with no spheres present were A, B, and C. These could be defined as $A = \left(\frac{\partial u_0}{\partial x} \right)$, etc., with $u_0 = Ax$, etc. With the spheres present, the principal dilatations at the origin are defined as

$$A^* = \left(\frac{\partial u}{\partial x} \right)_0 = A + \sum_\nu \left(\frac{\partial u_\nu}{\partial x} \right)_{x=0} = A - \sum_\nu \left(\frac{\partial u_\nu}{\partial x_\nu} \right)_{x=0}$$

with analogous expressions for B^* and C^*. Since the spheres are spaced far apart and their radius P is small compared to this distance, the alterations to the velocity components far from the spheres can be approximated from Eqs. (3.1), (3.3), and (3.26), keeping only terms to the third power in P/ρ_ν.

$$u_\nu = -\frac{5}{2}\frac{\mathrm{P}^3}{\rho_\nu^3}\frac{1}{\rho_\nu^2}\xi_\nu\left(A\xi_\nu^2 + B\eta_\nu^2 + C\zeta_\nu^2\right)$$

$$v_\nu = -\frac{5}{2}\frac{\mathrm{P}^3}{\rho_\nu^3}\frac{1}{\rho_\nu^2}\eta_\nu\left(A\xi_\nu^2 + B\eta_\nu^2 + C\zeta_\nu^2\right)$$

$$w_\nu = -\frac{5}{2}\frac{\mathrm{P}^3}{\rho_\nu^3}\frac{1}{\rho_\nu^2}\zeta_\nu\left(A\xi_\nu^2 + B\eta_\nu^2 + C\zeta_\nu^2\right)$$

At the origin, $\xi_\nu = x - x_\nu \to \xi_\nu = -x_\nu$ and $\rho_\nu = \sqrt{\xi_\nu^2 + \eta_\nu^2 + \zeta_\nu^2} \to r_\nu = \sqrt{x_\nu^2 + y_\nu^2 + z_\nu^2}$.

$$u_\nu = +\frac{5}{2}\frac{\mathrm{P}^3}{r_\nu^3}\frac{1}{r_\nu^2}x_\nu\left(Ax_\nu^2 + By_\nu^2 + Cz_\nu^2\right)$$

$$v_\nu = +\frac{5}{2}\frac{\mathrm{P}^3}{r_\nu^3}\frac{1}{r_\nu^2}y_\nu\left(Ax_\nu^2 + By_\nu^2 + Cz_\nu^2\right)$$

$$w_\nu = +\frac{5}{2}\frac{\mathrm{P}^3}{r_\nu^3}\frac{1}{r_\nu^2}z_\nu\left(Ax_\nu^2 + By_\nu^2 + Cz_\nu^2\right)$$

The expression for A^* becomes

$$A^* = A - \sum_\nu \left(\frac{\partial u_\nu}{\partial x_\nu}\right)_{x=0} \to A - \int ndV\left(\frac{\partial u_\nu}{\partial x_\nu}\right)$$

with $n = \#\ spheres/volume = $ constant,

$$A^* = A - n\int\left(\frac{\partial u_\nu}{\partial x_\nu}\right)_{x=0} dV$$

Writing $dV = dr_\nu\,dS$ and $\left(\frac{\partial u_\nu}{\partial x_\nu}\right)_{x=0} = \left(\frac{\partial u_\nu}{\partial r_\nu}\frac{\partial r_\nu}{\partial x_\nu}\right)_{x=0} = \left(\frac{\partial u_\nu}{\partial r_\nu}\frac{x_\nu}{r_\nu}\right)_{x=0}$ and canceling the dr_ν terms,

$$A^* = A - n\int du_\nu\frac{x_\nu}{r_\nu}ds = A - n\int u_\nu\frac{x_\nu}{r_\nu}ds$$

$$= A - n\frac{5}{2}\mathrm{P}^3\int dS\left[\frac{x_\nu^2}{r_\nu^6}\left(Ax_\nu^2 + By_\nu^2 + Cz_\nu^2\right)\right]$$

But $r_\nu = R$ on the surface, so

$$A^* = A - n\frac{5}{2}\frac{\mathrm{P}^3}{R^6}\int dS\left\{Ax_\nu^4 + By_\nu^2x_\nu^2 + Cz_\nu^2x_\nu^2\right\}$$

Using the integrals from the paper,[111]

$$A^* = A - n\frac{5}{2}\frac{\mathrm{P}^3}{R^6}\left[A\frac{4}{5}\pi R^6 + B\frac{4}{15}\pi R^6 + C\frac{4}{15}\pi R^6\right]$$

$$= A - n\frac{5}{2}\frac{\mathrm{P}^3}{R^6}\frac{4}{15}\pi R^6\left\{3A + B + C\right\}$$

But $3A + B + C = 2A + (A + B + C) = 2A$,

$$A^* = A - n\left(\frac{4}{3}\pi P^3\right) A = A\left(1 - n\Phi\right) = A\left(1 - \varphi\right)$$

Analogously, $B^* = B(1 - \varphi)$ and $C^* = C(1 - \varphi)$. Similar to $\delta^2 = A^2 + B^2 + C^2$, $(\delta^*)^2$ is defined as

$$\begin{aligned}(\delta^*)^2 &= (A^*)^2 + (B^*)^2 + (C^*)^2 \\ &= A^2\left(1 - \varphi\right)^2 + B^2\left(1 - \varphi\right)^2 + C^2\left(1 - \varphi\right)^2 \\ &= \delta^2\left(1 - \varphi\right)^2 = \delta^2\left(1 - 2\varphi\right)\end{aligned}$$

to first order in φ. With no spheres present, the heat generated per unit time per unit volume is $W = 2\delta^2 k$. If k^* denotes the coefficient of viscosity of the mixture, the heat generated per unit time per unit volume can be written as $W^* = 2(\delta^*)^2 k^*$. Equating this expression for W^* to the expression in Eq. (3.7b), Einstein obtains the desired expression for k^*:

$$2\delta^2 k\left(1 + \frac{\varphi}{2}\right) = 2\delta^2\left(1 - 2\varphi\right) k^*$$

$$k^* = k\left(\frac{1 + \varphi/2}{1 - 2\varphi}\right) = k\left(1 + \frac{\varphi}{2}\right)(1 + 2\varphi + \ldots) = k\left(1 + 2.5\varphi\right) \quad (3.29)$$

This is the corrected version of Eq. (3.9).

3.5.4 Determination of the Volume of a Dissolved Substance

From Section 3.2.2, using the corrected expressions,

$$\frac{k^*}{k} = 1 + 2.5\varphi$$

Using data for a 1% aqueous solution of sugar (1 gm of sugar in $100\,\text{cm}^3$ of water), the value of k^*/k was 1.0245. Thus $2.5\varphi = 0.0245$ and $\varphi = 0.0098 = \frac{volume\ spheres}{total\ volume} = \frac{volume\ spheres}{100\,\text{cm}^3}$. The total volume of the spheres in 100ml of water is 0.98 cm^3.

One gram of sugar has a volume of $0.61\,\text{cm}^3$. However, regarding viscosity, one gram of sugar has the effect of a volume of $0.98\,\text{cm}^3$. The ratio of $\frac{0.98}{0.61} \approx 1.6$ indicates the quantity of water bound to the sugar molecule has a volume about 60% of that of the sugar molecule.[112]

The molecular weight of sugar is 342 gm. If N is the number of molecules in 342 gm of sugar, there are $\frac{N}{342}$ molecules in one gm of sugar. The total volume of one molecule of dissolved sugar is $\frac{0.98\,\text{cm}^3}{N/342} = 0.98\,\text{cm}^3\frac{342}{N}$.

3.5.5 Derivation of the Expression for Entropy

This follows nearly identically the presentation in Albert Einstein's 1903 paper "A Theory of the Foundations of Thermodynamics,"[113] where Einstein addresses first the First Law of Thermodynamics,

$$dE = dW - dQ$$

Consider a system whose state is described completely by the state variables $(p_1, p_2, \ldots p_{3n}, q_1, q_2, \ldots q_{3n})$. The system is now allowed to undergo a small, very slow change. In addition to the state variables specifying the state of the system, there may be a set of parameters $(\lambda_1, \lambda_2, \ldots)$ describing the state of the system. The change in energy of the system in time dt is given by

$$dE = \sum \frac{\partial E}{\partial \lambda} d\lambda + \sum \left(\frac{\partial E}{\partial p_\nu} dp_\nu + \frac{\partial E}{\partial q_\nu} dq_\nu \right)$$

The first term is identified with the work done on the system, and the second term is identified with the heat dQ supplied to the system:

$$dE = \sum \frac{\partial E}{\partial \lambda} d\lambda + dQ$$

Consider a system undergoing this change adiabatically ($dQ = 0$). Before the change, the system was in a stationary state, with the probability of being in a given configuration given by[114]

$$dW = (constant = e^c)\, e^{-\frac{E}{kT}} dp_1 dp_2 \ldots dp_{3n} dq_1 dq_2 \ldots dq_{3n}$$

$$= e^{c - \frac{E}{kT}} dp_1 dp_2 \ldots dp_{3n} dq_1 dq_2 \ldots dq_{3n}$$

The constant c is determined from

$$\int dW = 1 = \int e^{c - 2\frac{E}{kT}} dp_1 dp_2 \ldots dp_{3n} dq_1 dq_2 \ldots dq_{3n}$$

$$= e^c \int e^{-\frac{E}{kT}} dp_1 dp_2 \ldots dp_{3n} dq_1 dq_2 \ldots dq_{3n}$$

$$c = -\ln \left\{ \int e^{-\frac{E}{kT}} dp_1 dp_2 \ldots dp_{3n} dq_1 dq_2 \ldots dq_{3n} \right\}$$

After the process, the system is again in a stationary state, with similar expressions for the probability dW for a given configuration, but with the possibility the energy and the constants c and $h(2h = \frac{1}{kT})$ will have changed slightly: $c \to c + dc, h \to h + dh$.

$$E \to E + dE = E + \sum \frac{\partial E}{\partial \lambda} d\lambda$$

$$\int dW = 1 = \int e^{(c+dc) - 2(h+dh)\left(E + \sum \frac{\partial E}{\partial \lambda} d\lambda \right)} dp_1 dp_2 dp_{3n} dq_1 dq_2 \ldots dq_{3n}$$

$$1 = \int e^{dc - 2\left(Edh + h \sum \frac{\partial E}{\partial \lambda} d\lambda + dh \sum \frac{\partial E}{\partial \lambda} d\lambda \right)} e^{c - \frac{E}{kT}} dp_1 dp_2 \ldots dp_{3n} dq_1 dq_2 \ldots dq_{3n}$$

Expanding the first exponential in a Taylor series, and neglecting terms of higher than first order:

$$1 = \int \left[1 + \left(dc - 2 \left(Edh + h \sum \frac{\partial E}{\partial \lambda} d\lambda + dh \sum \frac{\partial E}{\partial \lambda} d\lambda \right) \right) \right.$$
$$\left. + \frac{1}{2} \left(dc - 2 \left(\ldots \right) \right)^2 + \ldots \right] e^{c - \frac{E}{kT}} dp_1 dp_2 \ldots dp_{3n} dq_1 dq_2 \ldots dq_{3n}$$

$$1 = \int \left[1 + \left(dc - 2 \left(Edh + h \sum \frac{\partial E}{\partial \lambda} d\lambda + dh \sum \frac{\partial E}{\partial \lambda} d\lambda \right) \right) \right]$$
$$\times \ e^{c - \frac{E}{kT}} dp_1 dp_2 \ldots dp_{3n} dq_1 dq_2 \ldots dq_{3n}$$

$$0 = \int \left[\left(dc - 2 \left(Edh + h \sum \frac{\partial E}{\partial \lambda} d\lambda \right) \right) \right] e^{c - \frac{E}{kT}} dp_1 dp_2 \ldots dp_{3n} dq_1 dq_2 \ldots dq_{3n}$$

Since the term $e^{c - \frac{E}{kT}}$ is always positive (actually, non-negative), for the integral to equal zero the expression in $[\ldots]$ must be equal to zero:

$$dc - 2Edh - 2h \sum \frac{\partial E}{\partial \lambda} d\lambda = 0$$

But, from $dE = \sum \frac{\partial E}{\partial \lambda} d\lambda + dQ$, we have

$$-2hdE + 2h \sum \frac{\partial E}{\partial \lambda} d\lambda + 2hdQ = 0$$

Adding these two equations

$$2hdQ = d(2hE - c)$$

Substituting $2h = 1/(kT)$

$$\frac{dQ}{T} = d \left[\frac{E}{T} - kc \right] = dS$$

"This equation states that dQ/T is a total differential of a quantity that we will call the entropy S of the system. Taking into account the definition of c, one obtains,"[115]

$$S = \frac{E}{T} + k \ln \left\{ \int e^{-\frac{E}{kT}} dp_1 dp_2 \ldots dp_{3n} dq_1 dq_2 \ldots dq_{3n} \right\}$$

3.5.6 Derivation of $B = JV^{*n}$

The total volume of the n dissolved molecules or suspended particles in V^* is assumed to be small relative to the total volume V^*:

$$B = \int dB = \int e^{-2\frac{E}{kT}} dp_{x_1} dp_{y_1} \ldots dy_n dz_n$$

x_1, y_1, z_1 are the rectangular (Cartesian) coordinates of the center of mass of the first particle, and $\ldots, x_n, y_n, z_n$ the rectangular coordinates of the center of mass of the nth particle. Around each particle is an infinitesimally small parallelepiped region,

$dx_1 dy_1 dz_1, dx_2 dy_2 dz_2, \ldots, dx_n dy_n dz_n$, each of which lies in the volume V^*. It is claimed that the integral can be put into the form

$$dB = dx_1 dy_1 \ldots dz_n J$$

where J is independent of $dx_1 dy_1$, etc., as well as of $x_1, y_1, \ldots, z_n$, and V^*. To show this, consider the same n particles, but with different locations $x_1', y_1', \ldots, z_n'$ and with the small parallelepiped regions designated as $dx_1' dy_1' dz_1', dx_2' dy_2' dz_2', \ldots, dz_n' dy_n' dz_n'$, but with the volume of the parallelepiped region around each particle the same as in the original configuration., i.e. $dx_1 dx_2 \ldots dz_n = dx_1' dx_2' \ldots dz_n'$. In this second configuration, we have $dB' = dx_1' dy' \ldots dz_n' J'$. Dividing the two expressions for dB,

$$\frac{dB}{dB'}, \frac{J}{J'}$$

But the probability the particles are in the first configuration is dB/B, while the probability the particles are in the second configuration is dB'/B, since the integral over all configurations is the same, i.e., $B = B'$. However, the two probabilities must be the same if the motions of the particles are independent and the size of the regions $dx_1 dy_1 dz_1$, etc., are the same:

$$\frac{dB}{B} = \frac{dB'}{B}$$

But $dB = dB'$, from the prior equation, $\Rightarrow J = J'$. This shows that J does not depend on the location of the particles $x_1 y_1 z_1 \ldots z_n$, nor on V^*. Integrating the expression for B

$$B' \int dB = \int_{V^*} J dx_1 dy_1 dz_1 dx_2 \ldots dy_n dz_n$$

$$= J \int_{V^*} dx_1 dy_1 dz_1 dx_2 \ldots dy_n dz_n = JV^{*n}$$

3.5.7   Derivation of $\nu = f(x,t)$

From section 3 of Einstein's paper, $\varphi(\Delta)$ satisfies

$$dn = n\varphi(\Delta) d\Delta$$

$$\int_{-\infty}^{+\infty} \varphi(\Delta) d\Delta = 1$$

$$\varphi(\Delta) = \varphi(-\Delta)$$

where dn is the number of particles experiencing a displacement between Δ and $\Delta + d\Delta$ in the time interval τ. The total number of particles in the liquid is n, while ν is the number of particles per unit volume, $\nu = n/V$.

Designate by $\nu = f(x,t)$ the number of particles per unit volume at location x at time t. The number of particles per unit volume at location

$x + \Delta$ at time t is $f(x+\Delta, t)$. The number of these that will move a distance Δ to the location x in time τ is

$$df(x, t+\tau) = f(x+\Delta, t)\, \varphi(\Delta)\, d\Delta$$

The total number of particles per unit volume at location x at time $t + \tau$ is the sum (integral) of the df contributions. Since φ differs from zero only for very small values of Δ, for convenience of calculation, the limits of integration are extended to $\Delta = \pm\infty$. (In Section 3.5.8 this allows the use of Gaussian integrals.)

$$f(x, t+\tau) = \int_{\Delta=-\infty}^{\Delta=+\infty} f(x+\Delta, t)\, \varphi(\Delta)\, d\Delta$$

Since τ is small, $f(x, t+\tau)$ can be expanded in a Taylor series in τ,

$$f(x, t+\tau) = f(x, t) + \tau\frac{\partial f}{\partial t} + \frac{1}{2}\tau^2\frac{\partial^2 f}{\partial t^2} + \ldots \approx f(x, t) + \tau\frac{\partial f}{\partial t}$$

Since $\varphi(\Delta)$ differs from zero only for very small values of Δ, $f(x+\Delta, t)$ can be expanded in a Taylor series in Δ,

$$f(x+\Delta, t) = f(x, t) + \Delta\frac{\partial f}{\partial x} + \frac{1}{2}\Delta^2\frac{\partial^2 f}{\partial x^2} + \ldots \approx f(x, t) + \Delta\frac{\partial f}{\partial x} + \frac{1}{2}\Delta^2\frac{\partial^2 f}{\partial x^2}$$

Substituting into the above equation for $f(x, t)$

$$f(x, t) + \tau\frac{\partial f}{\partial t} = \int_{\Delta=-\infty}^{\Delta=+\infty}\left[f(x, t) + \Delta\frac{\partial f}{\partial t} + \frac{1}{2}\Delta^2\frac{\partial^2 f}{\partial t^2}\right]\varphi(\Delta)\, d\Delta$$

$$= f(x, t)\int_{\Delta=-\infty}^{\Delta=+\infty}\varphi(\Delta)\, d\Delta + \frac{\partial f}{\partial t}\int_{\Delta=-\infty}^{\Delta=+\infty}\Delta\varphi(\Delta)\, d\Delta$$

$$+ \frac{1}{2}\frac{\partial^2 f}{\partial t^2}\int_{\Delta=-\infty}^{\Delta=+\infty}\Delta^2\varphi(\Delta)\, d\Delta$$

On the right-hand side of the equation, the first integral is equal to unity by the definition of $\varphi(\Delta)$, and the second integral is zero since the integral of an odd function between symmetric limits is zero (Δ is an odd function of Δ, while $\varphi(\Delta)$ is an even function of Δ since $\varphi(\Delta) = \varphi(-\Delta)$). Rewriting,

$$\frac{\partial f}{\partial t} = \frac{1}{2\tau}\frac{\partial^2 f}{\partial t^2}\int_{\Delta=-\infty}^{\Delta=+\infty}\Delta^2\varphi(\Delta)d\Delta = D\frac{\partial^2 f}{\partial t^2}$$

This is the diffusion equation, with D the coefficient of diffusion, defined as

$$D = \frac{1}{2\tau} \int\limits_{\Delta=-\infty}^{\Delta=+\infty} \Delta^2 \varphi(\Delta) d\Delta$$

"The problem, which coincides with the problem of diffusion from one point ... is now completely determined mathematically; its solution is ..."[116]

$$f(x.t) = \frac{n}{\sqrt{4\pi D}} \frac{e^{-\frac{x^2}{4Dt}}}{\sqrt{t}}$$

3.5.8　Derivation of $\langle x^2 \rangle$

$f(x,t)$ is the expression for the number of particles with a displacement between x and $x + dx$ in time t. Dividing this expression by n, the number of particles in the volume, will give the probability that a single particle will experience a displacement between x and $x + dx$ in time t. Using the Gaussian integral $\int\limits_{-\infty}^{+\infty} e^{-ax^2} dx = \sqrt{\frac{\pi}{a}}$ one obtains

$\frac{1}{n} \int\limits_{-\infty}^{+\infty} f(x,t) dx = 1$. The average of the square of the displacements is,

noting $\int\limits_{-\infty}^{+\infty} x^2 e^{-ax^2} dx = \frac{1}{2} \sqrt{\frac{\pi}{a^3}}$,

$$\langle x^2 \rangle = \int\limits_{-\infty}^{+\infty} x^2 \frac{1}{n} f(x,t) dx = \frac{1}{\sqrt{4\pi D}} \frac{1}{\sqrt{t}} \int\limits_{-\infty}^{+\infty} x^2 e^{-\frac{x^2}{4Dt}} dx = 2Dt$$

The square root of this expression is the rms value, λ_x, for the average displacement

$$\lambda_x = \sqrt{\langle x^2 \rangle} = \sqrt{2Dt}$$

3.6　Notes

1. Einstein, Albert, A New Determination of Molecular Dimensions. Dissertation, University of Zürich, 1905; Stachel, John, editor, *The Collected Papers of Albert Einstein*, Volume 2, [CPAE2], Princeton University Press, Princeton, NJ, 1989, pp. 183–202; English translation by Anna Beck [CPAE2 ET], pp. 104–122. The original text contains a number of editorial comments and introductory comments (pp. 170–182) that are quite informative.
2. Einstein, Albert, On the Movement of Small Particles Suspended in Stationary Liquids Required by the Molecular-Kinetic Theory of Heat. *Annalen der Physik* 354 (7), (1905), pp. 549–560; [CPAE2, pp. 223–235;

CPAE2 ET, pp. 123–134]. The original text contains a number of editorial comments and introductory comments (pp. 206–222) that are quite informative.

3. Stachel, John, editor, *Einstein's Miraculous Year*, Princeton University Press, Princeton, NJ, 1998, p. 31.

4. Stachel, John, *Einstein's Miraculous Year*, pp. 31–32.

5. Stachel, John, *Einstein's Miraculous Year*, p. 33.

6. Einstein, Albert, Movement of Small Particles; [CPAE2, p. 223; CPAE2 ET, p. 123].

7. Sambur.sky, Shmuel, *Physical Thought from the Presocratics to the Quantum Physicists*, Pica Press, New York, 1975, p. 37.

8. Boorse, Henry A., and Motz, Lloyd, editors, *The World of the Atom*, Volume 1, Basic Books, Inc., New York, 1966, p. 4.

9. Boorse, Henry A., and Motz, Lloyd, *The World of the Atom*, p. 5.

10. Sambur.sky, *Physical Thought*, p. 56.

11. Sambur.sky, *Physical Thought*, pp. 83–84.

12. Dijksterhuis, E. J., *The Mechanization of the World Picture*, Princeton University Press, Princeton, NJ, 1986, p. 39.

13. Park, David, *The How and the Why*, Princeton University Press, Princeton, NJ, 1988, p. 45.

14. Park, David, *The How and the Why*, p. 195.

15. Park, David, *The How and the Why*, p. 195.

16. Boorse, Henry A., and Motz, Lloyd, *The World of the Atom*, pp. 47–51.

17. Boorse, Henry A., and Motz, Lloyd, *The World of the Atom*, p. 58; Newton, Isaac, *The Principia*, A new translation by I. Bernard Cohen and Anne Whitman, University of California Press, Berkeley, CA, 1999 (paperback version), Book 2, Proposition 23, Theorem 18, pp. 697–699.

18. Boorse, Henry A., and Motz, Lloyd, *The World of the Atom*, p. 111.

19. Boorse, Henry A., and Motz, Lloyd, *The World of the Atom*, p. 139.

20. Boorse, Henry A., and Motz, Lloyd, *The World of the Atom*, pp. 141, 143.

21. Boorse, Henry A., and Motz, Lloyd, *The World of the Atom*, p. 144.

22. Boorse, Henry A., and Motz, Lloyd, *The World of the Atom*, p. 144.

23. Maxwell, James Clerk, Illustrations of the Dynamical Theory of Gases, *Philosophical Magazine* 19 (1860), 19–32. Citation from Boorse, Henry A., and Motz, Lloyd, *The World of the Atom*, p. 271.

24. Boorse, Henry A., and Motz, Lloyd, *The World of the Atom*, pp. 271–273.

25. Kuhn, Thomas, *Black-Body Theory and the Quantum Discontinuity, 1894–1912*, The University of Chicago Press, Chicago, IL, 1978 (paperback edition), pp. 41–42.

26. Atkins, P. W., *The Second Law*, Scientific American Books, New York, 1984, p. 5.

27. Kuhn, Thomas, *The Quantum Discontinuity*, p. 21.

28. Maxwell, J. C., Tait's 'Thermodynamics', *Nature* 17 (1877–1878), 257–259, 278–280, with quotation from p. 279f.; *Scientific Papers*, II, 660–671, with quotation from p. 670; citation from Kuhn, Thomas, *The Quantum Discontinuity*, p. 24.

29. Cushing, James T., *Philosophical Concepts in Physics*, Cambridge University Press, Cambridge, 1998, p. 367.

30. Ostwald, W., *Verh. Ges. Deutsch. Naturf. Ärzte* 1, 155 (1895); citation from, and English translation of, Pais, Abraham, *Subtle is the Lord*, Oxford University Press, London, 1982, pp. 83, 106.

31. Boorse, Henry A., and Motz, Lloyd, *The World of the Atom*, p. 206.
32. Boorse, Henry A., and Motz, Lloyd, *The World of the Atom*, p. 211.
33. Boorse, Henry A., and Motz, Lloyd, *The World of the Atom*, p. 211.
34. Miller, Arthur, *Albert Einstein's Special Theory of Relativity*, Addison-Wesley Publishing Company, Reading, MA, 1981, pp. 45–47; Cushing, James T., *Philosophical Concepts*, p. 210, ref. 7. Citation to Goldberg 1970–1971, pp. 7–25, esp. 15.
35. Cushing, James T., *Philosophical Concepts*, p. 366. Citation 19: Lauden 1981, 9, 25, 34–44, ref. 20, Friedman 1992, 143, 171.
36. Boorse, Henry A., and Motz, Lloyd, *The World of the Atom*, p. 516.
37. Miller, Arthur, *Einstein's Special Theory*, p. 45.
38. Cushing, James T., *Philosophical Concepts*, p. 210.
39. Cushing, James T., *Philosophical Concepts*, p. 367.
40. Holton, Gerald, *Thematic Origins of Scientific Thought*, Harvard University Press, Cambridge, MA, (1973) 1988, p. 238.
41. Miller, Arthur, *Einstein's Special Theory*, p. 181, endnote 42-ii.
42. Einstein, Albert, Molecular Dimensions; [CPAE2, pp. 183–202; CPAE2 ET, pp. 104–122].
43. Einstein, Albert, Molecular Dimensions; [CPAE2, p. 186; CPAE2 ET, p. 105].
44. Einstein, Albert, Molecular Dimensions; [CPAE2, p. 187; CPAE2 ET, p. 106].
45. Einstein, Albert, Molecular Dimensions; [CPAE2, p. 188; CPAE2 ET, p. 107].
46. Einstein, Albert, Molecular Dimensions; [CPAE2, p. 188; CPAE2 ET, p. 107].
47. *McGraw Hill Encyclopedia of Science and Technology*, 9th edition, Volume 11, 2002, Navier–Stokes equations, p. 597.
48. Einstein, Albert, Molecular Dimensions; [CPAE2, pp. 189–191; CPAE2 ET, pp. 108–110]. In Eq. (6), there is an error. In the second term, the quantity ρ in the denominator should be raised to the fifth (5th) power, not to the sixth (6th) power. The equation as written has been corrected from Einstein's paper.
49. Einstein, Albert, Molecular Dimensions; [CPAE2, pp. 191–192; CPAE2 ET, p. 110].
50. Einstein, Albert, Molecular Dimensions; [CPAE2, p. 192; CPAE2 ET, p. 111].
51. Stachel, John, ed., [CPAE2, endnote 31, p. 204].
52. Einstein, Albert, Molecular Dimensions; [CPAE2, p. 194; CPAE2 ET, p. 113]. There are some arithmetical and mathematical errors in these equations, which were corrected in later versions of the same material. These corrections are contained in Stachel, John, ed., [CPAE2, endnotes, pp. 203–204].
53. Einstein, Albert, Molecular Dimensions; [CPAE2, p. 194; CPAE2 ET, p. 113].
54. The corrected expression shows the change in heat production to be an increase of $+\delta^2 k\Phi$, not a decrease of $2\delta^2 k\Phi$ as stated in the paper.
55. Einstein, Albert, Molecular Dimensions; [CPAE2, p. 194; CPAE2 ET, p. 113].
56. Einstein, Albert, Molecular Dimensions; [CPAE2, p. 196; CPAE2 ET, p. 115].

57. Einstein, Albert, Molecular Dimensions; [CPAE2, p. 196; CPAE2 ET, p. 115]. The corrected value is an increase of heat production by $2\delta^2 k\frac{\Phi}{2}$.
58. The corrected expression is $W = 2\delta^2 k\left(1 + \frac{\varphi}{2}\right)$.
59. The corrected expression is $W^* = 2\delta^2 k\left(1 + \frac{\varphi}{2}\right)$.
60. The corrected expression is $k^* = k\left(1 + 2.5\varphi\right)$.
61. Stachel, John, *Einstein's Miraculous Year*, p. 33.
62. Einstein, Albert, Molecular Dimensions; [CPAE2, p. 198; CPAE2 ET, p. 118].
63. Einstein, Albert, Molecular Dimensions; [CPAE2, p. 199; CPAE2 ET, pp. 118–119].
64. Einstein, Albert, Molecular Dimensions; [CPAE2, pp. 199–200; CPAE2 ET, p. 119].
65. Pais, Abraham, *Subtle is the Lord*, Oxford University Press, New York, 1982, p. 87.
66. Einstein, Albert, Molecular Dimensions; [CPAE2, pp. 199–200; CPAE2 ET, pp. 119-120].
67. Pais, Abraham, *Subtle is the Lord*, Oxford University Press, New York, 1982, p. 87.
68. Stachel, John, ed., [CPAE2, Brownian Motion, p. 211].
69. Stachel, John, ed., [CPAE2, endnote 60, p. 205]. Reference is to data in Landolt and Börnstein 1894, p. 306.
70. Einstein, Albert, Molecular Dimensions; [CPAE2, p. 202; CPAE2 ET, p. 122].
71. Using the corrected expressions for k and D, the values for P and for N are P $= 6.3 \times 10^{-8}$ and $N = 3 \times 10^{23}$.
72. Einstein, Albert, Movement of Small Particles; [CPAE2, pp. 223–235; CPAE2 ET, pp. 123–134].
73. Einstein, Albert, Movement of Small Particles; [CPAE2, p. 224; CPAE2 ET, p. 123].
74. Einstein, Albert, Movement of Small Particles; [CPAE2, p. 224; CPAE2 ET, p. 123].
75. Pais, Abraham, *Subtle is the Lord*, p. 87.
76. Einstein, Albert, Movement of Small Particles; [CPAE2, p. 225; CPAE2 ET, p. 124].
77. Einstein, Albert, Movement of Small Particles; [CPAE2, pp. 225–226; CPAE2 ET, p. 124].
78. Einstein, Albert, Movement of Small Particles; [CPAE2, p. 228; CPAE2 ET, p. 127].
79. The thermodynamic details can be found in Albert Einstein's 1903 paper, A Theory of the Foundations of Thermodynamics; [CPAE2, pp.76–94; CPAE2 ET, pp. 48–67].
80. Einstein used only p's as variables, representing both position and momentum coordinates. Consistent with Hamilton's equations the q variables (position) are expressed also.
81. Einstein, Albert, Movement of Small Particles; [CPAE2, p. 226; CPAE2 ET, p. 125]. Einstein uses the constant 2κ in the expression. Subsequently it was noted that $2\kappa = k$, Boltzmann's constant. For convenience, Boltzmann's constant will be used from the beginning.
82. Einstein, Albert, Movement of Small Particles; [CPAE2, p. 227; CPAE2 ET, p. 126].
83. Einstein, Albert, Movement of Small Particles; [CPAE2, p. 228; CPAE2 ET, pp. 126–127].

84. Einstein, Albert, Movement of Small Particles; [CPAE2, p. 228; CPAE2 ET, p. 127].
85. Einstein, Albert, Movement of Small Particles; [CPAE2, p. 229; CPAE2 ET, p. 128].
86. Einstein, Albert, Movement of Small Particles; [CPAE2, p. 229; CPAE2 ET, p. 128].
87. Einstein, Albert, Movement of Small Particles; [CPAE2, p. 229; CPAE2 ET, p. 128].
88. Einstein, Albert, Movement of Small Particles; [CPAE2, p. 230; CPAE2 ET, p. 129].
89. Einstein, Albert, Movement of Small Particles; [CPAE2, p. 231; CPAE2 ET, p. 130].
90. Einstein, Albert, Movement of Small Particles; [CPAE2, p. 231; CPAE2 ET, p. 130].
91. Einstein, Albert, Movement of Small Particles; [CPAE2, p. 234; CPAE2 ET, pp. 132–133].
92. Einstein, Albert, Movement of Small Particles; [CPAE2, p. 235; CPAE2 ET, p. 134].
93. Einstein, Albert, Movement of Small Particles; [CPAE2, p. 224; CPAE2 ET, p. 123].
94. Einstein, Albert, Movement of Small Particles; [CPAE2, p. 235; CPAE2 ET, p. 134].
95. Tipler, Paul A., Foundations of Modern Physics, Worth Publishers, New York, 1969, p. 83.
96. Pais, Abraham, *Subtle is the Lord*, p. 103.
97. Atkins, P. W., *The Second Law*, p. 5.
98. Stachel, John, *Einstein's Miraculous Year*, p. 33.
99. Pullam, Bernard, *The Atom in the History of Human Thought*, Oxford, 1998, p. 256; citation from Rigden, John S., *Einstein 1905, The Standard of Greatness*, Harvard University Press, Cambridge, MA, 2005, pp. 69, 157.
100. Brush, Stephen G., *The Kind of Motion We Call Heat: A History of the Kinetic theory of Gases in the 19th Century*, Vol. 2 (North Holland Publishers, Amsterdam, 1976), p. 699; citation from Rigden, John, *Einstein 1905, The Standard of Greatness*, pp. 71, 157.
101. Schilpp, Paul Arthur, Ed., *Albert Einstein: Philosopher–Scientist*, Cambridge University Press, London, 1970, p. 49.
102. Kirchhoff, Gustav, *Vorlesungen über Mechanik*, 26.Vorl [Lectures on Mechanics, Lecture 26], pp. 378–380, Section 4 of Lecture 26. Reference taken from [CPAE2, p. 189]. I thank Thomas Coffey for the translation of this article.
103. Einstein, Albert, Molecular Dimensions; [CPAE2, p. 203, endnote 15].
104. Stachel, John, ed., [CPAE2, endnote 21, p. 203]. $X\xi$ should have ξ as a subscript, X_ξ, and similar corrections for the other terms.
105. Kirchhoff, Gustav Robert, *Vorlesungen über Mathematische Physik. 1: Mechanik*, 4th edition, Wien, Wilhelm, editor, B. G. Teubner, Leipzig, 1897, p. 369; citation from Stachel, John, ed., [CPAE2, endnote 22, p. 203].
106. Stachel, John, ed., [CPAE2, corrections to these equations are given in endnotes 25, 26, p. 204].
107. Stachel, John, ed., [CPAE2, corrections given in endnote 27, p. 204].
108. Stachel, John, ed., [CPAE2, endnotes 27, 28, p. 204].

109. Stachel, John, ed., [CPAE2, endnotes 29, 30, p. 204].
110. Stachel, John, ed., [CPAE2, endnotes 31, 32, p. 204].
111. Stachel, John, ed., [CPAE2, p. 112].
112. Stachel, John, ed., [CPAE2, endnote 50, p. 205].
113. Stachel, John, ed., [CPAE2, pp. 76–94; CPAE2 ET, pp. 48–67].
114. In the exponent, Einstein uses the notation $-2hE$, and later identifies $h = 1/(4\kappa T)$. The κ used by Einstein is related to Boltzmann's constant k as $k = 2\kappa$. In this notation $2h = 1/(kT)$.
115. Stachel, John, ed., [CPAE2, p. 89; CPAE2 ET, p. 61].
116. Stachel, John, ed., [CPAE2, p. 233 ; CPAE2 ET, p. 132].

3.7 Bibliography

Atkins, P. W., *The Second Law*, Scientific American Books, New York, 1984.

Boorse, Henry A., and Motz, Lloyd, editors, *The World of the Atom*, Volume 1, Basic Books, Inc., New York, 1966.

Brush, Stephen G., *The Kind of Motion We Call Heat: A History of the Kinetic theory of Gases in the 19th Century*, Volume 2, North Holland Publishers, Amsterdam, 1976. [Reference from John Rigden.]

Cushing, James T., *Philosophical Concepts in Physics*, Cambridge University Press, Cambridge, 1998.

Dijksterhuis, E. J., *The Mechanization of the World Picture*, Princeton University Press, Princeton, NJ, 1986.

Einstein, Albert, A Theory of the Foundations of Thermodynamics, *Annalen der Physik* 11 (1903), 170–187. [CPAE2, pp. 76–97; CPAE2 ET, pp. 48–67.]

Einstein, Albert, *A New Determination of Molecular Dimensions*. Dissertation, University of Zurich, 1905. [CPAE2, pp. 183–202; CPAE2 ET, pp. 104–122.]

Einstein, Albert, On the Movement of Small Particles Suspended in Stationary Liquids Required by the Molecular-Kinetic Theory of Heat. *Annalen der Physik* 17 (1905), 549–560. [CPAE2, pp. 223–235; CPAE2ET, pp. 123–134.]

Holton, Gerald, *Thematic Origins of Scientific Thought*, Harvard University Press, Cambridge, MA, 1988.

Kirchhoff, Gustav, *Vorlesungen über Mechanik*, 26.Vorl [Lectures on Mechanics, Lecture 26].

Kirchhoff, Gustav Robert, *Vorlesungen über mathematische Physik. 1: Mechanik*, 4th edition, Wien, Wilhelm, editor, B. G. Teubner, Leipzig, 1897.

Kuhn, Thomas, *Black-Body Theory and the Quantum Discontinuity, 1894–1912*, The University of Chicago Press, Chicago, IL, 1978 (paperback edition).

Maxwell, James Clerk, Illustrations of the Dynamical Theory of Gases, *Philosophical Magazine* 19 (1860), 19–32. [Reference from Henry A. Boorse and Lloyd Motz, p. 271.]

Maxwell, J. C., Tait's 'Thermodynamics', *Nature* 17 (1877–1878). [Reference from Thomas Kuhn.]

McGraw Hill Encyclopedia of Science and Technology, 9th edition, Volume 11, McGraw Hill, New York, 2002.

Miller, Arthur, *Albert Einstein's Special Theory of Relativity*, Addison-Wesley Publishing Company, Reading, MA, 1981.

Newton, Isaac, *The Principia*, A new translation by I. Bernard Cohen and Anne Whitman, University of California Press, Berkeley, CA, 1999.

Ostwald, W., *Verh. Ges. Deutsch. Naturf. Ärzte* 1, 155 (1895). [Reference from Abraham Pais, p. 83, 106.]

Park, David, *The How and the Why*, Princeton University Press, Princeton, NJ, 1988.

Pais, Abraham, *Subtle is the Lord*, Oxford University Press, London, 1982.

Pullam, Bernard, *The Atom in the History of Human Thought*, Oxford, 1998. [Reference from John Rigden.]

Rigden, John S., *Einstein 1905*, Harvard University Press, Cambridge, MA, 2005.

Sambursky, Shmuel, *Physical Thought from the Presocratics to the Quantum Physicists*, Pica Press, New York, 1975.

Schilpp, Paul Arthur, *Albert Einstein: Philosopher–Scientist*, Cambridge University Press, London, 1970.

Stachel, John, editor, *The Collected Papers of Albert Einstein*, Volume 2, Princeton University Press, Princeton, NJ, 1989; English translation by Anna Beck, Princeton University Press, Princeton, NJ, 1989.

Stachel, John, editor, *Einstein's Miraculous Year*, Princeton University Press, Princeton, NJ, 1998.

Tipler, Paul A., *Modern Physics*, Worth Publishers, New York, 1969, 1978.

The Special Theory of Relativity

4

"On the Electrodynamics of Moving Bodies"[1]

> Received on 30 June, 1905, by *Annalen der Physik*

"Does the Inertia of a Body Depend on Its Energy Content?"[2]

> Received on 27 December, 1905, by *Annalen der Physik*

4.1	Historical Background	105
4.2	Albert Einstein's Paper, "On the Electrodynamics of Moving Bodies"	113
4.3	Albert Einstein's Paper, "Does the Inertia of a Body Depend Upon Its Energy Content?"	129
4.4	Discussion and Comments	131
4.5	Appendices	133
4.6	Notes	155
4.7	Bibliography	159

4.1 Historical Background

In *Einstein's Miraculous Year*,[3] John Stachel writes, "...[Albert] Einstein described the theory of relativity as arising from a specific problem: the apparent conflict between the principle of relativity and the Maxwell–Lorentz theory of electrodynamics. While the relativity principle asserts the physical equivalence of all inertial frames of reference, the Maxwell–Lorentz theory implies the existence of a privileged inertial frame."[4]

In these papers, Einstein sets down the foundation for the special theory of relativity and develops a number of consequences that follow from them, including the relation $E = mc^2$. He opens his paper "On the Electrodynamics of Moving Bodies"[5] with comments on "asymmetries [in electrodynamic induction] that do not seem to attach to the phenomena."[6] But he does not, as is often believed, base his work on the Michelson–Morley experiment, nor even directly mention it. Undoubtedly, the remark in the opening section of the paper about "...the failure of attempts to detect a motion of the earth relative to the 'light medium'..."[7] included, among others, the Michelson–Morley experiment.[8]

In these papers, Albert Einstein addresses the relativity theories of Galileo and Newton, the Maxwell–Lorentz theory of electrodynamics, and the electromagnetic induction experiment.

4.1.1 The Relativity of Galileo Galilei and of Isaac Newton

Galileo set for himself the goal to better describe the motion of falling bodies and projectiles. Considering the description of motion by different observers in different reference frames, Galileo arrived at the principle that "the motion of a system of bodies relative to each other does not change if the whole system is subjected to a common motion."[9] Although we know today this statement does not hold for a common motion that is accelerated or rotational, it does hold for uniform rectilinear motion. The point is not the absolute correctness, or incorrectness, of Galileo's statement but, rather, it is the initial formal statement of a principle of relativity.

Just as Galileo developed the kinematics of motion (i.e., its description, not its causes), Newton developed the dynamics of motion. Isaac Newton systematized the ideas of motion, many of which were yet of a "fragmentary and confused character."[10] These were summarized in Newton's three laws of motion:

Newton's First Law of Motion: Every body perseveres in its state of being at rest or of moving uniformly straight forward except as it is compelled to change its state by forces impressed.

Newton's Second Law of Motion: A change in motion is proportional to the motive force impressed and takes place along the straight line in which that force is impressed.

Newton's Third Law of Motion: To any action there is always an opposite and equal reaction; in other words, the actions of two bodies upon each other are always equal and always opposite in direction.[11]

Although the first law appears to be contained in the second law, it was necessary to counter the possibility that the momentum of a body might, as Dijksterhuis notes, "decrease spontaneously . . . This possibility is ruled out by the first [law]; the momentum, therefore, does not change of its own accord; for every change a force is required; the second [law] says something about the effect of this force. Or, in other words: the second [law] only asserts that for a change in momentum it is sufficient that a force be exerted; the first has already established that it is necessary."[12]

Newton's laws of motion are valid in an "inertial" frame of reference and, in fact, an inertial reference frame is defined as one in which Newton's laws of motion are valid. Once one inertial reference frame is established, any other reference frame in uniform rectilinear translational motion in relation to the first is also an inertial reference frame.[13] See Figure 4.1.

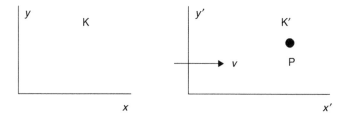

Fig. 4.1 Two reference frames.

Consider an object located at point P. If the point P is stationary in reference frame K′, it will be seen as moving with a velocity v in reference frame K, and vice versa (if stationary in K, it will be seen as moving with velocity $-v$ in K′). Of course, in many circumstances the point P will be observed moving both in reference frame K and in reference frame K′.

The location of point P as measured from reference frame K is (x, y, z), and the location of point P as measured from reference frame K′ is $(x′, y′, z′)$. The coordinates (x, y, z) are related to the coordinates $(x′, y′, z′)$ by the Galilean transformations. Starting the clocks at $t = 0$ when the origins of K and K′ are coincident, i.e., as K′ "passes" K, the Galilean transformations are:

$$x′ = x - vt$$
$$y′ = y$$
$$z′ = z$$
$$t′ = t$$

Taking the time derivative of each of these equations, and writing dx/dt as $\dot{x}$, etc.,

$$\dot{x}′ = \dot{x} - v$$
$$\dot{y}′ = \dot{y}$$
$$\dot{z}′ = \dot{z}$$

If the point P has a velocity $\dot{x}$ as measured in K, the velocity of point P as measured in K′ would be $\dot{x}′ = \dot{x} - v$, i.e., it differs from the measurement in K by v, the relative velocity between the two reference frames.

Taking the time derivative of these velocity transformations, since $v = constant$, the x-, y-, and z-components of the acceleration are the same in both the K and K′ reference frames. Thus, from Newton's Second Law, $F = ma$, no distinction can be made between the two reference frames.

Although Newton believed in absolute space, he was unable to determine a way to distinguish the inertial reference frame of absolute space from the inertial reference frame moving with uniform rectilinear translational motion relative to absolute space.[14] Newton raised this "inability" to a principle, today referred to as Newton's principle of relativity:

When bodies are enclosed in a given space, their motions in relation to one another are the same whether the space is at rest or whether it is moving uniformly straight forward without circular motion.[15]

Newton's principle of relativity stated it is impossible to determine our motion relative to absolute space using the equations of mechanics. Over the succeeding 200 years, it came to be accepted that Newton's principle of relativity applied to all areas of physics, not just to mechanics.

4.1.2   The Lorentz Transformations (from Lorentz)

Maxwell's equations of electromagnetism are

$$\nabla \times \boldsymbol{E} = -\frac{1}{c}\frac{\partial \boldsymbol{B}}{\partial t}$$

$$\nabla \times \boldsymbol{B} = \frac{1}{c}\frac{\partial \boldsymbol{E}}{\partial t} + \frac{4\pi}{c}\,\boldsymbol{j}$$

$$\nabla \cdot \boldsymbol{E} = 4\pi\rho$$

$$\nabla \cdot \boldsymbol{B} = 0$$

In empty space (no sources/sinks, i.e., $\boldsymbol{j} = 0$ and $\rho = 0$), and written in component form, Maxwell's equations are (the components of $\nabla \times \boldsymbol{E} = -\frac{1}{c}\frac{\partial \boldsymbol{B}}{\partial t}$ are Eqs. (a), (b), and (c) below, ... $\nabla \cdot \boldsymbol{B} = 0$ is Eq. (h))

(a) $\dfrac{\partial E_z}{\partial y} - \dfrac{\partial E_y}{\partial z} = -\dfrac{1}{c}\dfrac{\partial B_x}{\partial t}$
   (d) $\dfrac{\partial B_z}{\partial y} - \dfrac{\partial B_y}{\partial z} = \dfrac{1}{c}\dfrac{\partial E_x}{\partial t}$

(b) $\dfrac{\partial E_x}{\partial z} - \dfrac{\partial E_z}{\partial x} = -\dfrac{1}{c}\dfrac{\partial B_y}{\partial t}$
   (e) $\dfrac{\partial B_x}{\partial z} - \dfrac{\partial B_z}{\partial x} = \dfrac{1}{c}\dfrac{\partial E_y}{\partial t}$

(c) $\dfrac{\partial E_y}{\partial x} - \dfrac{\partial E_x}{\partial y} = -\dfrac{1}{c}\dfrac{\partial B_z}{\partial t}$
   (f) $\dfrac{\partial B_y}{\partial x} - \dfrac{\partial B_x}{\partial y} = \dfrac{1}{c}\dfrac{\partial E_z}{\partial t}$

(g) $\nabla \bullet \vec{E} = \dfrac{\partial E_x}{\partial x} + \dfrac{\partial E_y}{\partial y} + \dfrac{\partial E_z}{\partial z} = 0$

(h) $\nabla \bullet \vec{B} = \dfrac{\partial B_x}{\partial x} + \dfrac{\partial B_y}{\partial y} + \dfrac{\partial B_z}{\partial z} = 0$

Manipulating the empty space version of Maxwell's equations yields the wave equation for the electric and magnetic fields:[16]

$$\nabla^2 \vec{E} = \frac{1}{c^2}\frac{\partial^2 \vec{E}}{\partial t^2} \qquad \nabla^2 \vec{B} = \frac{1}{c^2}\frac{\partial^2 \vec{B}}{\partial t^2}$$

These electric and magnetic fields, based on the wave equation, travel at the speed of light, c, assumed relative to the luminiferous aether. Since the luminiferous aether was at rest relative to absolute space, one's speed relative to the aether would be the same as one's speed relative to absolute space. All one needed to do was measure the speed relative to the aether and the search begun by Newton would be completed.

By definition, Maxwell's equations are valid in absolute space (the aether rest frame). Transforming Maxwell's equations to a moving reference frame leaves them in almost the same form, differing only by an additional term proportional to v/c, where v is the speed of the moving reference frame through the aether (and, consequently, through absolute space) and c is the speed of light. Applying the Galilean transformation equations to Maxwell's equations, in the K′ reference frame they become (see Appendix 4.5.1.1 for the details of this derivation)[17],

(A) $\dfrac{\partial E_z}{\partial y'} - \dfrac{\partial E_y}{\partial z'} = -\dfrac{1}{c}\dfrac{\partial B_x}{\partial t} + \dfrac{v}{c}\dfrac{\partial B_x}{\partial x'}$

(D) $\dfrac{\partial B_z}{\partial y'} - \dfrac{\partial B_y}{\partial z'} = \dfrac{1}{c}\dfrac{\partial E_x}{\partial t} - \dfrac{v}{c}\dfrac{\partial E_x}{\partial x'}$

(B) $\dfrac{\partial E_x}{\partial z'} - \dfrac{\partial E_z}{\partial x'} = -\dfrac{1}{c}\dfrac{\partial B_y}{\partial t} + \dfrac{v}{c}\dfrac{\partial B_y}{\partial x'}$

(E) $\dfrac{\partial B_x}{\partial z'} - \dfrac{\partial B_z}{\partial x'} = \dfrac{1}{c}\dfrac{\partial E_y}{\partial t} - \dfrac{v}{c}\dfrac{\partial E_y}{\partial x'}$

(C) $\dfrac{\partial E_y}{\partial x'} - \dfrac{\partial E_x}{\partial y'} = -\dfrac{1}{c}\dfrac{\partial B_z}{\partial t} + \dfrac{v}{c}\dfrac{\partial B_z}{\partial x'}$

(F) $\dfrac{\partial B_y}{\partial x'} - \dfrac{\partial B_x}{\partial y'} = \dfrac{1}{c}\dfrac{\partial E_z}{\partial' t} - \dfrac{v}{c}\dfrac{\partial E_z}{\partial x'}$

(G) $\nabla' \bullet \vec{E} = \dfrac{\partial E_x}{\partial x'} + \dfrac{\partial E_y}{\partial y'} + \dfrac{\partial E_z}{\partial z'} = 0$

(H) $\nabla' \bullet \vec{B} = \dfrac{\partial B_x}{\partial x'} + \dfrac{\partial B_y}{\partial y'} + \dfrac{\partial B_z}{\partial z'} = 0$

It is clear that Eqs. (G) and (H) in reference frame K′ are of the same form as their counterparts, Eqs. (g) and (h), in reference frame K. Also, it is easily seen that Eqs. (A) through (F) are the same as their counterparts, Eqs. (a) through (f), in reference frame K, *except* there is an additional term proportional to v/c added to each equation. Since these terms are proportional to v/c, a measurement of these terms would allow one to determine the speed of the reference frame K′ relative to the stationary aether and, ultimately, to the absolute space of Newton. Experiments by Fizeau, and others, were unable to detect this first-order effect.[18]

Accepting the validity of the experimental results, Hendrik Lorentz set about to manipulate the form of Maxwell's equations in reference frame K′ to eliminate the "additional terms" that could not be found experimentally. In Eqs. (B) and (C), the "extra term" on the right-hand side could be moved to the left-hand side, and combined with the already existing derivative with respect to x'. For Eq. (A), it is necessary first to replace $(\partial B_x/\partial x')$ by $(\partial B_x/\partial x') = -(\partial B_y/\partial y') - (\partial B_z/\partial z')$ from Eq. (H). In like manner, Eqs. (D), (E), and (F) are manipulated using Eq. (G).

(A′) $\dfrac{\partial\left(E_z + \dfrac{v}{c}B_y\right)}{\partial y'} - \dfrac{\partial\left(E_y - \dfrac{v}{c}B_z\right)}{\partial z'} = -\dfrac{1}{c}\dfrac{\partial B_x}{\partial t}$

(B′) $\dfrac{\partial(E_x)}{\partial z'} - \dfrac{\partial\left(E_z + \dfrac{v}{c}B_y\right)}{\partial x'} = -\dfrac{1}{c}\dfrac{\partial B_y}{\partial t}$

(C′) $\dfrac{\partial\left(E_y - \dfrac{v}{c}B_z\right)}{\partial x'} - \dfrac{\partial(E_x)}{\partial y'} = -\dfrac{1}{c}\dfrac{\partial B_z}{\partial t}$

(D′) $\dfrac{\partial\left(B_z - \dfrac{v}{c}E_y\right)}{\partial y'} - \dfrac{\partial\left(B_y + \dfrac{v}{c}E_z\right)}{\partial z'} = \dfrac{1}{c}\dfrac{\partial E_x}{\partial t}$

(E′) $\dfrac{\partial(B_x)}{\partial z'} - \dfrac{\partial\left(B_z - \dfrac{v}{c}E_y\right)}{\partial x'} = \dfrac{1}{c}\dfrac{\partial E_y}{\partial t}$

(F′) $\dfrac{\partial\left(B_y + \dfrac{v}{c}E_z\right)}{\partial x'} - \dfrac{\partial(B_x)}{\partial y'} = \dfrac{1}{c}\dfrac{\partial E_z}{\partial' t}$

The subsequent development follows on the Eqs. (A′) through (F′). The expressions for Eqs. (G′) and (H′) are suppressed from this point forward. In K′, defining the electric and magnetic fields $E′$ and $B′$ in terms of the electric and magnetic fields E and B in K, as

$$E'_x = E_x \qquad\qquad B'_x = B_x$$

$$E'_y = E_y - \frac{v}{c}B_z \qquad B'_y = B_y + \frac{v}{c}E_z$$

$$E'_z = E_z + \frac{v}{c}B_y \qquad B'_z = B_z - \frac{v}{c}E_y$$

the equations become,

$$\text{(A″)} \quad \frac{\partial(E'_z)}{\partial y'} - \frac{\partial(E'_y)}{\partial z'} = -\frac{1}{c}\frac{\partial B_x}{\partial t} \qquad \text{(D″)} \quad \frac{\partial(B'_z)}{\partial y'} - \frac{\partial(B'_y)}{\partial z'} = \frac{1}{c}\frac{\partial E_x}{\partial t}$$

$$\text{(B″)} \quad \frac{\partial(E'_x)}{\partial z'} - \frac{\partial(E'_z)}{\partial x'} = -\frac{1}{c}\frac{\partial B_y}{\partial t} \qquad \text{(E″)} \quad \frac{\partial(B'_x)}{\partial z'} - \frac{\partial(B'_z)}{\partial x'} = \frac{1}{c}\frac{\partial E_y}{\partial t}$$

$$\text{(C″)} \quad \frac{\partial(E'_y)}{\partial x'} - \frac{\partial(E'_x)}{\partial y'} = -\frac{1}{c}\frac{\partial B_z}{\partial t} \qquad \text{(F″)} \quad \frac{\partial(B'_y)}{\partial x'} - \frac{\partial(B'_x)}{\partial y'} = \frac{1}{c}\frac{\partial E_z}{\partial 't}$$

At first glance, each of these equations appears to be in the proper form. However, the terms on the left-hand side of each equation contain only quantities referred to in the K′ system, while on the right-hand side of the equations, the electric and magnetic fields, E and B, are still those measured in reference frame K. Lorentz then made another change of variables, defining a new time variable $t' = t - \frac{vx'}{c^2}$, yielding (see Appendix 4.5.1.1 for the details of this derivation),[19]

$$\text{(A‴)} \quad \frac{\partial(E'_z)}{\partial y'} - \frac{\partial(E'_y)}{\partial z'} = -\frac{1}{c}\frac{\partial B'_x}{\partial t'}$$

$$\text{(B‴)} \quad \frac{\partial(E'_x)}{\partial z'} - \frac{\partial(E'_z)}{\partial x'} = -\frac{1}{c}\frac{\partial B'_y}{\partial t'} - \frac{1}{c}\left(\frac{v}{c}\right)^2\frac{\partial B_y}{\partial t'}$$

$$\text{(C‴)} \quad \frac{\partial(E'_y)}{\partial x'} - \frac{\partial(E'_x)}{\partial y'} = -\frac{1}{c}\frac{\partial B'_z}{\partial t'} - \frac{1}{c}\left(\frac{v}{c}\right)^2\frac{\partial B_z}{\partial t'}$$

$$\text{(D‴)} \quad \frac{\partial(B'_x)}{\partial z'} - \frac{\partial(B'_z)}{\partial x'} = \frac{1}{c}\frac{\partial E'_y}{\partial t'}$$

$$\text{(E‴)} \quad \frac{\partial(B'_x)}{\partial z'} - \frac{\partial(B'_z)}{\partial x'} = \frac{1}{c}\frac{\partial E'_y}{\partial t'} + \frac{1}{c}\left(\frac{v}{c}\right)^2\frac{\partial E_y}{\partial t'}$$

$$\text{(F‴)} \quad \frac{\partial(B'_y)}{\partial x'} - \frac{\partial(B'_x)}{\partial y'} = \frac{1}{c}\frac{\partial E'_z}{\partial t'} + \frac{1}{c}\left(\frac{v}{c}\right)^2\frac{\partial E_z}{\partial t'}$$

In this last transformation only the time variable has been changed, to what was called "local time" because it was dependent on the location x', where the time measurement was being made. This transformation was made expressly to rid the equations of the terms in first order in (v/c). Introducing the local time achieved this purpose, leaving the Maxwell equations in the same form in the moving reference frame K′ as they were in the rest frame – at least to first order in (v/c).

Galilean Transformations		Modified Transformations
$x' = x - vt$		$x' = x - vt$
$y' = y$	$\Rightarrow$	$y' = y$
$z' = z$		$z' = z$
$t' = t$		$t' = t - \frac{vx}{c^2}$

At this time, the general belief was these transformations were simply a mathematical ploy to save the form of Maxwell's equations in the moving reference frame. As Robert Rynasiewicz writes, "Nowhere (prior to 1906) does Lorentz indicate he intends E' and $[B']$ to represent the fields relative to a moving frame.... The pre-relativistic understanding took the electromagnetic field to constitute an intrinsic, objective, non-relational state of the aether. To ask about the electric or magnetic field *relative* to a moving frame would have made as much sense as asking about the atomic number of an element relative to a moving frame."[20]

The introduction of a local time to remove difficulties associated specifically with the Galilean transformations was successful in that it removed the difficulties – at least to order v/c. The introduction of a specific hypothesis to address a specific issue was the method employed by Lorentz as he constructed his theory of electrons. This led to his great success with the theory of electrons – as well as to his failure to recognize the significance of Einstein's special theory of relativity, let alone be the primary author of the theory of relativity. In Lorentz's words:

The chief cause of my failure was my clinging to the idea that the variable t only can be considered as the true time and that my local time t' must be regarded as no more than an auxiliary mathematical quantity. In Einstein's theory, on the contrary, t' plays the same part as t; if we want to describe phenomena in terms of x', y', z', and t', we must work with these variables exactly as we could do with x, y, z, t.[21]

The introduction of the local time removed the first order terms as intended, but it also introduced second order terms which, at the time, were too small to be measured. Subsequent experiments, notably the Michelson–Morley experiment,[22] were unable to detect the second order effects. The result consistently was $v = 0$. Another adjustment of the transformation equations became necessary.

The Michelson–Morley experiment, performed by Albert Michelson and Edward Morley in 1887, was ingeniously designed to measure these second order effects. James Cushing describes the experiment in these words:

The basic idea of the experiment is to send light down to a mirror and back (a distance L), once along the line of motion of the earth and once at right angles to that motion and to measure the time difference for these two round trip journeys....

This time difference is not measured directly in the experiment, but an interference effect is looked for. Since the wave trains are both in phase when they begin their trips along these different paths and since the times of flight of the light signals along these two paths are different, the two beams will be slightly out of phase when they are recombined. This phase difference can produce an interference fringe pattern....[23]

In the Michelson–Morley experiment, the time to travel down and back the arm parallel to the motion of the earth is calculated to be (see Appendix 4.5.1.2 for the details of these calculations)

$$t_{parallel} = \frac{2l}{c} \left(\frac{1}{1 - \frac{v^2}{c^2}} \right)$$

whereas, in the direction perpendicular to the motion of the earth, the time is calculated to be

$$t_{perpendicular} = \frac{2l}{c} \frac{1}{\sqrt{1 - \frac{v^2}{c^2}}}$$

It is the difference between these two times, $t_{parallel}$ and $t_{perpendicular}$, that would give rise to interference fringes. When the apparatus was rotated by $\pi/2$, the roles played by the arms parallel to, and perpendicular to, the motion of the earth through the aether would be exchanged, producing a shift in the fringe pattern. However, no fringe shift was detected. In 1892, the physicists George FitzGerald in Ireland and Hendrik Lorentz independently arrived at the same explanation: if the arm of the apparatus traveling parallel to the earth's motion contracts by exactly the right amount (while the arm perpendicular to the motion does not contract), the $t_{parallel}$ and $t_{perpendicular}$ would be exactly equal to one another. The amount of contraction would be $\sqrt{1 - (v/c)^2}$. This, today, is known as the Lorentz–FitzGerald contraction.

As he had done with local time, Lorentz introduced the contraction hypothesis to address a specific issue – this time the null result of the Michelson–Morley experiment. Lorentz considered the contraction to be a real physical contraction of the apparatus as it moved through the aether, not a result simply of relative motion of the two reference frames.

In 1904, Lorentz published a set of modified Galilean transformations, including the local time and the contraction adjustments.[24] These transformations were labeled the Lorentz transformations by Poincaré:

$$t' = \gamma \left(t - \frac{v}{c^2} x \right)$$

$$x' = \gamma(x - vt)$$

$$y' = y$$

$$z' = z$$

with $\gamma = \frac{1}{\sqrt{1 - (\frac{v}{c})^2}}$

Modern convention uses the symbol γ to represent this relativistic factor, whereas Einstein used the symbol β (modern convention uses the symbol β for the factor v/c).

Lorentz arrived at the Lorentz transformations from hypotheses postulated to address specific issues, a result that was to be arrived at from

a more general point of view by Einstein's special theory of relativity.[25]
It needs to be noted that the use of the Lorentz transformations rather
than the Galilean transformations leaves Maxwell's equations in exactly
the same form in the moving reference frame as they were in the frame
stationary in the aether. Consequently, no electromagnetic experiments
(such as the Michelson–Morley experiment) could be expected to detect
motion through the aether.[26]

In response to this continuing trend of modifications Lorentz says:

> Poincaré has objected to the existing theory of electric and optic phenomena
> in moving bodies that, in order to explain Michelson's negative result, the
> introduction of a new hypothesis has been required, and that the same
> necessity may occur each time new facts may be brought to light. Surely
> this course of inventing special hypotheses for each new experimental result is
> somewhat artificial.[27]

By 1904, prior to Einstein's paper on the special theory of relativity,
Cushing points out that Lorentz's "theory was, in its predictions, indistinguishable from Einstein's special theory of relativity."[28] Thus we see,
at this stage, that Lorentz has all the same results that Albert Einstein
is to present in 1905. Yet Lorentz was unable to change from a dynamical
foundation for these results to the kinematical foundation that Einstein
would provide one year later.

4.2 Albert Einstein's Paper, "On the Electrodynamics of Moving Bodies"[29]

Albert Einstein described his theory of relativity as arising from an
apparent conflict between the principle of relativity of Newton and the
theory of electrodynamics of Maxwell, as modified by Lorentz, i.e., the
Lorentz transformations. Newton's principle of relativity showed it is
impossible to determine the motion of an inertial system relative to
absolute space by mechanical experiments. Maxwell's theory of electrodynamics, in principle, should allow one to determine the motion of an
inertial system relative to the aether by optical experiments, but it failed
to do so – leading to a number of additional assumptions, such as length
contraction and local time.

Einstein took this failure of optical experiments to detect the motion
of the earth through the aether as verification of the principle of relativity and set about the task of making the Maxwell–Lorentz theory of
electrodynamics consistent with the principle of relativity. He took what
had been a problem in the Maxwell–Lorentz theory – the constancy of
the speed of light – accepted it as empirical fact, and added it to the
principle of relativity as his second postulate. The problem Einstein
was addressing in this paper was the description of the electrodynamics
of bodies in a reference frame K' moving through the aether, i.e.,
the electrodynamics of moving bodies.[30] Rather than focusing on the
theory of electrodynamics as others were, Einstein showed how a deeper

understanding of space and time would lead naturally to the Lorentz transformations. Such a deeper understanding would have an impact on all of physics, not just electrodynamics. Pursuing this line of thought led Einstein to reexamine the concept of time and, in particular, the concept of simultaneity (Section 4.2.1).

In the electromagnetic induction experiment, a moving magnet induces an emf in a stationary coil of wire, producing a current. However, if the situation is reversed, with the magnet stationary and the coil of wire moving, a current again is established in the coil because the charges in the wire experience a magnetic force as they move through the magnetic field. See figure 4.2.

Einstein's observation was that, if the *relative motion* between the magnet and coil of wire was the same in the two cases, the resulting current was the same. But, as he further noted, depending on which is at rest, in the two cases the explanation is different (see Section 4.4 for how the theory of relativity resolved this). This, along with "the failure of attempts to detect a motion of the earth relative to the 'light medium [the aether],' lead to the conjecture that not only in mechanics, but in electrodynamics as well, the phenomena do not have any properties corresponding to the concept of absolute rest.... We shall raise this conjecture (whose content will be called 'the principle of relativity' hereafter) to the status of a postulate and shall introduce, in addition, the postulate, only seemingly incompatible with the former one, that in empty space light is always propagated with a definite velocity $[c]$[31] that is independent of the state of motion of the emitting body."[32] It is here that Einstein comments that "[t]he introduction of a 'light ether' will prove superfluous..."[33] The now common use of the symbol c for the speed of light is said to be from the Latin word "celeritas," meaning "swiftness."[34]

(a) Motion by the magnet produces a current in the
      stationary loop of wire.

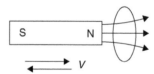

(b) Motion by the loop of wire in the vicinity of a stationary
      magnet produces a current in the loop of wire.

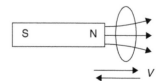

Fig. 4.2 Electromagnetic induction.

4.2.1 Definition of Simultaneity

In Section 1, it may appear an inordinate amount of time was expended on the idea of simultaneity. But, it should be remembered, this concern was a new focus in 1905. For this reason, and since it was to be the foundation on which Einstein's entire paper was built, first to obtain the Lorentz transformations, and then to obtain a variety of results following from the Lorentz transformations, it is crucial to establish with certainty, and with clarity, just what is meant by the concept of simultaneity.

As noted by John Stachel, once Einstein began to pursue the line of thought that the apparent contradiction of the two postulates was based on the Newtonian addition of velocities, he began to focus on the concept of velocity, noting the "concept of the velocity of an object with respect to an inertial frame depends on time readings made at two different places in that inertial frame.... How do we know that time readings at two such distant places are properly correlated? Ultimately this boils down to the question: how do we decide when events at two different places in the same frame of reference occur at the same time, i.e., simultaneously?... [This] led to the radically novel idea that, once one physically defines simultaneity of two distant events relative to one inertial frame of reference, it by no means follows that these two events will be simultaneous when the *same* definition is used relative to another inertial frame moving with respect to the first."[35]

Thus, at the heart of things, one needed to determine very clearly the concept of simultaneity. This clarification is Section 1 of the paper.

Simultaneity is defined for clocks *in the same reference frame*, i.e., at this stage there is no "relativity" of measurements made in one reference frame relative to measurements made in another reference frame. In the coordinate system considered, Newton's laws of mechanics are assumed to be valid.[36]

Time is defined in terms of a collection of clocks at different locations, each clock remaining stationary at its given location. For a clock at location A, an observer can record the time, t_A, of events occurring in the "immediate vicinity of A." Likewise, an observer at location B can record the time, t_B, of events occurring in the "immediate vicinity of B." But, as Einstein noted, "it is not possible to compare the time of an event at A with one at B without a further stipulation,"[37] i.e., some way to correlate the time of clock A with the time of clock B. See Figure 4.3.

Einstein states, by definition, the time for light to travel from point A to point B is the same as the time for light to travel from point B to point A (isotropy of space), and uses this to synchronize the two clocks. Consider a ray of light that leaves point A at time t_A, arrives at point B at time t_B, is reflected back to point A, arriving back at point A at time t_A^*, where t_A is the time measured on the clock at point A and t_B is the time measured on the clock at point B. "The two clocks are synchronous by definition if"[38]

$$t_B - t_A = t_A^* - t_B \qquad (4.1)$$

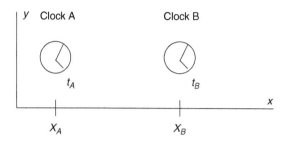

Fig. 4.3 Simultaneity and two stationary clocks.

Two stationary clocks, one located at X_A, the other at X_B.

Note that all times are defined in terms of clocks at rest in the reference frame. The concern of synchronization is because the clocks are in locations separate from one another.

The "time" of an event, as defined by Einstein, "... is the reading obtained simultaneously with the event from a clock at rest that is located at the place of the event and that for all time determinations is in synchrony with a specified clock at rest."[39] He then continues, postulating that, based on experience, $\frac{2(X_B - X_A)}{t_A^* - t_A} = c$, a universal constant.

4.2.2   On the Relativity of Lengths and Times

In Section 2 of his paper, Einstein states the two principles from which all else will flow:[40]

1. The laws governing the changes of the state of any physical system do not depend on which of the two coordinate systems in uniform translational motion relative to each other these changes of the state are referred to.
2. Each ray of light moves in the coordinate system "at rest" with the definite velocity [c][41], independent of whether this ray of light is emitted by a body at rest or a body in motion. Here,

$$velocity = \frac{light\ path}{time\ interval}$$

where "time interval" should be understood in the sense of the definition given in [Section 1 of Einstein's paper, Section 4.2.1 of this chapter].

John Stachel notes, "When combined with the relativity principle, this [c = constant] leads to an apparently paradoxical conclusion: the velocity of light must be the same in all inertial frames. This result conflicts with the Newtonian law of addition of velocities, forcing a revision of the kinematical foundations underlying all of physics."[42]

All time measurements are to be made in terms of the clocks mentioned in Section 4.2.1 and all length measurements are to be made in terms of rigid rods of fixed length (meter sticks, etc.). Two coordinate systems are considered, one called the rest system K, with x-, y-, and z-axes, and a second one called the moving system K′, with x'-. y'-, and

z'-axes. [N.B. This notation differs from that used by Einstein. Instead of designating the rest reference frame by Latin letter coordinates, and the moving reference frame by Greek letter coordinates (as done by Einstein), the rest system is designated as system K with coordinates x, y, z, and time t (the same notation as Einstein), while the moving system is designated as system K$'$ with coordinates x', y', z' and time t' (Einstein's designated the moving system as system k with coordinates ξ, η, ζ, and time τ).] The K$'$ system is moving at a constant velocity v relative to the K system, with the relative velocity along the $x - x'$ axes. The y'- and z'-axes are parallel to the y- and z- axes.[43] The clocks in each reference frame are set at time $t = 0$, $t' = 0$, when the origins of the two coordinate systems are coincident. See Figure 4.1.

When the rod is at rest in the rest frame it has some length ℓ, when measured by a person at rest in the rest frame. The axis of the rod is now set along the x-axis and the rod is set into uniform translational motion along the x-axis with speed v. See Figure 4.4.

As measured by a person moving with the rod, the length of the rod will be ℓ. But, Einstein noted, "We will determine... 'the length of the (moving) rod in the system at rest,' on the basis of our two principles, and will find it to be different from ℓ."[44]

On the moving rod, consider two clocks, one at each end, i.e., one at end A and one at end B, of the rod. These two clocks run synchronously *with the clocks in the rest frame*, i.e., each time clock A is adjacent to a clock in the rest frame, the reading on clock A and the reading on the rest clock are the same. The same is true for clock B. Thus these two moving clocks are "synchronous in the system at rest."[45]

A light signal is sent from one end of the moving rod to the other, i.e., from point A to point B. In the rest frame, the signal is seen approaching point B at the speed $c-v$. The length of the moving rod as measured from the rest frame K is designated r_{AB}, where (as will be shown later) r_{AB} is not equal to ℓ. The time to reach point B, as measured in the rest reference frame, will be $t_B - t_A = r_{AB}/(c - v)$. At point B, the signal is reflected back to point A. The time to reach point A from point B, as measured in the rest reference frame, will be $t_A^* - t_B = r_{AB}/(c + v)$. Since the time from A to B is different than the time from B to A, the observers "co-moving with the moving rod would thus find that the two clocks do not run synchronously while the observers in the system at rest would declare them synchronous.... Thus we see that we must not ascribe *absolute* meaning to the concept of simultaneity..."[46]

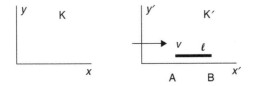

Fig. 4.4 Rod of length ℓ at rest in reference frame K$'$.

4.2.3   Theory of Transformation of Coordinates and Time from a System at Rest to a System in Uniform Translational Motion Relative to It

In Section 3 of his paper, Einstein uses the condition of clock synchronization to obtain the Lorentz transformations. Again we consider the two reference frames K and K′ introduced in Section 4.2.2. For the sake of convenience, we consider the reference frame K to be at rest, with the reference frame K′ moving with velocity v parallel to the x-axis of K. All of the clocks at rest in K are synchronized using the method outlined in Section 4.2.1. In a similar manner, all of the clocks at rest in K′ (the moving system) are synchronized within K′. The time in K is denoted as t, while the time in K′ is denoted as t'. An event in K occurring at location (x, y, z) and time t, occurs at a corresponding location (x', y', z') and time t' in reference frame K′.

The transformations between the quantities x, y, z, and t and the quantities x', y', z' and t' are the focus of this section of the paper. Einstein begins by stating, "First of all, it is clear that these equations must be *linear* because of the properties of homogeneity that we attribute to space and time."[47]

The derivation of the Lorentz transformations flows from the definition of synchronicity, Eq. (4-1) in Section 1 of the paper (Section 4.2.1 of this chapter),

$$t_B - t_A = t_A^* - t_B$$

Consider a point at location (x', y', z') at time t' in reference frame K′. The coordinates of this point in reference frame K at time t are (x, y, z). As time t' continues to change, there will be some corresponding change in the time t. If the point is at a fixed value of x', in K the value of x will be changing or, vice versa, if at rest in K at location x, the variable x' will be changing or, even possibly, both x and x' could be changing. And this is for motion parallel to the $x - x'$ axes (no change in the y and y' or z and z' directions). Thus there will be *at least* three variables changing in any transformation equations (x or x', t and t').

Einstein introduced a new variable X to describe a point at rest in reference frame K′ (but moving in reference frame K). As observed from reference frame K the point is at a fixed location X from the origin of K′. See Figure 4.1.

In reference frame K, the location of the point fixed in K′ is given by the equation $x = X + vt$, where v is the velocity of the origin of K′ relative to K. The fixed point has a "definite time-independent system of values $[X]$, y, and z assigned to it." Subsequently, in the calculation, X will be taken to be infinitesimally small.[48] Einstein then described the location of the point in reference frame K by the coordinate X rather than x, where $X = x - vt$. Since X is constant this removes one of the variables in the potential transformation equations, i.e., we can write $t' = t'(X, t) = t'(t)$.

The first set of transformation equations found are between the coordinates (X, y, z, t) and (x', y', z', t'). Once these have been obtained, X is replaced by $X = x - vt$ in the equations, resulting in the Lorentz transformation equations.

In reference frame K, consider a point P moving parallel to the x-axis with a velocity v. The "x-position" of the point P is denoted as $X = x - vt$, with all of the quantities measured in reference frame K. The coordinates of point P can be written (X, y, z), with $X = $ constant. In reference frame K' the point P is stationary, remaining at a fixed value of x'. In K', the coordinates of point P can be written (x', y', z').

Consider the criterion of synchronization in K'. Changing our notation slightly so that the light ray leaves the origin of K' at time t_0', is reflected from the point P at time t_1', and arrives back at the origin at time t_2', the condition of synchronicity, Eq. (4.1), becomes

$$t_1' - t_0' = t_2' - t_1'$$

$$\frac{1}{2}(t_0' + t_2') = t_1'$$

The transformation of t', expressing t' in terms of the coordinates of reference frame K, is written as a function of (X, y, z), rather than as a function of (x, y, z), i.e., $t' = t'(X, y, z, t)$. Once the transformation is obtained in terms of X, X is replaced by $x-vt$. Einstein arrived at the following three differential equations for t' (see Appendix 4.5.2.1 for the details of this derivation):

$$\frac{\partial t'}{\partial X} + \frac{v}{c^2 - v^2}\frac{\partial t'}{\partial t} = 0$$

$$\frac{\partial t'}{\partial y} = 0$$

$$\frac{\partial t'}{\partial z} = 0$$

Using the linearity condition, Einstein shows the solutions to these equation are of the form (see Appendices 4.5.2.2, 4.5.2.3, and 4.5.2.4 for the details of this derivation):

$$t' = \varphi(v)\gamma\left(t - \frac{v}{c^2}x\right)$$

$$x' = \varphi(v)\gamma(x - vt)$$

$$y' = \varphi(v)y$$

$$z' = \varphi(v)z$$

with $\gamma = 1/\sqrt{1 - v^2/c^2}$. $\varphi(v)$ is an unspecified function of the relative velocity v, but is independent of the spatial and time coordinates. The reverse transformations, from K' to K, are obtained by exchanging $(x, y, z, t) \leftrightarrow (x', y', z', t')$ and substituting $-v$ for v.

To verify these results are consistent with the two postulates, using these transformations Einstein showed a spherical wave emitted from the origin has the same form in both the rest frame K and the moving frame K′. Consider the two reference frames K and K′, and consider their origins to coincide at $t = 0$, $t' = 0$. At this instant, a spherical wave is emitted from the origin of K, propagating outward with the velocity c. In K, the equation of the wave front is

$$x^2 + y^2 + z^2 = c^2 t^2$$

Using the above transformations, Einstein transformed this equation into the moving coordinate system, K′,

$$x^2 + y^2 + z^2 = c^2 t^2$$

$$[\phi(-v)\gamma(x' + vt')]^2 + [\phi(-v)y']^2 + [\phi(-v)z']^2 = c^2 \left[\phi(-v)\gamma\left(t' + \frac{vx}{c^2}\right)\right]^2$$

Cancelling the common factor, $\phi^2(-v)$,

$$\gamma^2(x' + vt')^2 + (y')^2 + (z')^2 = c^2\gamma^2\left(t' + \frac{vx}{c^2}\right)^2$$

$$\gamma^2 x'^2\left(1 - \frac{v^2}{c^2}\right) + y'^2 + z'^2 = \gamma^2 t'^2(c^2 - v^2) = \gamma^2 t'^2 c^2\left(1 - \frac{v^2}{c^2}\right)$$

$$x'^2 + y'^2 + z'^2 = c^2 t'^2$$

Thus, a wave front that is spherical in K, also is a spherical wave front in K′. As Einstein then notes, "This proves that our two fundamental principles are compatible."[49]

He then proceeds to show $\varphi(v) = 1$, resulting in what today are called the Lorentz transformations (see Appendices 4.5.2.5 and 4.5.2.6 for the details of this derivation):

$$t' = \gamma\left(t - \frac{v}{c^2}x\right)$$

$$x' = \gamma(x - vt)$$

$$y' = y$$

$$z' = z$$

with $\gamma = \frac{1}{\sqrt{1 - (\frac{v}{c})^2}}$

4.2.4   The Physical Meaning of the Equations Obtained Concerning Moving Rigid Bodies and Moving Clocks

In his Section 4, Einstein applies the transformation equations to some examples.

Example 4.1 – *The Sphere:*
Consider a rigid sphere of radius R that is at rest in reference frame K′ (the moving frame) and centered on the origin of K′. The equation of its surface is

$$x'^2 + y'^2 + z'^2 = R^2$$

Transforming to reference frame K, this equation of this surface at time t = 0 is

$$\frac{x^2}{\left(\sqrt{1-\left(\frac{v}{c}\right)^2}\right)^2} + y^2 + z^2 = R^2$$

A spherical body at rest is an ellipsoid when viewed in motion, with axes

$$R\sqrt{1-\left(\frac{v}{c}\right)^2}, R, R$$

"The x-dimension appears to be contracted in the ratio 1 : $\sqrt{1-(v/c)^2}$ .... At v = [c], all moving objects – observed from the system 'at rest' – shrink into plane structures."[50]

Example 4.2 – *Time:*
Consider the two reference frames K and K′. At t = 0, t′ = 0, their origins coincide. A clock is at rest at the origin of K′. K′ is moving in the positive x-direction with velocity v. In the K frame, after time t the origin of K′ will be at x = vt. Using the time transformation equation

$$t' = \frac{1}{\sqrt{1-\left(\frac{v}{c}\right)^2}}\left(t - \frac{v}{c^2}x\right) = \frac{1}{\sqrt{1-\left(\frac{v}{c}\right)^2}}\left(t - \frac{v}{c^2}vt\right)$$

$$t' = t\sqrt{1-\left(\frac{v}{c}\right)^2} = t - \left(1 - \sqrt{1-\left(\frac{v}{c}\right)^2}\right)t$$

"which shows that the clock (observed in the system at rest) is retarded each second by $(1 - \sqrt{1-(v/c)^2})$... "[51]

4.2.5 The Addition Theorem of Velocities

Consider a particle moving with constant velocity motion in the K′ reference frame. In Section 5 of his paper, Einstein determines the velocity of the particle as seen from the rest frame K. In the reference frame K′, the particle starts from the origin at $t' = 0$, moving with constant velocity w in the $x' - y'$ plane. After a time t', the particle will have the coordinates

$$x' = w_x t'$$

$$y' = w_y t'$$

$$z' = 0$$

Using the transformation equations for the first equation

$$x' = w_x t'$$

$$\gamma(x - vt) = w_x \gamma \left(t - \frac{v}{c^2} x \right)$$

$$x = \left(\frac{w_x + v}{1 + \frac{v w_x}{c^2}} \right) t$$

since, in K, $x = u_x t \Rightarrow u_x = \dfrac{w_x + v}{1 + \frac{v w_x}{c^2}}$

where u_x is the x component of the velocity in K. In a similar manner, the second equation gives

$$y' = w_y t'$$

$$y = w_y \gamma \left(t - \frac{v}{c^2} x \right)$$

and substituting the previously obtained equation for x into this equation

$$y = w_y \gamma \left(t - \frac{v}{c^2} \left(\frac{w_x + v}{1 + \frac{v w_x}{c^2}} \right) t \right)$$

$$y = \frac{w_y \sqrt{1 - \left(\frac{v}{c} \right)^2}}{1 + \frac{v w_x}{c^2}} t$$

since, in K, $y = u_y t \Rightarrow u_y = \dfrac{w_y \sqrt{1 - \left(\frac{v}{c} \right)^2}}{1 + \frac{v w_x}{c^2}}$

4.2.6   Transformation of the Maxwell–Hertz Equations for Empty Space. On the Nature of the Electromotive Forces that Arise upon Motion in a Magnetic Field

It should be pointed out that, in Maxwell's equations, the terms we today call the electric and magnetic fields, Einstein referred to as the electric and magnetic force vectors.[52]

One of the major concerns that led Einstein to develop his special theory of relativity was that Maxwell's equations appeared to select out a preferred frame of reference, that of the stationary aether, in which the speed of light was $c = 3 \times 10^8$ m/s. In other inertial reference frames, the speed of light should be greater/smaller by the relative speed of the two reference frames. In Section 4.1.2 we showed, using the Galilean transformations to transform Maxwell's equations to a "moving reference frame," Maxwell's equations take almost the same form, with small additional terms proportional to the relative speed of the two reference frames. By measuring these "additional" terms to Maxwell's equations in the moving reference frame, one could thereby determine the speed of the moving reference frame relative to the stationary aether which, in turn, was assumed to be at rest relative to the absolute space of Newton. But these terms were very small, on the order of (v/c) (taking v to be the orbital speed of the earth in orbit, (v/c) was on the order of 10^{-4}). Experiments by Fizeau, and others, were unable to detect the presence of these first order effects.[53] In response to this, Lorentz showed that the introduction of local time, and defining the electric and magnetic fields in the moving reference frame in terms of appropriate linear combinations of the electric and magnetic fields from the rest frame, one could make the first order terms vanish (but, at the same time, introducing second order terms in (v/c), i.e., $(v/c)^2$, which were too small to be measured). The relation between the electric and magnetic fields in the rest frame and in the moving frame were obtained for the specific purpose of making the first order terms in the transformed Maxwell equations vanish, so that Maxwell's equations would have the same form (to first order in (v/c)) in the rest frame and in the moving frame, i.e., that Maxwell's equations would be consistent with Newton's principle of relativity.

Einstein reversed this process. Starting with the principle of relativity, he obtained the Lorentz transformations. In Section 6 of his paper, he applies the Lorentz transformation equations to Maxwell's equations. Requiring Maxwell's equations to have the same form in the moving reference frame as in the rest frame, Einstein obtains the same transformations of the electric and magnetic fields as were obtained by Lorentz, not in response to a particular experimental result and correct to first order, but as a natural and exact result of the principle of relativity.

In empty space, since there are no sources/sinks for the fields, Maxwell's equations are:

(a) $\dfrac{1}{c}\dfrac{\partial E_x}{\partial t} = \dfrac{\partial B_z}{\partial y} - \dfrac{\partial B_y}{\partial z}$ (d) $\dfrac{1}{c}\dfrac{\partial B_x}{\partial t} = \dfrac{\partial E_y}{\partial z} - \dfrac{\partial E_z}{\partial y}$

(b) $\dfrac{1}{c}\dfrac{\partial E_y}{\partial t} = \dfrac{\partial B_x}{\partial z} - \dfrac{\partial B_z}{\partial x}$ (e) $\dfrac{1}{c}\dfrac{\partial B_y}{\partial t} = \dfrac{\partial E_z}{\partial x} - \dfrac{\partial E_x}{\partial z}$

(c) $\dfrac{1}{c}\dfrac{\partial E_z}{\partial t} = \dfrac{\partial B_y}{\partial x} - \dfrac{\partial B_x}{\partial y}$ (f) $\dfrac{1}{c}\dfrac{\partial B_z}{\partial t} = \dfrac{\partial E_x}{\partial y} - \dfrac{\partial E_y}{\partial x}$

(g) $\nabla \bullet \vec{E} = \dfrac{\partial E_x}{\partial x} + \dfrac{\partial E_y}{\partial y} + \dfrac{\partial E_z}{\partial z} = 0$

(h) $\nabla \bullet \vec{B} = \dfrac{\partial B_x}{\partial x} + \dfrac{\partial B_y}{\partial y} + \dfrac{\partial B_z}{\partial z} = 0$

These equations are valid in the rest system K. Applying the Lorentz transformation equations obtained in Section 4.2.3, these equation in the K′ reference frame become[54] (see Appendices 4.5.3.1 and 4.5.3.2 for the details of this derivation):

(A) $$\frac{1}{c}\frac{\partial E_x}{\partial t'} = \frac{\partial \left[\gamma \left(B_z - \frac{v}{c}E_y\right)\right]}{\partial y'} - \frac{\partial \left[\gamma \left(B_y + \frac{v}{c}E_z\right)\right]}{\partial z'}$$

(B) $$\frac{1}{c}\frac{\partial \left[\gamma \left(E_y - \frac{v}{c}B_z\right)\right]}{\partial t'} = \frac{\partial B_x}{\partial z'} - \frac{\partial \left[\gamma \left(B_z - \frac{v}{c}E_y\right)\right]}{\partial x'}$$

(C) $$\frac{1}{c}\frac{\partial \left[\gamma \left(E_z + \frac{v}{c}B_y\right)\right]}{\partial t'} = \frac{\partial \left[\gamma \left(B_y + \frac{v}{c}E_z\right)\right]}{\partial x'} - \frac{\partial B_x}{\partial y'}$$

(D) $$\frac{1}{c}\frac{\partial B_x}{\partial t'} = \frac{\partial \left[\gamma \left(E_y - \frac{v}{c}B_z\right)\right]}{\partial z'} - \frac{\partial \left[\gamma \left(E_z + \frac{v}{c}B_y\right)\right]}{\partial z'}$$

(E) $$\frac{1}{c}\frac{\partial \left[\gamma \left(B_y + \frac{v}{c}E_z\right)\right]}{\partial t'} = \frac{\partial \left[\gamma \left(E_z + \frac{v}{c}B_y\right)\right]}{\partial x'} - \frac{\partial E_x}{\partial y'}$$

(F) $$\frac{1}{c}\frac{\partial \left[\gamma \left(B_z - \frac{v}{c}E_y\right)\right]}{\partial t'} = \frac{\partial E_x}{\partial y'} - \frac{\partial \left[\gamma \left(E_y - \frac{v}{c}B_z\right)\right]}{\partial x'}$$

But, Einstein notes, if Maxwell's equations are valid in reference frame K, the principle of relativity requires that they also be valid in K'.[55] Thus, by inspection of the above equations, and the form Maxwell's equations would have in K', Einstein obtains[56]:

$$E'_x = E_x \qquad\qquad B'_x = B_x$$

$$E'_y = \gamma \left(E_y - \frac{v}{c}B_z\right) \qquad B'_y = \gamma \left(B_y + \frac{v}{c}E_z\right)$$

$$E'_z = \gamma \left(E_z + \frac{v}{c}B_y\right) \qquad B'_z = \gamma \left(B_z - \frac{v}{c}E_y\right)$$

Einstein noted that, on the right-hand side of each relation, there could be a multiplicative factor that depended on velocity v, $\Psi(v)$. He then showed such a factor would be equal to unity. (See Appendix 5.5.3.3 for the details of this derivation.)[57] Einstein's approach gave the same Lorentz transformations and the same transformations for the electric and magnetic fields as Lorentz had obtained, but from a simple, coherent principle, the principle of relativity.

4.2.7   Theory of Doppler's Principle and of Aberration

"Imagine in the system K, very far from the coordinate origin, a source of electrodynamic waves, which in a part of space containing the coordinate origin is represented with sufficient accuracy by..." plane waves.[58] See Figure 4.5.

In reference frame K, the plane waves are represented as

$$E_x = E_{x0}\sin\Phi \quad B_x = B_{x0}\sin\Phi$$

$$E_y = E_{y0}\sin\Phi \quad B_y = B_{x0}\sin\Phi \quad \text{with} \quad \Phi = \omega\left(t - \frac{a_x x + a_y y + a_z z}{c}\right)$$

$$E_z = E_{z0}\sin\Phi \quad B_z = B_{x0}\sin\Phi$$

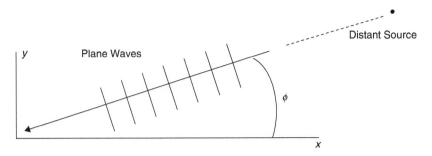

Fig. 4.5 Plane waves from a distant source.

where a_x, a_y, a_z, are the direction cosines of the normal to the wave front. From Figure 4.5, $a_x = \cos\phi$, etc. (The notation has been altered from that used by Einstein. Einstein used the symbols a, b, and c as the direction cosines, and the symbol V for the velocity of light. To avoid any confusion with c representing the speed of light, the direction cosines are written as a_x, a_y, and a_z.)

Consider now the reference frame K$'$, moving at a constant speed v along the X-axis of reference frame K. Using the transformation equations just obtained for the electric and magnetic fields, and applying the Lorentz transformations to the coordinate and time variables, the above equations become (see Appendix 4.5.4)

$$E'_x = E_{x0}\sin\Phi' \qquad\qquad B'_x = B_{x0}\sin\Phi'$$
$$E'_y = \gamma\left(E_{y0} - \tfrac{v}{c}B_{z0}\right)\sin\Phi' \;\; B'_y = \gamma\left(B_{y0} + \tfrac{v}{c}E_{z0}\right)\sin\Phi'$$
$$E'_z = \gamma\left(E_{z0} + \tfrac{v}{c}B_{y0}\right)\sin\Phi' \;\; B'_z = \gamma\left(B_{z0} - \tfrac{v}{c}E_{y0}\right)\sin\Phi'$$

with $\Phi' = \omega'\left(t' \frac{a'_x x' + a'_y y' + a'_z z'}{c}\right)$

and with

$$\omega' = \omega\gamma\left(1 - a_x\frac{v}{c}\right)$$

$$a'_x = \frac{a_x - \frac{v}{c}}{1 - a_x\frac{v}{c}}$$

$$a'_y = \frac{a_y}{\gamma\left(1 - a_x\frac{v}{c}\right)}$$

$$a'_z = \frac{a_z}{\gamma\left(1 - a_x\frac{v}{c}\right)}$$

From the above relation between ω and ω', with the direction cosine a_x expressed as $\cos\phi$, the frequencies of the wave as observed in the two reference frames are related as

$$\nu' = \nu\frac{1 - \frac{v}{c}\cos\phi}{\sqrt{1 - \frac{v^2}{c^2}}}$$

For $\phi = 0$, this equation reduces to the simple Doppler shift formula:

$$\nu' = \nu \sqrt{\frac{1 - \frac{v}{c}}{1 + \frac{v}{c}}}$$

"If A and A$'$ denote the electric or magnetic [field] in the system at rest and in motion, respectively, we get..." (See Appendix 4.5.4.)[59]

$$A'^2 = A^2 \frac{\left(1 - \frac{v}{c}\cos\phi\right)^2}{1 - \left(\frac{v}{c}\right)^2} \tag{4.2}$$

where ϕ is the angle between the observer's velocity and a line from the observer to the light source, i.e. between the x-axis and the wave normal.

4.2.8   Transformation of the Energy of Light Rays. Theory of the Radiation Pressure Exerted on Perfect Mirrors

Einstein considers the same physical situation as in Section 4.2.7, a source of electromagnetic waves so distant from the origin of K that the waves can be approximated as plane waves in the vicinity of the origin of K.

Consider a spherical surface moving parallel to the motion of the plane waves, and at speed c in reference frame K. See Figure 4.6.

"If $[a_x, a_y, a_z]$ are the direction cosines of the wave normal of the light in the system at rest [K], then the surface elements of the spherical surface

$$(x - cta_x)^2 + (y - cta_y)^2 + (z - cta_z)^2 = R^2$$

which moves with the velocity of light, are not traversed by any energy; we may therefore say that this surface permanently encloses the same light complex."[60] The sphere is traveling with the wave front at the speed of light so that radiation energy neither enters nor leaves the sphere, i.e., the amount of radiation energy within the surface remains constant. Einstein transforms this spherical surface seen in K into the surface as seen from K$'$ (an ellipsoid). Using the Lorentz transformations,

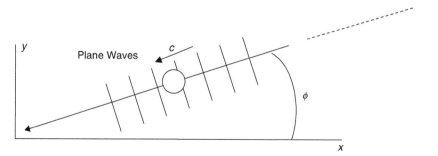

Fig. 4.6 Spherical surface moving parallel to the plane waves.

the equation of this ellipsoidal surface as seen in the K′ reference frame is

$$\left[\gamma\left(x'+vt'\right)-ca_x\gamma\left(t'+\frac{v}{c^2}x'\right)\right]^2+\left[y'-ca_y\gamma\left(t'+\frac{v}{c^2}x'\right)\right]^2$$
$$+\left[z'-ca_z\gamma\left(t'+\frac{v}{c^2}x'\right)\right]=R^2$$

At time $t'=0$, this equation becomes

$$\gamma^2\left(1-\frac{v}{c}a_x\right)^2x'^2+\left(y'-\gamma\frac{v}{c}a_yx'\right)^2+\left(z'-\gamma\frac{v}{c}a_zx'\right)^2=R^2$$

At $t'=0$, this is the equation of an ellipsoid centered at $x'=0, y'=\gamma\frac{v}{c}a_yx', z'=\gamma\frac{v}{c}a_zx'$, with semi-axes of length $X'=\frac{R}{\gamma(1-\frac{v}{c}a_x)}; Y'=R; Z'=R$.

The volume of an ellipsoid is $\frac{4}{3}\pi XYZ$. The ratio of the volume of the ellipsoid in K′ to the volume of the sphere in K is

$$\frac{Vol'_{ellipsoid}}{Vol_{Sphere}}=\frac{\frac{4}{3}\pi X'Y'Z'}{\frac{4}{3}\pi R^3}=\frac{\sqrt{1-\frac{v^2}{c^2}}}{1-\frac{v}{c}a_x}=\frac{\sqrt{1-\frac{v^2}{c^2}}}{1-\frac{v}{c}\cos\phi}$$

The energy of light per unit volume is equal to $A^2/8\pi$. Using the relation between A and A′ obtained in Section 4.2.7, Eq. (4.2), "If the energy of the light enclosed by the surface under consideration is denoted by E when measured in the system at rest and by E' when measured in the moving system, we obtain"[61]

$$\frac{E'}{E}=\frac{(A'^2/8\pi)}{(A^2/8\pi)}\left(\frac{Vol'_{ellipsoid}}{Vol_{sphere}}\right)=\frac{A'^2}{A^2}\frac{\sqrt{1-\frac{v^2}{c^2}}}{1-\frac{v}{c}\cos\phi}$$

$$=\frac{\left(1-\frac{v}{c}\cos\phi\right)^2}{1-\left(\frac{v}{c}\right)^2}\frac{\sqrt{1-\frac{v^2}{c^2}}}{1-\frac{v}{c}\cos\phi}=\frac{1-\frac{v}{c}\cos\phi}{\sqrt{1-\frac{v^2}{c^2}}}$$

This relation between E and E' is the same as the relation between the frequencies of a wave, ν and ν', in Section 4.2.7. Einstein commented that he found it noteworthy that "the energy and the frequency of a light complex vary with the observer's state of motion according to the same law."[62] Three months earlier (April, 2005), in his paper "On a Heuristic Point of View Concerning the Production and Transformation of Light," Einstein had presented the idea that the energy of a light wave is proportional to its frequency, $E=h\nu$, yet he did not point out that relation in this context.[63]

This relation between E and E' will be used in the subsequent paper, "Does the Inertia of a Body Depend on Its Energy Content?"[64]

4.2.9   Transformation of the Maxwell–Hertz Equations when Convection Currents Are Taken into Consideration

Einstein shows that Maxwell's equations transform properly under the Lorentz transformation when convection currents are present, i.e., "...with our kinematic principles taken as a basis, the electrodynamic foundation of Lorentz's theory of the electrodynamics of moving bodies is in agreement with the principle of relativity."[65]

Einstein states that it "can easily be deduced"[66] that the electrical charge will have the same value whether it is viewed at rest or moving.

4.2.10   Dynamics of the (Slowly Accelerated) Electron

Again consider the same two reference frames, K and K′, with K "at rest" and K′ moving with constant velocity v in the positive x-direction. Let an electron be at rest at the origin of K′ (the moving reference frame). In K′, the electromagnetic force on the electron will be purely electric. In K′, Newton's second law is

$$m\frac{d^2x'}{dt'^2} = qE'_x$$

$$m\frac{d^2y'}{dt'^2} = qE'_y$$

$$m\frac{d^2z'}{dt'^2} = qE'_z$$

In K, the electron is moving at speed v. Transforming the above equations to the reference frame K (see Appendix 4.5.5 for the details of this derivation),[67]

$$(m\gamma^3)\frac{d^2x}{dt^2} = qE_x$$

$$(m\gamma)\frac{d^2y}{dt^2} = q\left(E_y - \frac{v}{c}B_z\right)$$

$$(m\gamma)\frac{d^2z}{dt^2} = q\left(E_z - \frac{v}{c}B_y\right)$$

It should be noted that the right-hand side of these equations gives the Lorentz force on a charged particle moving in an electromagnetic field as a consequence of the principle of relativity, obtained in a straightforward and natural way, not as a separate postulate as was done by Lorentz. The left-hand side of these equations indicates the mass will be a function of the speed v. Consistent with the terminology of the day, $m\gamma^3$ was termed the longitudinal mass and $m\gamma$ was termed the transverse mass. In the terminology of today, m is the rest mass.

When an electron starting from rest is accelerated by an electric field along the x-axis, the kinetic energy of the electron is equal to the work done on it by the electric field, $W = \int_0^v F dx$ with $F = qE_x$. As the electron is accelerated, Newton's second law for the electron is $(m\gamma^3)\frac{d^2x}{dt^2} = qE_x$. Inverting this, $E_x = \frac{m\gamma^3}{q}\frac{d^2x}{dt^2}$,

$$KE = W = \int_0^v F dx = \int_0^v qE_x \, dx$$

$$= \int q\left(\frac{m\gamma^3}{q}\frac{dv}{dt}\right) dx$$

$$= m\int_0^v (\gamma^3) dv \frac{dx}{dt}$$

$$= m\int_0^v \left(\frac{1}{\left(1 - \frac{v^2}{c^2}\right)^{3/2}}\right) v dv$$

$$= mc^2\left(\frac{1}{\sqrt{1 - \frac{v^2}{c^2}}} - 1\right)$$

$$= \gamma mc^2 - mc^2$$

From the second to last expression for the kinetic energy of the electron, it can be seen the work supplied to the electron becomes infinitely large when $v = c$. This shows the impossibility of reaching superluminal speeds.

4.3   Albert Einstein's Paper, "Does the Inertia of a Body Depend Upon Its Energy Content?"[68]

This is the paper in which the famous relation $E = mc^2$ is obtained. However, in the paper, this relation does not appear in this form. In the paper, Einstein considers a body at rest that emits two light rays (in opposite directions to conserve momentum), each with energy $L/2$. Viewing the process from a moving reference frame, it appears the kinetic energy of the body is reduced by an amount $\frac{L}{c^2}\frac{v^2}{2}$. Equating this to $\frac{1}{2}mv^2$, and designating the energy L as E, one obtains the relation $E = mc^2$.

Consider a system of plane light waves having the energy E in system K. The normal to the wave front makes an angle ϕ with the X-axis. See Figure 4.7.

In system K' (in uniform translational motion along the x-axis of K), this system of plane light waves has energy E'. From Section 8 of "On the Electrodynamics of Moving Bodies,"[69] it was shown that E and E' are related as [70]

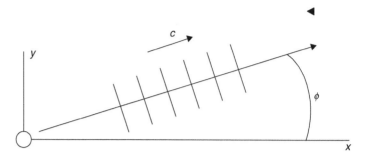

Fig. 4.7 Body at rest at the origin
emits plane light waves.

$$E' = E \frac{1 - \frac{v}{c} \cos \phi}{\sqrt{1 - \frac{v^2}{c}}}$$

In K, a body at rest has the energy E_0. Relative to the reference frame
K′, the body has energy H_0. As measured in the reference frame K, the
body emits plane light waves of energy $L/2$ at an angle ϕ relative to
the x-axis. Simultaneously it emits "an equal amount of light in the
opposite direction."[71] After the emission of the plane light waves the
body has energy E_1. From conservation of energy,

$$E_0 = E_1 + \left[\frac{L}{2} + \frac{L}{2} \right] = E_1 + L$$

In K′, the energy after the emission of the plane light waves is H_1.
The principle of relativity requires that energy conservation also hold in
K′. Using the energy relation above,

$$H_0 = H_1 + \left[\frac{L}{2} \frac{1 - \frac{v}{c} \cos \phi}{\sqrt{1 - \frac{v^2}{c^2}}} + \frac{L}{2} \frac{1 + \frac{v}{c} \cos \phi}{\sqrt{1 - \frac{v^2}{c^2}}} \right]$$

$$H_0 = H_1 + \frac{L}{\sqrt{1 - \frac{v^2}{c^2}}}$$

H and E are the energies of the same body as measured from two
different reference frames, K′ and K. The body is at rest in reference
frame K, and is moving relative to reference frame K′. E is the energy of
the body at rest (in K), while H is the energy of the body as seen moving
(in K′). The difference H–E will then be the kinetic energy of the body,
H_0–E_0 being the kinetic energy before the emission of the plane light
waves, and H_1–E_1 being the kinetic energy of the body after the emission
of the plane light waves. Subtracting the above equations for H and
for E,

$$(H_0 - E_0) - (H_1 - E_1) = L \left\{ \frac{1}{\sqrt{1 - \frac{v^2}{c^2}}} - 1 \right\}$$

$$(KE_0) - (KE_1) = L \left\{ \frac{1}{\sqrt{1 - \frac{v^2}{c^2}}} - 1 \right\}$$

The kinetic energy of the body, as viewed from reference frame K′, "decreases as a result of the emission of light by an amount that is independent of the body's characteristics."[72]

Expanding the right-hand side of the equation, and keeping only the lowest order non-vanishing term,

$$(KE_0) - (KE_1) = L \left\{ 1 + \frac{1}{2} \left(\frac{v^2}{c^2} \right) + \ldots - 1 \right\}$$

$$(KE_0) - (KE_1) = \frac{L}{c^2} \frac{v^2}{2} = \frac{1}{2} \left(\frac{L}{c^2} \right) v^2 = \frac{1}{2} (\Delta m) v^2$$

Einstein concludes:

1.   If a body emits the energy L in the form of radiation, its mass decreases by L/c^2.
2.   The mass of a body is a measure of its energy content; if the energy changes by L, the mass changes in the same sense by $L/9 \cdot 10^{20} [\text{i.e.} L/c^2]$.[73]

In more current terms, $\Delta mass = \frac{\Delta E}{c^2}$ or, rearranging, $E = mc^2$.

4.4   Discussion and Comments

By 1900, Newton's principle of relativity, in practice, was believed by many scientists to extend beyond mechanics to all of physics – that no experiment would enable one to determine one's speed relative to absolute space. But Maxwell's theory of electromagnetism predicted a definite speed for light, presumably in the "at rest" reference frame of the luminiferous aether. If true, the speed of light relative to the earth would be greater (or lesser) than the speed of light in the aether by an amount equal to the speed of the earth through the aether, thus enabling one to determine the speed of the earth relative to the aether which was at rest relative to absolute space. However, experiment was unable to detect any such adjustment to the speed of light. To address this incompatibility with experiment, Lorentz devised modifications to Maxwell's equations to remove the offending terms, once to be rid of first order terms in (v/c) and then, later, to be rid of second order terms in (v/c). But Newton's principle of relativity and the predictions of Maxwell's theory that the speed of light has a fixed value remained at odds.

In the words of John Stachel, "When combined with the relativity principle, this $[c = \text{constant}]$ leads to an apparently paradoxical conclusion: the velocity of light must be the same in all inertial frames. This result conflicts with the Newtonian law of addition of velocities, forcing a revision of the kinematical foundations underlying all of physics. Einstein showed that the simultaneity of distant events is only defined physically relative to a particular inertial frame, leading to kinematical transformations between the spatial and temporal coordinates of two inertial frames that agree formally with the transformations that Lorentz had introduced in 1904."[74]

The concern with the electromagnetic induction is resolved in the joining of the electric and magnetic fields into one electromagnetic field. The difference in the explanations whether the magnet is moving or the coil is moving is one of perspective, i.e., there is but one electromagnetic phenomenon, but the description of it is dependent upon which reference frame one is viewing it from.

Because both Newton's principle of relativity and Maxwell's electromagnetic theory had so many other successes, Einstein accepted both of them and, from them, proposed the two apparently conflicting postulates from which he developed the special theory of relativity. Once Einstein had narrowed the area of incompatibility to the Galilean–Newtonian addition of velocities, he hit upon the concept of time being at the root of the problem. From this all else followed.

Lorentz's modification of Maxwell's equations was a careful, methodical, systematic development of a theory based on experimental results – a "theory of construction."

Lorentz was looking to discover the basic constituents of the physical world, i.e., of what the world is "constructed." He was tied to experiment to obtain greater and deeper insights. Lorentz had "constructed" the Lorentz transformations and the transformations of the electric and magnetic fields in response to experiments showing the first order terms were zero, and then in response to experiments showing the second order terms to be zero. On the other hand, Einstein was focused on a theory of principle, a principle based on a minimum of postulates that had application to a wide range of phenomena and possessed inner consistency, inner perfection, and simplicity. From Einstein's "theory of principle," the principle of relativity, he obtained directly the Lorentz transformations and the same transformations for the electric and magnetic fields.

From 1900 to 1905, Walter Kaufmann had conducted a number of experiments on high-speed electrons, showing the mass of electrons increased with increasing speed.[75] In 1905, after a series of such experiments, Kaufmann wrote that, "The prevalent results decidedly speak against the correctness of Lorentz's assumptions as well as Einstein's."[76] Einstein did not reply immediately to this assertion of Kaufmann. In Einstein's 1907 paper, "On the Relativity Principle and the Conclusions Drawn from It," he writes, "It should also be mentioned that ... [other] ... theories of the motion of the electron yield curves that are significantly closer to the observed curve than the curve obtained from the theory of relativity. However, the probability that their theories are correct is rather small, in my opinion, because their basic assumptions concerning the dimensions of the moving electron are not suggested by theoretical systems that encompass larger complexes of phenomena."[77] Planck, after two years of investigation of the Kaufmann results, found an inconsistency in Kaufmann's data that shifted the conclusions to slightly in favor of Einstein's theory.[78]

4.5 Appendices

4.5.1 Lorentz and the Transformed Maxwell Equations

4.5.1.1 Transforming Maxwell's Equations

The Galilean transformations are

$$x' = x - vt$$

$$y' = y$$

$$z' = z$$

$$t' = t$$

In empty space Maxwell's equations are

(a) $\dfrac{\partial E_z}{\partial y} - \dfrac{\partial E_y}{\partial z} = -\dfrac{1}{c}\dfrac{\partial B_x}{\partial t}$
    (d) $\dfrac{\partial B_z}{\partial y} - \dfrac{\partial B_y}{\partial z} = \dfrac{1}{c}\dfrac{\partial E_x}{\partial t}$

(b) $\dfrac{\partial E_x}{\partial z} - \dfrac{\partial E_z}{\partial x} = -\dfrac{1}{c}\dfrac{\partial B_y}{\partial t}$
    (e) $\dfrac{\partial B_x}{\partial z} - \dfrac{\partial B_z}{\partial x} = \dfrac{1}{c}\dfrac{\partial E_y}{\partial t}$

(c) $\dfrac{\partial E_y}{\partial x} - \dfrac{\partial E_x}{\partial y} = -\dfrac{1}{c}\dfrac{\partial B_z}{\partial t}$
    (f) $\dfrac{\partial B_y}{\partial x} - \dfrac{\partial B_x}{\partial y} = \dfrac{1}{c}\dfrac{\partial E_z}{\partial t}$

(g) $\nabla \bullet \vec{E} = \dfrac{\partial E_x}{\partial x} + \dfrac{\partial E_y}{\partial y} + \dfrac{\partial E_z}{\partial z} = 0$

(h) $\nabla \bullet \vec{B} = \dfrac{\partial B_x}{\partial x} + \dfrac{\partial B_y}{\partial y} + \dfrac{\partial B_z}{\partial z} = 0$

By definition, Maxwell's equations are valid in the rest system K. The Galilean transformation equations are applied to the Maxwell equations to determine the form they will assume in the K$'$ reference frame (for convenience, we keep the designation of time in reference frame K$'$ as t, not as t').

Working first with Eq. (g),

$$\frac{\partial E_x}{\partial x} = \frac{\partial E_x}{\partial t}\left(\frac{\partial t}{\partial x}\right) + \frac{\partial E_x}{\partial x'}\left(\frac{\partial x'}{\partial x}\right) + \frac{\partial E_x}{\partial y'}\left(\frac{\partial y'}{\partial x}\right) + \frac{\partial E_x}{\partial z'}\left(\frac{\partial z'}{\partial x}\right)$$

$$= \frac{\partial E_x}{\partial t}(0) + \frac{\partial E_x}{\partial x'}(1) + \frac{\partial E_x}{\partial y'}(0) + \frac{\partial E_x}{\partial z'}(0)$$

$$= \frac{\partial E_x}{\partial x'}$$

Similarly, $\dfrac{\partial E_y}{\partial y} = \dfrac{\partial E_y}{\partial y'}$ and $\dfrac{\partial E_z}{\partial z} = \dfrac{\partial E_z}{\partial z'}$. In K', Eq. (g) is

$$\Rightarrow \frac{\partial E_x}{\partial x'} + \frac{\partial E_y}{\partial y'} + \frac{\partial E_z}{\partial z'} = 0 \quad \text{Eq. G}$$

and, in a similar manner, in K', Eq. (h) is

$$\frac{\partial B_x}{\partial x'} + \frac{\partial B_y}{\partial y'} + \frac{\partial B_z}{\partial z'} = 0 \quad \text{Eq. H}$$

Working with Eq. (a),

$$\frac{\partial E_z}{\partial y} = \frac{\partial E_z}{\partial t}\left(\frac{\partial t}{\partial y}\right) + \frac{\partial E_z}{\partial x'}\left(\frac{\partial x'}{\partial y}\right) + \frac{\partial E_z}{\partial y'}\left(\frac{\partial y'}{\partial y}\right) + \frac{\partial E_z}{\partial z'}\left(\frac{\partial z'}{\partial y}\right)$$

$$= \frac{\partial E_z}{\partial t}(0) + \frac{\partial E_z}{\partial x'}(0) + \frac{\partial E_z}{\partial y'}(1) + \frac{\partial E_z}{\partial z'}(0)$$

$$= \frac{\partial E_z}{\partial y'}$$

And, in like manner,

$$\frac{\partial E_y}{\partial z} = \frac{\partial E_y}{\partial z'}$$

$$\frac{\partial E_x}{\partial z} = \frac{\partial E_x}{\partial z'}$$

$$\frac{\partial E_z}{\partial x} = \frac{\partial E_z}{\partial x'}$$

$$\frac{\partial E_y}{\partial x} = \frac{\partial E_y}{\partial x'}$$

$$\frac{\partial E_x}{\partial y} = \frac{\partial E_x}{\partial y'}$$

The right-hand side of Eq. (a) becomes

$$\frac{-1}{c}\frac{\partial B_x}{\partial t} = \frac{-1}{c}\left\{\frac{\partial B_x}{\partial t}\left(\frac{\partial t}{\partial t}\right) + \frac{\partial B_x}{\partial x'}\left(\frac{\partial x'}{\partial t}\right) + \frac{\partial B_x}{\partial y'}\left(\frac{\partial y'}{\partial t}\right) + \frac{\partial B_x}{\partial z'}\left(\frac{\partial z'}{\partial t}\right)\right\}$$

$$= \frac{-1}{c}\left\{\frac{\partial B_x}{\partial t}(1) + \frac{\partial B_x}{\partial x'}(-v) + \frac{\partial B_x}{\partial y'}(0) + \frac{\partial B_x}{\partial z'}(0)\right\}$$

$$= \frac{-1}{c}\left\{\frac{\partial B_x}{\partial t} - v\frac{\partial B_x}{\partial x'}\right\}$$

In like manner,

$$\frac{-1}{c}\frac{\partial B_y}{\partial t} = \frac{-1}{c}\left\{\frac{\partial B_y}{\partial t} - v\frac{\partial B_y}{\partial x'}\right\}$$

$$\frac{-1}{c}\frac{\partial B_z}{\partial t} = \frac{-1}{c}\left\{\frac{\partial B_z}{\partial t} - v\frac{\partial B_z}{\partial x'}\right\}$$

In reference frame K′, Eqs. (a), (b) and (c) are

(A) $\dfrac{\partial E_z}{\partial y'} - \dfrac{\partial E_y}{\partial z'} = -\dfrac{1}{c}\dfrac{\partial B_x}{\partial t} + \dfrac{v}{c}\dfrac{\partial B_x}{\partial x'}$

(B) $\dfrac{\partial E_x}{\partial z'} - \dfrac{\partial E_z}{\partial x'} = -\dfrac{1}{c}\dfrac{\partial B_y}{\partial t} + \dfrac{v}{c}\dfrac{\partial B_y}{\partial x'}$

(C) $\dfrac{\partial E_y}{\partial x'} - \dfrac{\partial E_x}{\partial y'} = -\dfrac{1}{c}\dfrac{\partial B_z}{\partial t} + \dfrac{v}{c}\dfrac{\partial B_z}{\partial x'}$

In a similar manner, in reference frame K′, Eqs. (d), (e) and (f) are

(D) $\dfrac{\partial B_z}{\partial y'} - \dfrac{\partial B_y}{\partial z'} = \dfrac{1}{c}\dfrac{\partial E_x}{\partial t} - \dfrac{v}{c}\dfrac{\partial E_x}{\partial x'}$

(E) $\dfrac{\partial B_x}{\partial z'} - \dfrac{\partial B_z}{\partial x'} = \dfrac{1}{c}\dfrac{\partial E_y}{\partial t} - \dfrac{v}{c}\dfrac{\partial E_y}{\partial x'}$

(F) $\dfrac{\partial B_y}{\partial x'} - \dfrac{\partial B_x}{\partial y'} = \dfrac{1}{c}\dfrac{\partial E_z}{\partial' t} - \dfrac{v}{c}\dfrac{\partial E_z}{\partial x'}$

From equations G and H,

(G) $\dfrac{\partial E_x}{\partial x'} + \dfrac{\partial E_y}{\partial y'} + \dfrac{\partial E_z}{\partial z'} = 0$

(H) $\dfrac{\partial B_x}{\partial x'} + \dfrac{\partial B_y}{\partial y'} + \dfrac{\partial B_z}{\partial z'} = 0$

It is easily seen that Eqs. (A) through (F) are the same as their counterparts, Eqs. (a) through (f), in reference frame K, *except* there is an additional term proportional to v/c added to each equation. Hendrik Lorentz set about to eliminate the "additional terms" that could not be found experimentally.[79] In Eqs. (B) and (C), the "extra term" on the right-hand side can be moved to the left-hand side, and combined with the already existing derivative with respect to x'. For Eq. (A), it is necessary first to replace $(\partial B_x/\partial x')$ by $(\partial B_x/\partial x') = -(\partial B_y/\partial y') - (\partial B_y/\partial y')$ from Eq. (H).

Equation (A) becomes

$\dfrac{\partial E_z}{\partial y'} - \dfrac{\partial E_y}{\partial z'} = -\dfrac{1}{c}\dfrac{\partial B_x}{\partial t} + \dfrac{v}{c}\left(\dfrac{-\partial B_y}{\partial y'}\dfrac{-\partial B_z}{\partial z'}\right)$

(A′) $\dfrac{\partial}{\partial y'}\left(E_z + \dfrac{v}{c}B_y\right) - \dfrac{\partial}{\partial z'}\left(E_y - \dfrac{v}{c}B_z\right) = -\dfrac{1}{c}\dfrac{\partial B_x}{\partial t}$

Equations (B) and (C) become

(B′) $\dfrac{\partial E_x}{\partial z'} - \dfrac{\partial}{\partial x'}\left(E_z + \dfrac{v}{c}B_y\right) = -\dfrac{1}{c}\dfrac{\partial B_y}{\partial t}$

(C′) $\dfrac{\partial}{\partial x'}\left(E_y - \dfrac{v}{c}B_z\right) - \dfrac{\partial E_x}{\partial y'} = -\dfrac{1}{c}\dfrac{\partial B_z}{\partial t}$

Defining the electric field in K′ as

$$E'_x = E_x$$

$$E'_y = E_y - \frac{v}{c}B_z$$

$$E'_z = E_z + \frac{v}{c}B_y$$

(A″) $$\frac{\partial\left(E'_z\right)}{\partial y'} - \frac{\partial\left(E'_y\right)}{\partial z'} = -\frac{1}{c}\frac{\partial B_x}{\partial t}$$

(B″) $$\frac{\partial(E'_x)}{\partial z'} - \frac{\partial\left(E'_z\right)}{\partial x'} = -\frac{1}{c}\frac{\partial B_y}{\partial t}$$

(C″) $$\frac{\partial(E'_y)}{\partial x'} - \frac{\partial(E'_x)}{\partial y'} = -\frac{1}{c}\frac{\partial B_z}{\partial t}$$

In like manner, Eqs. (D), (E), and (F) are transformed, using Eq. (G), to obtain

(D″) $$\frac{\partial(B'_z)}{\partial y'} - \frac{\partial\left(B'_y\right)}{\partial z'} = \frac{1}{c}\frac{\partial E_x}{\partial t}$$

(E″) $$\frac{\partial(B'_x)}{\partial z'} - \frac{\partial\left(B'_z\right)}{\partial x'} = \frac{1}{c}\frac{\partial E_y}{\partial t}$$

(F″) $$\frac{\partial(B'_y)}{\partial x'} - \frac{\partial(B'_x)}{\partial y'} = \frac{1}{c}\frac{\partial E_z}{\partial' t}$$

With the components of the magnetic field in K′ given by

$$B'_x = B_x$$

$$B'_y = B_y + \frac{v}{c}E_z$$

$$B'_z = B_z - \frac{v}{c}E_y$$

Each of these equations is in the desired form. However, each of the terms on the left-hand side of the equations (properly) contains only quantities referred to the K′ system, whereas on the right-hand side of the equations, the electric and magnetic fields, E and B, are still those measured in reference frame K. Working first with Eq. (B″):

$$\frac{\partial(E'_x)}{\partial z'} - \frac{\partial(E'_z)}{\partial x'} = -\frac{1}{c}\frac{\partial}{\partial t}\left(B'_y - \frac{v}{c}E_z\right)$$

$$\frac{\partial(E'_x)}{\partial z'} - \left\{\frac{\partial}{\partial x'}\left(E_z + \frac{v}{c}B_y\right) + \frac{1}{c}\left(\frac{v}{c}\right)\frac{\partial E_z}{\partial t}\right\} = -\frac{1}{c}\frac{\partial B'_y}{\partial t}$$

$$\frac{\partial(E'_x)}{\partial z'} - \left\{\frac{\partial E_z}{\partial x'} + \frac{1}{c}\left(\frac{v}{c}\right)\frac{\partial E_z}{\partial t} + \frac{v}{c}\frac{\partial B_y}{\partial x'}\right\} = -\frac{1}{c}\frac{\partial B'_y}{\partial t}$$

Lorentz introduced a new time variable, $t' = t - \dfrac{vx'}{c^2}$, yielding:

$$\frac{\partial E_z}{\partial x'} = \frac{\partial E_z}{\partial t'}\frac{\partial t'}{\partial x'} + \frac{\partial E_z}{\partial x'}\frac{\partial x'}{\partial x'} + \frac{\partial E_z}{\partial y'}\frac{\partial y'}{\partial x'} + \frac{\partial E_z}{\partial z'}\frac{\partial z'}{\partial x'}$$

$$= \frac{\partial E_z}{\partial t'}\left(\frac{-v}{c^2}\right) + \frac{\partial E_z}{\partial x'}(1) + \frac{\partial E_z}{\partial y'}(0) + \frac{\partial E_z}{\partial z'}(0)$$

$$= \frac{-v}{c^2}\frac{\partial E_z}{\partial t'} + \frac{\partial E_z}{\partial x'}$$

$$\frac{\partial E_z}{\partial t} = \frac{\partial E_z}{\partial t'}\frac{\partial t''}{\partial t'} + \frac{\partial E_z}{\partial x'}\frac{\partial x'}{\partial t} + \frac{\partial E_z}{\partial y'}\frac{\partial y'}{\partial t} + \frac{\partial E_z}{\partial z'}\frac{\partial z'}{\partial t}$$

$$= \frac{\partial E_z}{\partial t'}(1) + \frac{\partial E_z}{\partial x}(0) + \frac{\partial E_z}{\partial y}(0) + \frac{\partial E_z}{\partial z}(0)$$

$$= \frac{\partial E_z}{\partial t'}$$

(similarly $\dfrac{\partial B'_y}{\partial t} = \dfrac{\partial B'_y}{\partial t'}$)

$$\frac{\partial B_y}{\partial x'} = \frac{\partial B_y}{\partial t'}\frac{\partial t'}{\partial x'} + \frac{\partial B_y}{\partial x'}\frac{\partial x'}{\partial x'} + \frac{\partial B_y}{\partial y'}\frac{\partial y'}{\partial x'} + \frac{\partial B_y}{\partial z'}\frac{\partial z'}{\partial x'}$$

$$= \frac{\partial B_y}{\partial t'}\left(\frac{-v}{c^2}\right) + \frac{\partial B_y}{\partial x'}(1) + \frac{\partial B_y}{\partial y'}(0) + \frac{\partial B_y}{\partial z'}(0)$$

$$= \frac{-v}{c^2}\frac{\partial B_y}{\partial t'} + \frac{\partial B_y}{\partial x'}$$

Putting these together Eq. (B″) becomes

$$\frac{\partial(E'_x)}{\partial z'} - \left\{\frac{-v}{c^2}\frac{\partial E_z}{\partial t''} + \frac{\partial E_z}{\partial x'} + \frac{1}{c}\left(\frac{v}{c}\right)\frac{\partial E_z}{\partial t'} - \frac{v}{c^2}\left(\frac{v}{c}\right)\frac{\partial B_y}{\partial t'} + \frac{v}{c}\frac{\partial B_y}{\partial x'}\right\} = -\frac{1}{c}\frac{\partial B'_y}{\partial t''}$$

$$\frac{\partial(E'_x)}{\partial z'} - \left\{\frac{\partial E_z}{\partial x'} - \frac{v}{c^2}\left(\frac{v}{c}\right)\frac{\partial B_y}{\partial t'} + \frac{v}{c}\frac{\partial B_y}{\partial x'}\right\} = -\frac{1}{c}\frac{\partial B'_y}{\partial t'}$$

$$\frac{\partial(E'_x)}{\partial z'} - \left\{\frac{\partial}{\partial x'}\left(E_z + \frac{v}{c}B_y\right)\right\} = -\frac{1}{c}\frac{\partial B'_y}{\partial t'} - \frac{1}{c}\left(\frac{v}{c}\right)^2\frac{\partial B_y}{\partial t'''}$$

$$\frac{\partial(E'_x)}{\partial z'} - \frac{\partial(E'_z)}{\partial x'} = -\frac{1}{c}\frac{\partial B'_y}{\partial t'} - \frac{1}{c}\left(\frac{v}{c}\right)^2\frac{\partial B_y}{\partial t'}$$

Equations (C″), (E″), and (F″) are obtained in similar fashion. Consider Eq. (A″):

$$\frac{\partial(E'_z)}{\partial y'} - \frac{\partial(E'_y)}{\partial z'} = -\frac{1}{c}\frac{\partial B_x}{\partial t}$$

Similar to the above calculations, $\frac{\partial E'_z}{\partial y'} = \frac{\partial E'_z}{\partial y'}$, $\frac{\partial E'_y}{\partial z'} = \frac{\partial E'_y}{\partial z'}$, and $\frac{\partial B'_x}{\partial t} = \frac{\partial B'_x}{\partial t'}$,

$$\frac{\partial (E'_z)}{\partial y'} - \frac{\partial (E'_y)}{\partial z'} = -\frac{1}{c}\frac{\partial B_x}{\partial t'}$$

and $\frac{\partial (B'_z)}{\partial y'} - \frac{\partial (B'_y)}{\partial z'} = -\frac{1}{c}\frac{\partial E_x}{\partial t'}$

With this series of transformations, in reference frame K′, Maxwell's equations are

(A‴) $$\frac{\partial (E'_z)}{\partial y'} - \frac{\partial (E'_y)}{\partial z'} = -\frac{1}{c}\frac{\partial B'_x}{\partial t'}$$

(B‴) $$\frac{\partial (E'_x)}{\partial z'} - \frac{\partial (E'_z)}{\partial x'} = -\frac{1}{c}\frac{\partial B'_y}{\partial t'} - \frac{1}{c}\left(\frac{v}{c}\right)^2\frac{\partial B_y}{\partial t'}$$

(C‴) $$\frac{\partial (E'_y)}{\partial x'} - \frac{\partial (E'_x)}{\partial y'} = -\frac{1}{c}\frac{\partial B'_z}{\partial t'} - \frac{1}{c}\left(\frac{v}{c}\right)^2\frac{\partial B_z}{\partial t'}$$

(D‴) $$\frac{\partial (B'_x)}{\partial z'} - \frac{\partial (B'_z)}{\partial x'} = \frac{1}{c}\frac{\partial E'_y}{\partial t'}$$

(E‴) $$\frac{\partial (B'_x)}{\partial z'} - \frac{\partial (B'_z)}{\partial x'} = \frac{1}{c}\frac{\partial E'_y}{\partial t'} + \frac{1}{c}\left(\frac{v}{c}\right)^2\frac{\partial E_y}{\partial t'}$$

(F‴) $$\frac{\partial (B'_y)}{\partial x'} - \frac{\partial (B'_x)}{\partial y'} = \frac{1}{c}\frac{\partial E'_z}{\partial t'} + \frac{1}{c}\left(\frac{v}{c}\right)^2\frac{\partial E_z}{\partial t'}$$

In this last transformation only the time variable has been changed, to what was called "local time," because it was dependent on the location x' where the time measurement was being made. This transformation was made expressly to rid the equations of the terms in first order in (v/c). Introducing the local time achieved this purpose, leaving the Maxwell equations in the same form in the moving reference frame K′ as they were in the rest frame – at least to first order in (v/c). Although these manipulations rid the transformed Maxwell equations of the terms to first order in (v/c), at the same time they introduced terms of second order in (v/c), i.e., terms of the order $(v/c)^2$. At this time, the general belief was these transformations were simply a mathematical ploy to save the form of Maxwell's equations in the moving reference frame.[80]

Subsequent experiments, notably the Michelson–Morley experiment,[81] were unable to detect these second order effects, and another adjustment of the transformation equations became necessary.

4.5.1.2   The Michelson–Morley Experiment

Consider the universe filled completely by an aether that is at rest. This aether is the medium through which light waves travel at the speed of light c. Now consider a person traveling through this aether at a speed v. We designate the direction of motion to be the x-axis, and the y-axis to be perpendicular to the x-axis. See Figure 4.1. The aether is at rest in reference frame K. K$'$ is the reference frame of the person traveling through the aether at a speed v.

In the Michelson–Morley experiment, a beam of light is sent along the negative x-axis toward the origin. At the origin, the beam of light is split into two beams, one beam continuing along the positive x-axis a distance L, where it is reflected from a mirror back toward the origin; the total round trip from the origin and back again is length $2L$. The second beam of light travels along the y-axis a distance L, where it is reflected from a mirror back toward the origin; the total round trip from the origin and back again is length $2L$. At the origin when the two beams of light recombine, if the travel time for the two beams is not exactly the same, interference fringes will result. See Figure 4.8.

For the beam of light traveling along the x-axis from the origin to $x = L$, the speed of the beam in the aether is c. As seen by the person traveling with the apparatus, the light will be measured to travel at the speed $c - v$ relative to the apparatus. The time to travel the distance L is $t_{x1} = L/(c - v)$. In a like manner, the time to return from the mirror at distance L to the origin is $t_{x2} = L/(c + v)$. The time for the round trip along the x-axis from the origin to $x = L$ and back to the origin is

$$t_{parallel} = t_{x1} + t_{x2} = \frac{L}{c - v} + \frac{L}{c + v} = \frac{2Lc}{c^2 - v^2} = \frac{2L}{c}\left(\frac{1}{1 - \frac{v^2}{c^2}}\right)$$

For the beam of light traveling along the y-axis from the origin to $y = L$, the speed of the beam *in the aether* is c. As seen by the person traveling with the apparatus, the light will be measured to travel at a speed u where, as viewed from the aether reference frame, u, v, and c are related as $c^2 = u^2 + v^2$, or $u = \sqrt{c^2 - v^2}$. See Figure 4.9.

The time to travel the distance L from the origin to $y = L$ is $t_{y1} = L/u = L/\sqrt{c^2 - v^2}$. The time for the return trip is the same. The time for the round trip along the y-axis from the origin to $y = L$ and back to the origin is

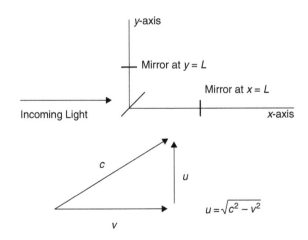

Fig. 4.8 Configuration for the Michelson–Morley experiment.

Fig. 4.9 Speed through the aether as viewed from a moving earth.

$$t_{perpendicular} = t_{y1} + t_{y2} = \frac{L}{\sqrt{c^2 - v^2}} + \frac{L}{\sqrt{c^2 - v^2}} = \frac{2L}{\sqrt{c^2 - v^2}}$$

$$= \frac{2L}{c}\left(\frac{1}{\sqrt{1 - \frac{v^2}{c^2}}}\right)$$

As can be seen, the time parallel to the motion through the aether differs from the time perpendicular to the motion through the aether by a factor of $\sqrt{1 - (v/c)^2}$. This time difference would give rise to a phase shift between the two beams of light and, subsequently, to fringes when the two beams came together. Since a phase shift could be due to other effects, such as the two arms not being exactly the same value of L, Michelson and Morley rotated the experiment so the arm of the apparatus parallel to the motion of the earth through the aether became the arm perpendicular to the motion (and the previously perpendicular arm became the parallel arm) and looked for a shift in the fringes. But none was found.

In 1892, the physicists George FitzGerald in Ireland and Hendrik Lorentz independently arrived at the same explanation: if the arm of the apparatus traveling parallel to the earth's motion contracts by exactly the right amount (while the arm perpendicular to the motion does not contract) the $t_{parallel}$ and $t_{perpendicular}$ would be exactly equal to one another. The amount of contraction would be $\sqrt{1 - (v/c)^2}$. In 1904, Lorentz summarized his modifications of the Galilean transformations, including the local time and the contraction adjustments.[82] These transformations were labeled the Lorentz transformations by Poincaré:

$$t' = \gamma\left(t - \frac{v}{c^2}x\right)$$

$$x' = \gamma\left(x - vt\right)$$

$$y' = y$$

$$z' = z$$

with $\gamma = \frac{1}{\sqrt{1 - (\frac{v}{c})^2}}$

4.5.2   Derivation of the Lorentz Transformation Equations

4.5.2.1   The Differential Equations for t'

In Section 4.2.3 (Section 3 of the paper "On the Electrodynamics of Moving Bodies"), Einstein obtained three differential equations for the time t' in the moving reference frame, beginning with the condition for synchronicity,

$$\frac{1}{2}(t'_0 + t'_2) = t'_1$$

Remembering the transformation equations are linear (Section 4.2.3), we can rewrite this equation with the variables explicitly shown, noting t' is expressed in terms of the coordinates (X, y, z) rather than the coordinates (x, y, z) (once the transformation is obtained in terms of X, X is replaced by $x-vt$):

$$\frac{1}{2}\left[t'(0,0,0,t) + t'\left(0,0,0,t + \frac{X}{c-v} + \frac{X}{c+v}\right)\right] = t'\left(X,0,0,t + \frac{X}{c-v}\right)$$

If X is chosen to be infinitesimally small, this can be expressed as a differential equation:[83]

$$\frac{1}{2}\left[t'(0,0,0,t) + t'(0,0,0,t) + \frac{\partial t'}{\partial t}\Delta t\right] = t'(0,0,0,t) + \frac{\partial t'}{\partial t}\Delta t + \frac{\partial t'}{\partial X}\Delta X$$

$$\frac{1}{2}\left[t'(0,0,0,t) + t'(0,0,0,t) + \frac{\partial t'}{\partial t}\left(\frac{X}{c-v} + \frac{X}{c+v}\right)\right]$$

$$= t'(0,0,0,t) + \frac{\partial t'}{\partial t}\left(\frac{X}{c-v}\right) + \frac{\partial t'}{\partial X}(X)$$

$$\frac{1}{2}\left[\frac{\partial t'}{\partial t}\left(\frac{X}{c-v} + \frac{X}{c+v}\right)\right] = \frac{\partial t'}{\partial t}\left(\frac{X}{c-v}\right) + \frac{\partial t'}{\partial X}X$$

$$\frac{\partial t'}{\partial X} + \frac{v}{c^2 - v^2}\frac{\partial t'}{\partial t} = 0$$

In an analogous manner, in K' a light signal is sent along the y'-axis to a point distance Y fixed in K'. As seen from the K reference frame, the light travels at some angle to the y-axis with speed c along the diagonal, with speed v along the x-axis, and with speed $\sqrt{c^2 - v^2}$ along the y'-axis. See Figure 4.10.

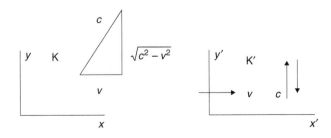

Fig. 4.10 Light beam along the y'-axis, as seen from reference frame K and from reference frame K'.

Again, using the condition of synchronicity,

$$\frac{1}{2}\left[t'(0,0,0,t)+t'\left(0,0,0,t+\frac{Y}{\sqrt{c^2-v^2}}+\frac{Y}{\sqrt{c^2-v^2}}\right)\right]$$

$$=t'\left(0,Y,0,t+\frac{Y}{\sqrt{c^2-v^2}}\right)$$

$$\frac{1}{2}\left[t'(0,0,0,t)+t'(0,0,0,t)+\frac{\partial t'}{\partial t}\Delta t\right]=t'(0,0,0,t)+\frac{\partial t'}{\partial t}\Delta t+\frac{\partial t'}{\partial y}\Delta Y$$

$$\frac{1}{2}\left[\frac{\partial t'}{\partial t}\left(\frac{2Y}{\sqrt{c^2-v^2}}\right)\right]=\frac{\partial t'}{\partial t}\left(\frac{Y}{\sqrt{c^2-v^2}}\right)+\frac{\partial t'}{\partial y}Y$$

$$\frac{\partial t'}{\partial y}=0$$

Similarly, $\frac{\partial t'}{\partial z}=0$

4.5.2.2 The Solution for t'

As noted previously, the transformations must be linear. Thus, the solution for t' will be of the form

$$t'=at+bX+cy+dz$$

The above conditions on the derivatives of t' with respect to y and z being zero gives

$$\frac{\partial t'}{\partial y}=c=0$$

$$\frac{\partial t'}{\partial z}=d=0$$

The first equation for t', relating the derivatives with respect to X and to t, gives

$$b+\frac{v}{c^2-v^2}a=0\quad\Rightarrow b=\frac{-v}{c^2-v^2}a$$

and the transformation for the time t' becomes

$$t'=a\left[t-\frac{v}{c^2-v^2}X\right]$$

where a is a quantity not dependent on X, y, z, or t, but could be dependent on the velocity v.

4.5.2.3 The Solution for the Spatial Coordinates

With the transformation for t' known, Einstein proceeded to determine the transformation equations for the spatial coordinates.

In the moving system K$'$, a ray of light is emitted in the direction of increasing x'. The light ray, as measured in K$'$, travels at the speed c. After a time t', it will have traveled some distance x', with

$$x' = ct' = c\left[a\left(t - \frac{v}{c^2 - v^2}X\right)\right]$$

It should be noted all of the quantities on the left-hand side of the equation are measured in reference frame K′, while all of the quantities on the right-hand side are measured in reference frame K. But, Einstein continues, "as measured in the system at rest [K], the light ray propagates with velocity $c-v$ relative to the origin of K′, so that"[84]

$$\frac{X}{c - v} = t$$

Substituting into the above equation for t,

$$x' = c\left[a\left(\frac{X}{c - v} - \frac{v}{c^2 - v^2}X\right)\right] = a\frac{c^2}{c^2 - v^2}X = a\frac{c^2}{c^2 - v^2}(x - vt)$$

Similarly, along the y'-axis, noting X is zero along the y'-axis,

$$y' = ct' = c\left[a\left(t - \frac{v}{c^2 - v^2}X\right)\right] = cat$$

In the rest frame K,

$$y = \sqrt{c^2 - v^2}t \quad \Rightarrow t = \frac{y}{\sqrt{c^2 - v^2}}$$

Substituting,

$$y' = ca\frac{y}{\sqrt{c^2 - v^2}} = a\left(\frac{1}{\sqrt{1 - \frac{v^2}{c^2}}}\right)y$$

In the same manner,

$$z' = ca\frac{z}{\sqrt{c^2 - v^2}} = a\left(\frac{1}{\sqrt{1 - \frac{v^2}{c^2}}}\right)z$$

4.5.2.4 Simplifying the Transformations

Noting there is common factor of $a\left(\frac{1}{\sqrt{1 - \frac{v^2}{c^2}}}\right)$ in these equations, and that the quantity a may depend on the velocity, these are combined into one factor $\varphi(v)$. Rewriting the equations,

$$t' = a\left[t - \frac{v}{c^2 - v^2}X\right] = a\left[t - \frac{v}{c^2 - v^2}(x - vt)\right] = \frac{a}{1 - \frac{v^2}{c^2}}\left(t - \frac{v}{c^2}x\right)$$

$$= \varphi(v)\,\gamma\left(t - \frac{v}{c^2}x\right)$$

$$x' = a\frac{c^2}{c^2 - v^2}(x - vt) = a\frac{1}{1 - \frac{v^2}{c^2}}(x - vt) = \varphi(v)\,\gamma\,(x - vt)$$

$$y' = a\left(\frac{1}{\sqrt{1 - \frac{v^2}{c^2}}}\right)y = \varphi(v)\,y$$

$$z' = a\left(\frac{1}{\sqrt{1 - \frac{v^2}{c^2}}}\right)z = \varphi(v)\,z$$

with $\gamma = \dfrac{1}{\sqrt{1 - \frac{v^2}{c^2}}}$

4.5.2.5   The Form of $\varphi(v)$

It is left now only to determine the unknown function $\varphi(v)$. A third coordinate system, K″, is introduced. The axes of K″ are parallel to the axes of K′. K″ is moving along the x'-axis with velocity $-v$, i.e., in the negative direction relative to K′. The origins of K, K′, and K″ are coincident at $t = 0$, $t' = 0$, and $t'' = 0$ (it will be noted subsequently that K″ is coincident with K).Transforming first from K′ to K″, then from K to K′, we have

$$t'' = \varphi(-v)\gamma(-v)\left\{t' + \frac{v}{c^2}x'\right\} = \varphi(-v)\gamma(-v)\left\{\left[\varphi(v)\gamma(v)\left(t - \frac{v}{c^2}x\right)\right]\right.$$

$$\left. + \frac{v}{c^2}\left[\varphi(v)\gamma(v)(x - vt)\right]\right\}$$

$$t'' = \varphi(-v)\varphi(v)\gamma(-v)\gamma(v)\left\{\left[\left(t - \frac{v}{c^2}x\right)\right] + \frac{v}{c^2}\left[(x - vt)\right]\right\} = \varphi(-v)\,\varphi(v)\,t$$

$$t'' = \varphi(-v)\varphi(v)t$$

Similarly, for the coordinate x'', we obtain

$$x'' = \varphi(-v)\gamma(-v)\left\{x' + vt'\right\}$$

$$= \varphi(-v)\gamma(-v)\left\{\varphi(v)\,\gamma(v)\,(x - vt) + v\varphi(v)\gamma(v)\left(t - \frac{v}{c^2}x\right)\right\}$$

$$x'' = \varphi(-v)\varphi(v)\gamma(-v)\gamma(v)\left\{(x - vt) + \left(vt - \frac{v^2}{c^2}x\right)\right\}$$

$$x'' = \varphi(-v)\varphi(v)\gamma(-v)\gamma(v)\left\{1 - \frac{v^2}{c^2}\right\}x = \varphi(-v)\varphi(v)x$$

$$x'' = \varphi(-v)\varphi(v)x$$

In a similar manner, for the coordinates y'' and z'', we obtain

$$y'' = \varphi(-v)y' = \varphi(-v)\varphi(v)y$$
$$z'' = \varphi(-v)z' = \varphi(-v)\varphi(v)z$$

Einstein then notes, "Since the relations between $[x'', y'', z'']$ and x, y, z do not contain the time t, the systems K and $[K'']$ are at rest relative to each other, and it is clear that the transformation from K to $[K'']$ must be the identity transformation. Hence, $\varphi(v)\varphi(-v) = 1.$"[85]
It remains now only to determine the precise form of $\varphi(v)$. Consider a rod of length ℓ at rest in the K$'$ reference frame (the moving reference frame). The rod is aligned with the y'-axis, with one end (designated as A) at the origin of K$'$ ($x' = 0$, $y' = 0$, $z' = 0$) and the other end (designated as B) located at $x' = 0$, $y' = \ell$, $z' = 0$. Using the above transformations, in K the ends of the rod are located at

$$x_A = vt \qquad y_A = 0 \qquad\qquad z_A = 0$$
$$x_B = vt \qquad \ell = \varphi(v)y \Rightarrow y = \tfrac{\ell}{\varphi(v)} \qquad z_B = 0$$

Einstein continues, "The length of the rod, measured in K, is thus $\ell/\varphi(v)$;... For reasons of symmetry it is obvious that the length of a rod measured in the system at rest and moving perpendicular to its own axis can depend only on its velocity and not on the direction and sense of its motion. Thus, the length of the moving rod measured in the system at rest does not change when v is replaced by $-v$. From this we arrive at:

$$\frac{l}{\varphi(v)} = \frac{l}{\varphi(-v)}$$

or

$$\varphi(v) = \varphi(-v)$$

It follows from this relation and the one found before $[\varphi(v)\varphi(-v) = 1]$ that $\varphi(v)$ must equal 1."[86]

4.5.2.6 The Lorentz Transformation Equations
Since $\varphi(v) = 1$, the transformation equations become

$$t' = \gamma\left(t - \frac{v}{c^2}x\right)$$
$$x' = \gamma\left(x - vt\right)$$
$$y' = y$$
$$z' = z$$

with $\gamma = \dfrac{1}{\sqrt{1 - (\frac{v}{c})^2}}$

These are the Lorentz transformations.

4.5.3 The Electromagnetic Field Transformations

In empty space (no sources/sinks) the Maxwell equations are:

(a) $\dfrac{1}{c}\dfrac{\partial E_x}{\partial t} = \dfrac{\partial B_z}{\partial y} - \dfrac{\partial B_y}{\partial z}$ 　　(d) $\dfrac{1}{c}\dfrac{\partial B_x}{\partial t} = \dfrac{\partial E_y}{\partial z} - \dfrac{\partial E_z}{\partial y}$

(b) $\dfrac{1}{c}\dfrac{\partial E_y}{\partial t} = \dfrac{\partial B_x}{\partial z} - \dfrac{\partial B_z}{\partial x}$ 　　(e) $\dfrac{1}{c}\dfrac{\partial B_y}{\partial t} = \dfrac{\partial E_z}{\partial x} - \dfrac{\partial E_x}{\partial z}$

(c) $\dfrac{1}{c}\dfrac{\partial E_z}{\partial t} = \dfrac{\partial B_y}{\partial x} - \dfrac{\partial B_x}{\partial y}$ 　　(f) $\dfrac{1}{c}\dfrac{\partial B_z}{\partial t} = \dfrac{\partial E_x}{\partial y} - \dfrac{\partial E_y}{\partial x}$

In empty space, since there are no sources/sinks for the fields,

(g) $\quad \nabla \bullet \vec{E} = \dfrac{\partial E_x}{\partial x} + \dfrac{\partial E_y}{\partial y} + \dfrac{\partial E_z}{\partial z} = 0$

(h) $\quad \nabla \bullet \vec{B} = \dfrac{\partial B_x}{\partial x} + \dfrac{\partial B_y}{\partial y} + \dfrac{\partial B_z}{\partial z} = 0$

4.5.3.1 The Transformed Equations for $\nabla \bullet \vec{E}$ and $\nabla \bullet \vec{B}$

First, we express Eq. (g) and Eq. (h) in reference frame K′. Transforming each term of Eq. (g), using the Lorentz transformation equations,

$$\frac{\partial E_x}{\partial x} = \frac{\partial E_x}{\partial t'}\frac{\partial t'}{\partial x} + \frac{\partial E_x}{\partial x'}\frac{\partial x'}{\partial x} + \frac{\partial E_x}{\partial y'}\frac{\partial y'}{\partial x} + \frac{\partial E_x}{\partial z'}\frac{\partial z'}{\partial x}$$

$$= \frac{\partial E_x}{\partial t'}\left(-\gamma\frac{v}{c^2}\right) + \frac{\partial E_x}{\partial x'}(\gamma) + \frac{\partial E_x}{\partial y'}(0) + \frac{\partial E_x}{\partial z'}(0)$$

$$= -\gamma\frac{v}{c^2}\frac{\partial E_x}{\partial t'} + \gamma\frac{\partial E_x}{\partial x'}$$

$$\frac{\partial E_y}{\partial y} = \frac{\partial E_y}{\partial t'}\frac{\partial t'}{\partial y} + \frac{\partial E_y}{\partial x'}\frac{\partial x'}{\partial y} + \frac{\partial E_y}{\partial y'}\frac{\partial y'}{\partial y} + \frac{\partial E_y}{\partial z'}\frac{\partial z'}{\partial y}$$

$$= \frac{\partial E_y}{\partial t'}(0) + \frac{\partial E_y}{\partial x'}(0) + \frac{\partial E_y}{\partial y'}(1) + \frac{\partial E_y}{\partial z'}(0)$$

$$= \frac{\partial E_y}{\partial y'}$$

$$\frac{\partial E_z}{\partial z} = \frac{\partial E_z}{\partial t'}\frac{\partial t'}{\partial z} + \frac{\partial E_z}{\partial x'}\frac{\partial x'}{\partial z} + \frac{\partial E_z}{\partial y'}\frac{\partial y'}{\partial z} + \frac{\partial E_z}{\partial z'}\frac{\partial z'}{\partial z}$$

$$= \frac{\partial E_z}{\partial t'}(0) + \frac{\partial E_z}{\partial x'}(0) + \frac{\partial E_z}{\partial y'}(0) + \frac{\partial E_z}{\partial z'}(1)$$

$$= \frac{\partial E_z}{\partial z'}$$

Thus $\nabla' \bullet \vec{E} = -\gamma\dfrac{v}{c^2}\dfrac{\partial E_x}{\partial t'} + \gamma\dfrac{\partial E_x}{\partial x'} + \dfrac{\partial E_y}{\partial y'} + \dfrac{\partial E_z}{\partial z'} = 0$

In a similar manner, Eq. (h) becomes

$$\nabla' \bullet \vec{B} = -\gamma \frac{v}{c^2} \frac{\partial B_x}{\partial t'} + \gamma \frac{\partial B_x}{\partial x'} + \frac{\partial B_y}{\partial y'} + \frac{\partial B_z}{\partial z'} = 0$$

4.5.3.2 The Transformed Maxwell Equations

Working first with Eq. (a),

$$\text{(a)} \quad \frac{1}{c} \frac{\partial E_x}{\partial t} = \frac{\partial B_z}{\partial y} - \frac{\partial B_y}{\partial z}$$

Transforming each term, using the Lorentz transformation equations,

$$\frac{1}{c} \frac{\partial E_x}{\partial t} = \frac{1}{c} \left\{ \frac{\partial E_x}{\partial t'} \frac{\partial t'}{\partial t} + \frac{\partial E_x}{\partial x'} \frac{\partial x'}{\partial t} + \frac{\partial E_x}{\partial y'} \frac{\partial y'}{\partial t} + \frac{\partial E_x}{\partial z'} \frac{\partial z'}{\partial t} \right\}$$

$$= \frac{1}{c} \left\{ \frac{\partial E_x}{\partial t'} (\gamma) + \frac{\partial E_x}{\partial x'} (-\gamma v) + \frac{\partial E_x}{\partial y'} (0) + \frac{\partial E_x}{\partial z'} (0) \right\}$$

$$= \frac{1}{c} \left\{ \gamma \frac{\partial E_x}{\partial t'} - \gamma v \frac{\partial E_x}{\partial x'} \right\}$$

$$\frac{\partial B_z}{\partial y} = \left\{ \frac{\partial B_z}{\partial t'} \frac{\partial t'}{\partial y} + \frac{\partial B_z}{\partial x'} \frac{\partial x'}{\partial y} + \frac{\partial B_z}{\partial y'} \frac{\partial y'}{\partial y} + \frac{\partial B_z}{\partial z'} \frac{\partial z'}{\partial y} \right\}$$

$$= \left\{ \frac{\partial B_z}{\partial t'} (0) + \frac{\partial B_z}{\partial x'} (0) + \frac{\partial B_z}{\partial y'} (1) + \frac{\partial B_z}{\partial z'} (0) \right\}$$

$$= \frac{\partial B_z}{\partial y'}$$

$$\frac{\partial B_y}{\partial z} = \left\{ \frac{\partial B_y}{\partial t'} \frac{\partial t'}{\partial z} + \frac{\partial B_y}{\partial x'} \frac{\partial x'}{\partial z} + \frac{\partial B_y}{\partial y'} \frac{\partial y'}{\partial z} + \frac{\partial B_y}{\partial z'} \frac{\partial z'}{\partial z} \right\}$$

$$= \left\{ \frac{\partial B_y}{\partial t'} (0) + \frac{\partial B_y}{\partial x'} (0) + \frac{\partial B_y}{\partial y'} (0) + \frac{\partial B_y}{\partial z'} (1) \right\}$$

$$= \frac{\partial B_y}{\partial z'}$$

In reference frame K', Eq. (a) is

$$\frac{1}{c} \left\{ \gamma \frac{\partial E_x}{\partial t'} - \gamma v \frac{\partial E_x}{\partial x'} \right\} = \frac{\partial B_z}{\partial y'} - \frac{\partial B_y}{\partial z'}$$

But from Eq. (g), above,

$$-\gamma \frac{v}{c^2} \frac{\partial E_x}{\partial t'} + \gamma \frac{\partial E_x}{\partial x'} + \frac{\partial E_y}{\partial y'} + \frac{\partial E_z}{\partial z'} = 0$$

Substituting $\quad \gamma\dfrac{\partial E_x}{\partial x'} = \gamma\dfrac{v}{c^2}\dfrac{\partial E_x}{\partial t'} - \dfrac{\partial E_y}{\partial y'} - \dfrac{\partial E_z}{\partial z'}\quad$ into the previous equation,

$$\frac{1}{c}\left\{\gamma\frac{\partial E_x}{\partial t'} - v\left(\gamma\frac{v}{c^2}\frac{\partial E_x}{\partial \tau'} - \frac{\partial E_y}{\partial y'} - \frac{\partial E_z}{\partial z'}\right)\right\} = \frac{\partial B_z}{\partial y'} - \frac{\partial B_y}{\partial z'}$$

$$\frac{1}{c}\left\{\gamma\left(1 - \frac{v^2}{c^2}\right)\frac{\partial E_x}{\partial t'} + v\frac{\partial E_y}{\partial y'} + v\frac{\partial E_z}{\partial z'}\right\} = \frac{\partial B_z}{\partial y'} - \frac{\partial B_y}{\partial z'}$$

$$\frac{1}{c}\frac{\partial E_x}{\partial t'} = \frac{\partial\left[\gamma\left(B_z - \frac{v}{c}E_y\right)\right]}{\partial y'} - \frac{\partial\left[\gamma\left(B_y + \frac{v}{c}E_z\right)\right]}{\partial z'}$$

This is Eq. (A).

In a similar manner, Eq. (D) is obtained:

$$\frac{1}{c}\frac{\partial B_x}{\partial t'} = \frac{\partial\left[\gamma\left(E_y - \frac{v}{c}B_z\right)\right]}{\partial z'} - \frac{\partial\left[\gamma\left(E_z + \frac{v}{c}B_y\right)\right]}{\partial y'}$$

For Eq. (b),

$$\text{(b)}\quad \frac{1}{c}\frac{\partial E_y}{\partial t} = \frac{\partial B_x}{\partial z} - \frac{\partial B_z}{\partial x}$$

Transforming each term, using the Lorentz transformation equations,

$$\frac{1}{c}\frac{\partial E_y}{\partial t} = \frac{1}{c}\left\{\frac{\partial E_y}{\partial t'}\frac{\partial t'}{\partial t} + \frac{\partial E_y}{\partial x'}\frac{\partial x'}{\partial t} + \frac{\partial E_y}{\partial y'}\frac{\partial y'}{\partial t} + \frac{\partial E_y}{\partial z'}\frac{\partial z'}{\partial t}\right\}$$

$$= \frac{1}{c}\left\{\frac{\partial E_y}{\partial t'}(\gamma) + \frac{\partial E_y}{\partial x'}(-\gamma v) + \frac{\partial E_y}{\partial y'}(0) + \frac{\partial E_y}{\partial z'}(0)\right\}$$

$$= \frac{1}{c}\left\{\gamma\frac{\partial E_y}{\partial t'} - \gamma v\frac{\partial E_y}{\partial x'}\right\}$$

$$\frac{\partial B_x}{\partial z} = \left\{\frac{\partial B_x}{\partial t'}\frac{\partial t'}{\partial z} + \frac{\partial B_x}{\partial x'}\frac{\partial x'}{\partial z} + \frac{\partial B_x}{\partial y'}\frac{\partial y'}{\partial z} + \frac{\partial B_x}{\partial z'}\frac{\partial z'}{\partial z}\right\}$$

$$= \left\{\frac{\partial B_x}{\partial t'}(0) + \frac{\partial B_x}{\partial x'}(0) + \frac{\partial B_x}{\partial y'}(0) + \frac{\partial B_x}{\partial z'}(1)\right\}$$

$$= \frac{\partial B_x}{\partial z'}$$

$$\frac{\partial B_z}{\partial x} = \left\{\frac{\partial B_z}{\partial t'}\frac{\partial t'}{\partial x} + \frac{\partial B_z}{\partial x'}\frac{\partial x'}{\partial x} + \frac{\partial B_z}{\partial y'}\frac{\partial y'}{\partial x} + \frac{\partial B_z}{\partial z'}\frac{\partial z'}{\partial x}\right\}$$

$$= \left\{\frac{\partial B_z}{\partial t'}\left(-\gamma\frac{v}{c^2}\right) + \frac{\partial B_z}{\partial x'}(\gamma) + \frac{\partial B_z}{\partial y'}(0) + \frac{\partial B_z}{\partial z'}(0)\right\}$$

$$= \left\{-\gamma\frac{v}{c^2}\frac{\partial B_z}{\partial t'} + \gamma\frac{\partial B_z}{\partial x'}\right\}$$

In reference frame K', Eq. (b) is

$$\frac{1}{c}\left\{\gamma\frac{\partial E_y}{\partial t'} - \gamma v\frac{\partial E_y}{\partial x'}\right\} = \frac{\partial B_x}{\partial z'} - \left\{-\gamma\frac{v}{c^2}\frac{\partial B_z}{\partial t'} + \gamma\frac{\partial B_z}{\partial x'}\right\}$$

$$\frac{1}{c}\frac{\partial\left[\gamma\left(E_y - \frac{v}{c}B_z\right)\right]}{\partial t'} = \frac{\partial B_x}{\partial z'} - \frac{\partial\left[\gamma\left(B_z - \frac{v}{c}E_y\right)\right]}{\partial x'}$$

This is Eq. (B). Eqs. (C), (E), and (F) are obtained in a manner similar to that for obtaining Eq. (B).

4.5.3.3 The Electromagnetic Field Transformations

In empty space, in reference frame K, Maxwell's equations are:

(a) $\dfrac{1}{c}\dfrac{\partial E_x}{\partial t} = \dfrac{\partial B_z}{\partial y} - \dfrac{\partial B_y}{\partial z}$ (d) $\dfrac{1}{c}\dfrac{\partial B_x}{\partial t} = \dfrac{\partial E_y}{\partial z} - \dfrac{\partial E_z}{\partial y}$

(b) $\dfrac{1}{c}\dfrac{\partial E_y}{\partial t} = \dfrac{\partial B_x}{\partial z} - \dfrac{\partial B_z}{\partial x}$ (e) $\dfrac{1}{c}\dfrac{\partial B_y}{\partial t} = \dfrac{\partial E_z}{\partial x} - \dfrac{\partial E_x}{\partial z}$

(c) $\dfrac{1}{c}\dfrac{\partial E_z}{\partial t} = \dfrac{\partial B_y}{\partial x} - \dfrac{\partial B_x}{\partial y}$ (f) $\dfrac{1}{c}\dfrac{\partial B_z}{\partial t} = \dfrac{\partial E_x}{\partial y} - \dfrac{\partial E_y}{\partial x}$

These equations, transformed to the moving reference frame K' are:

(A) $\dfrac{1}{c}\dfrac{\partial E_x}{\partial t'} = \dfrac{\partial\left[\gamma\left(B_z - \frac{v}{c}E_y\right)\right]}{\partial y'} - \dfrac{\partial\left[\gamma\left(B_y + \frac{v}{c}E_z\right)\right]}{\partial z'}$

(B) $\dfrac{1}{c}\dfrac{\partial\left[\gamma\left(E_y - \frac{v}{c}B_z\right)\right]}{\partial t'} = \dfrac{\partial B_x}{\partial z'} - \dfrac{\partial\left[\gamma\left(B_z - \frac{v}{c}E_y\right)\right]}{\partial x'}$

(C) $\dfrac{1}{c}\dfrac{\partial\left[\gamma\left(E_z + \frac{v}{c}B_y\right)\right]}{\partial t'} = \dfrac{\partial\left[\gamma\left(B_y + \frac{v}{c}E_z\right)\right]}{\partial x'} - \dfrac{\partial B_x}{\partial y'}$

(D) $\dfrac{1}{c}\dfrac{\partial B_x}{\partial t'} = \dfrac{\partial\left[\gamma\left(E_y - \frac{v}{c}B_z\right)\right]}{\partial z'} - \dfrac{\partial\left[\gamma\left(E_z + \frac{v}{c}B_y\right)\right]}{\partial z'}$

(E) $\dfrac{1}{c}\dfrac{\partial\left[\gamma\left(B_y + \frac{v}{c}E_z\right)\right]}{\partial t'} = \dfrac{\partial\left[\gamma\left(E_z + \frac{v}{c}B_y\right)\right]}{\partial x'} - \dfrac{\partial E_x}{\partial y'}$

(F) $\dfrac{1}{c}\dfrac{\partial\left[\gamma\left(B_z - \frac{v}{c}E_y\right)\right]}{\partial t'} = \dfrac{\partial E_x}{\partial y'} - \dfrac{\partial\left[\gamma\left(E_y - \frac{v}{c}B_z\right)\right]}{\partial x'}$

But, Einstein notes, if Maxwell's equations are valid in reference frame K, the principle of relativity requires that they also be valid in K'.[87] In reference frame K', Maxwell's equations are, with (E'_x, E'_y, E'_z) and (B'_x, B'_y, B'_z), the components of the electric and magnetic fields measured in reference frame K',

$$\text{(a')} \quad \frac{1}{c}\frac{\partial E'_{x'}}{\partial t'} = \frac{\partial B'_{x'}}{\partial y'} - \frac{\partial B'_{y'}}{\partial z'} \qquad \text{(d')} \quad \frac{1}{c}\frac{\partial B'_{x'}}{\partial t'} = \frac{\partial E'_{y'}}{\partial z'} - \frac{\partial E'_{z'}}{\partial y'}$$

$$\text{(b')} \quad \frac{1}{c}\frac{\partial E'_{y'}}{\partial t'} = \frac{\partial B'_{x'}}{\partial z'} - \frac{\partial B'_{z'}}{\partial x'} \qquad \text{(e')} \quad \frac{1}{c}\frac{\partial B'_{y'}}{\partial t'} = \frac{\partial E'_{z'}}{\partial x'} - \frac{\partial E'_{x'}}{\partial z'}$$

$$\text{(c')} \quad \frac{1}{c}\frac{\partial E'_{z'}}{\partial t'} = \frac{\partial B'_{y'}}{\partial x'} - \frac{\partial B'_{x'}}{\partial y'y} \qquad \text{(f')} \quad \frac{1}{c}\frac{\partial B'_{z'}}{\partial t'} = \frac{\partial E'_{x'}}{\partial y'} - \frac{\partial E'_{y'}}{\partial x'}$$

Thus, by inspection of the above Maxwell equations in K', and the form of Maxwell's equations after being transformed from reference frame K to reference frame K', Einstein obtained[88]

$$E'_x = \psi(v)E_x \qquad\qquad B'_x = \psi(v)B_x$$

$$E'_y = \psi(v)\gamma\left(E_y - \tfrac{v}{c}B_z\right) \qquad B'_y = \psi(v)\gamma\left(B_y + \tfrac{v}{c}E_z\right)$$

$$E'_z = \psi(v)\gamma\left(E_z + \tfrac{v}{c}B_y\right) \qquad B'_z = \psi(v)\gamma\left(B_z - \tfrac{v}{c}E_y\right)$$

where the factor $\psi(v)$ is an arbitrary function of the velocity v, but not a function of the spatial or time coordinates. To determine the form of $\psi(v)$, each of the above equations is inverted, going first from K to K' and then from K' to K using $\psi(-v)$. Consider, for example the y-component of the electric field:

$$E_y = \psi(-v)\gamma\left\{E'_y + \frac{v}{c}B'_z\right\}$$

$$= \psi(-v)\gamma\left\{\left[\psi(v)\gamma\left(E_y - \frac{v}{c}B_z\right)\right] + \frac{v}{c}\left[\psi(v)\gamma\left(B_z - \frac{v}{c}E_y\right)\right]\right\}$$

$$= \psi(-v)\psi(v)\gamma^2\left\{E_y\left(1 - \frac{v^2}{c^2}\right) - \frac{v}{c}B_z(1-1)\right\} = \psi(-v)\psi(v)E_y$$

$$\Rightarrow \psi(-v)\psi(v) = 1$$

A similar result is obtained for each of the other components of the electric and magnetic fields. From symmetry $\psi(v) = \psi(-v)$ and, thus, $\psi(v) = 1$. The electromagnetic field transformations are:

$$E'_x = E_x \qquad\qquad B'_x = B_x$$

$$E'_y = \gamma\left(E_y - \tfrac{v}{c}B_z\right) \qquad B'_y = \gamma\left(B_y + \tfrac{v}{c}E_z\right)$$

$$E'_z = \gamma\left(E_z + \tfrac{v}{c}B_y\right) \qquad B'_z = \gamma\left(B_z - \tfrac{v}{c}E_y\right)$$

4.5.4   The Doppler Principle

In transforming from the reference frame K to the reference frame K', with K' moving at a constant velocity v along the x-axis of K, the electric and magnetic fields transform as given in the Section 4.5.3.3:

$$E'_x = E'_{x0} \sin \Phi' = E_{x0} \sin \Phi'$$

$$B'_x = B'_{x0} \sin \Phi' = B_{x0} \sin \Phi'$$

$$E'_y = E'_{y0} \sin \Phi' = \gamma \left(E_{y0} - \frac{v}{c} B_{z0} \right) \sin \Phi'$$

$$B'_y = B'_{y0} \sin \Phi' = \gamma \left(B_{y0} + \frac{v}{c} E_{z0} \right) \sin \Phi'$$

$$E'_z = E'_{z0} \sin \Phi' = \gamma \left(E_{z0} + \frac{v}{c} B_{y0} \right) \sin \Phi'$$

$$B'_z = B'_{z0} \sin \Phi' = \gamma \left(B_{z0} - \frac{v}{c} E_{y0} \right) \sin \Phi'$$

with Φ' the phase angle measured in the K$'$ reference frame,

$$\Phi' = \omega \left(t - \frac{a_x x + a_y y + a_z z}{c} \right) = \omega t - \frac{\omega}{c} a_x x - \frac{\omega}{c} a_y y - \frac{\omega}{c} a_z z$$

$$= \omega \gamma \left(t' + \frac{vx'}{c^2} \right) - \frac{\omega}{c} a_x \gamma (x' + vt') - \frac{\omega}{c} a_y y - \frac{\omega}{c} a_z z$$

$$= \omega \gamma \left(1 - \frac{1}{c} a_x v \right) t' - \omega \gamma \left(\frac{1}{c} a_x - \frac{v}{c^2} \right) x' - \frac{\omega}{c} a_y y - \frac{\omega}{c} a_z z$$

Identifying the first term as $\omega' t'$ determines ω' to be

$$\omega' = \omega \gamma \left(1 - a_x \frac{v}{c} \right)$$

Using the reciprocal relation $\omega = \frac{\omega'}{\gamma (1 - a_x \frac{v}{c})}$, the second term becomes

$$-\omega \gamma \left(\frac{1}{c} a_x - \frac{v}{c^2} \right) x' = - \left[\frac{\omega'}{\gamma \left(1 - a_x \frac{v}{c} \right)} \right] \gamma \left(\frac{1}{c} a_x - \frac{v}{c^2} \right) x' = -\frac{\omega'}{c} a'_x x'$$

This defines the direction cosine a'_x as

$$a'_x = \frac{a_x - \frac{v}{c}}{1 - a_x \frac{v}{c}}$$

Using again the reciprocal relation, the third term becomes

$$-\frac{\omega}{c} a_y y = -\frac{\omega'}{\gamma \left(1 - a_x \frac{v}{c} \right)} a_y y' = -\frac{\omega'}{c} a'_y y'$$

This defines the direction cosine a'_y as

$$a'_y = \frac{a_y}{\gamma \left(1 - a_x \frac{v}{c} \right)}$$

In a similar manner, the direction cosine a'_z is determined to be

$$a'_z = \frac{a_z}{\gamma \left(1 - a_x \frac{v}{c} \right)}$$

Einstein notes that the normal to the wave front, the electric field, and the magnetic field are mutually perpendicular in both reference frame K and reference frame K$'$, and that, to satisfy Maxwell's equations, the electric field and magnetic field are equal in magnitude (in empty space, i.e., a vacuum,$\rho = 0$, and from Eqs. (a') through (f') of Section 4.5.3.3, with the exception of a minus sign, the relation between the electric field E and the magnetic field B is the same). He then proceeds to develop the relation between the magnitude of the electromagnetic wave in reference frame K and in reference frame K'.[89]

Consider the same situation as in Section 4.2.7. See Figure 4.5.

(a) The unit normal to the wave front will be designated $\hat{n}$ and points along the line from the distant source of the waves toward the origin of K.
(b) The x-axis is the direction of motion of K' relative to K.
(c) The y-axis is perpendicular to the x-axis, and is selected such that the x–y plane contains the wave normal $\hat{n}$.
(d) The z-axis is perpendicular to both the x- and y-axes, with the direction determined by the right-hand rule.

Consider first the situation of the electric field parallel to the z-axis. The components of the electric field in reference frame K will be

$$E_{x0} = 0$$

$$E_{y0} = 0$$

$$E_{z0} = A$$

Since the magnetic field is perpendicular to the electric field,$B_{z0} = 0$. Thus, the magnetic field will lie in the x–y plane in a direction perpendicular to $\hat{n}$, with magnitude A. Just as the angle between the x-axis and $\hat{n}$ is the angle φ, the angle between $\vec{B}$ and the y-axis is φ. The components of the magnetic field will be

$$B_{x0} = A \sin \varphi$$

$$B_{y0} = -A \cos \varphi$$

$$B_{z0} = 0$$

Using the results of Section 4.5.3, and remembering $E_x = E_{x0} \sin \Phi$, etc., in reference frame K' the electric and magnetic fields are

$$E'_x = E_x = 0$$

$$E'_y = \gamma \left(E_y - \frac{v}{c} B_z \right) = 0$$

$$E'_z = \gamma \left(E_z + \frac{v}{c} B_y \right) = \gamma \left(1 - \frac{v}{c} \cos \varphi \right) A \sin \Phi'$$

$$B'_x = B_x = A \sin \varphi \sin \Phi'$$

$$B'_y = \gamma \left(B_y + \frac{v}{c} E_z \right) = \gamma \left(-\cos \varphi + \frac{v}{c} \right) A$$

$$B'_z = \gamma \left(B_z - \frac{v}{c} E_y \right) = 0$$

Denoting the amplitude of the electric field in K' as A' , we obtain

$$A' = \gamma \left(1 - \frac{v}{c} \cos \varphi\right) A = A \frac{1 - \frac{v}{c} \cos \varphi}{\sqrt{1 - \frac{v^2}{c^2}}}$$

Starting with the magnetic field perpendicular to the direction of the wave motion and the relative motion, the same results are obtained. Since any general case can be viewed as a superposition of these two cases ($\vec{E} \perp z$-axis and $\vec{B} \perp z$-axis) the relation is valid in general.

4.5.5 The Electrodynamic Lorentz Force

Consider two reference frames, K and K', with K "at rest" and K' moving with constant velocity v in the positive x-direction. Let an electron be at rest at the origin of K' (the moving reference frame). In K', the electromagnetic force on the electron will be purely electric. In K', Newton's second law is

$$m\frac{d^2x'}{dt'^2} = qE'_x$$

$$m\frac{d^2y'}{dt'^2} = qE'_y$$

$$m\frac{d^2z'}{dt'^2} = qE'_z$$

In K, the electron is moving at speed v. The Lorentz transformation and the electromagnetic field transformations are used to transform the above equations to the reference frame K.[90] This derivation follows closely that given by Einstein in his 1907 review paper on the relativity principle.[91]

The Lorentz Transformations *The Electromagnetic Field Transformations*

$t' = \gamma \left(t - \frac{v}{c^2}x\right)$

$x' = \gamma (x - vt)$ $E'_x = E_x$ $B'_x = B_x$

$y' = y$ $E'_y = \gamma \left(E_y - \frac{v}{c}B_z\right)$ $B'_y = \gamma \left(B_y + \frac{v}{c}E_z\right)$

$z' = z$ $E'_z = \gamma \left(E_z + \frac{v}{c}B_y\right)$ $\gamma'_z = \beta \left(B_z - \frac{v}{c}E_y\right)$

with $\gamma = \frac{1}{\sqrt{1-(\frac{v}{c})^2}}$

Setting $dx/dt = \dot{x}$, etc., and working first with the x-component of Newton's second law, we have

$$\frac{d^2 x'}{dt'^2} = \frac{d}{dt'}\left(\frac{dx'}{dt'}\right)$$

$$\frac{dx'}{dt'} = \left(\frac{dx'}{dt}\right)\left(\frac{dt}{dt'}\right) = \frac{dx'/dt}{dt'/dt} = \frac{d\left[\gamma(x-vt)\right]/dt}{d\left[\gamma\left(t-\frac{v}{c^2}x\right)\right]dt}$$

$$= \frac{\gamma\left(\dot{x}-v\right)}{\gamma\left(1-\frac{v}{c^2}\dot{x}\right)} = \frac{\left(\dot{x}-v\right)}{\left(1-\frac{v}{c^2}\dot{x}\right)}$$

$$\frac{d}{dt'}\left(\frac{dx'}{dt'}\right) = \frac{d}{dt'}\left(\frac{\left(\dot{x}-v\right)}{\left(1-\frac{v}{c^2}\dot{x}\right)}\right) = \frac{d\left(\frac{\left(\dot{x}-v\right)}{\left(1-\frac{v}{c^2}\dot{x}\right)}\right)/dt}{dt'\,dt}$$

$$= \frac{1}{\gamma}\frac{1}{\left(1-\frac{v}{c^2}\dot{x}\right)}\left\{\frac{\ddot{x}}{\left(1-\frac{v}{c^2}\dot{x}\right)} - \frac{\left(\dot{x}-v\right)\left(-\frac{v}{c^2}\ddot{x}\right)}{\left(1-\frac{v}{c^2}\dot{x}\right)^2}\right\}$$

$$= \frac{1}{\gamma}\frac{1}{\left(1-\frac{v}{c^2}\dot{x}\right)^3}\left\{\ddot{x}\left(1-\frac{v}{c^2}\dot{x}\right) + \left(\dot{x}-v\right)\frac{v}{c^2}\ddot{x}\right\}$$

Noting that $\dot{x} = v$, this becomes

$$\frac{d^2 x'}{dt'^2} = \frac{1}{\gamma}\frac{1}{\left(1-\frac{v^2}{c^2}\right)^3}\left\{\ddot{x}\left(1-\frac{v^2}{c^2}\right)\right\} = \gamma^3 \ddot{x}$$

Substituting into the x-component of Newton's second law,

$$m\frac{d^2 x'}{dt'^2} = qE'_x \rightarrow m\gamma^3\frac{d^2 x}{dt^2} = qE_x$$

Working now with the y-component of Newton's second law, we have

$$\frac{d^2 y'}{dt'^2} = \frac{d}{dt'}\left(\frac{dy'}{dt'}\right)$$

$$\frac{dy'}{dt'} = \left(\frac{dy'}{dt'}\right)\left(\frac{dt}{dt'}\right) = \frac{dy'/dt}{dt'/dt} = \frac{d\left[y\right]/dt}{d\left[\gamma\left(t-\frac{v}{c^2}x\right)\right]/dt} = \frac{\dot{y}}{\gamma\left(1-\frac{v}{c^2}\dot{x}\right)}$$

$$\frac{d}{dt'}\left(\frac{dy'}{dt'}\right) = \frac{d}{dt'}\left(\frac{\dot{y}}{\gamma\left(1-\frac{v}{c^2}\dot{x}\right)}\right) = \frac{d\left(\frac{\dot{y}}{\gamma\left(1-\frac{v}{c^2}\dot{x}\right)}\right)/dt}{dt'/dt}$$

$$= \frac{1}{\gamma}\frac{1}{\left(1-\frac{v}{c^2}\dot{x}\right)}\left\{\frac{\ddot{y}}{\gamma\left(1-\frac{v}{c^2}\dot{x}\right)} - \frac{\dot{y}\left(-\frac{v}{c^2}\ddot{x}\right)}{\gamma\left(1-\frac{v}{c^2}\dot{x}\right)^2}\right\}$$

Noting $\ddot{x} = 0$, and that $\dot{x} = v$, this becomes

$$\frac{d^2 y'}{dt'^2} = \frac{1}{\gamma^2}\frac{1}{\left(1-\frac{v^2}{c^2}\right)^2}\ddot{y} = \gamma^2 \ddot{y}$$

Substituting into the y-component of Newton's second law,

$$m\frac{d^2 y'}{dt'^2} = qE'_y \rightarrow m\gamma^2\frac{d^2 y}{dt^2} = q\gamma\left(E_y - \frac{v}{c}B_z\right) \rightarrow m\gamma\frac{d^2 y}{dt^2} = q\left(E_y - \frac{v}{c}B_z\right)$$

and, in a manner similar to the y-component, the z-component of Newton's second law becomes

$$m\gamma\frac{d^2z}{dt^2} = q\left(E_z + \frac{v}{c}B_y\right)$$

In summary,

$$m\frac{d^2x'}{dt'^2} = qE'_x \quad \rightarrow m\gamma^3\frac{d^2x}{dt^2} = qE_x$$

$$m\frac{d^2y'}{dt'^2} = qE'_y \quad \rightarrow m\gamma\frac{d^2y}{dt^2} = q\left(E_y - \frac{v}{c}B_z\right)$$

$$m\frac{d^2z'}{dt'^2} = qE'_z \quad \rightarrow m\gamma\frac{d^2z}{dt^2} = q\left(E_z + \frac{v}{c}B_y\right)$$

4.6   Notes

1. Einstein, Albert, On the Electrodynamics of Moving Bodies, *Annalen der Physik* 322 (10), (1905), pp. 891–921; Stachel, John, editor, *The Collected Papers of Albert Einstein*, Volume 2 [CPAE2], Princeton University Press, Princeton, NJ, 1989, pp. 275–306; English translation by Anna Beck [CPAE2 ET], pp. 140–171. The original text contains a number of editorial comments and introductory comments (pp. 253–274) that are quite informative.
2. Einstein, Albert, Does the Inertia of a Body Depend Upon Its Energy Content? *Annalen der Physik* 323 (13), (1905), pp. 639–641; Stachel, John, ed., [CPAE2, pp. 311–314; CPAE2 ET, pp. 172–174].
3. Stachel, John, editor, *Einstein's Miraculous Year*, Princeton University Press, Princeton, NJ, 1998.
4. Stachel, John, ed., *Einstein's Miraculous Year*, pp. 102–103.
5. Einstein, Albert, Electrodynamics of Moving Bodies; [CPAE2, pp. 275–306; CPAE2ET, pp. 140–171].
6. Einstein, Albert, Electrodynamics of Moving Bodies; [CPAE2, p. 276; CPAE2 ET, p. 140].
7. Einstein, Albert, Electrodynamics of Moving Bodies; [CPAE2, p. 276; CPAE2 ET, p. 140].
8. Howard, Don, and Stachel, John, editors, *Einstein: The Formative Years, 1879–1909*, Birkhäuser, Boston, MA, 2000, article by Robert Rynasiewitz, p. 165, regarding Einstein's reading of Wien's article listing the Michelson–Morley experiment; also see Miller, Arthur, *Albert Einstein's Special Theory of Relativity*, Addison-Wesley Publishing Company, Reading, MA, 1981, p. 91; Buchwald, Diana Kormos, Rosenkranz, Zéev, Sauer, Tilman, Illy, József, and Holmes, Virginia Iris, editors, *The Collected Papers of Albert Einstein*, Volume 12 [CPAE12], Princeton University Press, Princeton, NJ, 2009, p. xxxvii.
9. Dijksterhuis, E. J., *The Mechanization of the World Picture*, Princeton University Press, Princeton, NJ, 1986, p. 354.
10. Dijksterhuis, E. J., *The Mechanization of the World Picture*, p. 464.
11. Newton, Isaac, *The Principia*, A new translation by I. Bernard Cohen and Anne Whitman, University of California Press, Berkeley, CA, 1999 (paperback version), pp. 416–417.

12. Dijksterhuis, E. J., *Mechanization of the World Picture*, p. 474.
13. Dijksterhuis, E. J., *Mechanization of the World Picture*, p. 466.
14. Dijksterhuis, E. J., *Mechanization of the World Picture*, p. 468.
15. Newton, Isaac, *The Principia*, Corollary 5, p. 423.
16. Cushing, James T., *Philosophical Concepts in Physics*, Cambridge University Press, Cambridge, 1988, p. 206.
17. Nienkamp, Paul K., *Lorentz' Contributions to the Theory of Relativity*, M.S. Thesis, Creighton University, 1999. Many of the details of the recreation of Lorentz' derivation of the Lorentz equations are based on the M.S. thesis of Paul Nienkamp.
18. Miller, Arthur, *Einstein's Special Theory*, p. 24.
19. Lorentz, H. A., *The Theory of Electrons*, 2nd edition, Dover Publications, New York, 1952, pp. 57–58.
20. Rynasiewicz, Robert, *Lorentz's Local Time and the Theorem of Corresponding States*, Volume 1, Fine, Arthur, and Leplin, Jarrett, editors, Philosophy of Science Association, East Lansing, MI, 1938, pp. 68–69.
21. Lorentz, H. A., *The Theory of Electrons*, p. 321.
22. Cushing, James T., *Philosophical Concepts*, pp. 199–202.
23. Cushing, James T., *Philosophical Concepts*, pp. 199, 201.
24. Lorentz, H. A., *The Theory of Electrons*, pp. 57–58, 198.
25. Boorse, Henry A., and Motz, Lloyd, editors, *The World of the Atom*, Basic Books, Inc., Publishers, New York, 1966, p. 519.
26. Hoffmann, Banesh, *Relativity and Its Roots*, Scientific American Books, W. H. Freeman and Company, New York, 1983, p. 87.
27. Lorentz, Hendrik, *The Principle of Relativity*, pp. 12–13. Citation from Holton, Gerald, *Thematic Origins of Scientific Thought*, revised edition, Harvard University Press, Cambridge, MA, 1973, 1988, p. 323.
28. Cushing, James T., *Philosophical Concepts*, p. 209.
29. Einstein, Albert, Electrodynamics of Moving Bodies; [CPAE2, pp. 275–306; CPAE2 ET, pp. 140–171].
30. Stachel, John, ed., [CPAE2, pp. 254–257].
31. Einstein used V to represent the speed of light. For consistency with modern notation c will be used in all cases to represent the speed of light.
32. Einstein, Albert, Electrodynamics of Moving Bodies; [CPAE2, pp. 276–277; CPAE2 ET, p. 140].
33. Einstein, Albert, Electrodynamics of Moving Bodies; [CPAE2, p. 277; CPAE2 ET, p. 141].
34. In the late 1800s the speed of light was designated sometimes by the upper case letter L (used by Einstein, CPAE2 ET, p. 89), sometimes by an upper case letter V (used by Einstein [CPAE2 ET, p. 143]), and sometimes by the letter c (most likely from the Latin word "celeritas" meaning speed or swiftness, but possibly from Weber's constant in electrodynamics. Reference: math.ucr.edu/home/baez/physics/Relativity/SpeedOfLight/c.html). Shortly after 1900 the letter c became the standard designation of the speed of light (used by Einstein [CPAE2 ET, p. 256]).
35. Stachel, John, *Einstein from 'B' to 'Z'*, Birkhäuser, Boston, 2002, p. 166.
36. Einstein, Albert, Electrodynamics of Moving Bodies; [CPAE2, p. 277; CPAE2 ET, p. 141].
37. Einstein, Albert, Electrodynamics of Moving Bodies; [CPAE2, p. 279; CPAE2 ET, p. 142].

38. Einstein, Albert, Electrodynamics of Moving Bodies; [CPAE2, p. 279; CPAE2 ET, p. 142].
39. Einstein, Albert, Electrodynamics of Moving Bodies; [CPAE2, p. 279; CPAE2 ET, p. 143].
40. Einstein, Albert, Electrodynamics of Moving Bodies; [CPAE2, p. 280; CPAE2 ET, pp. 143, 144].
41. See endnote 31.
42. Stachel, John, ed., *Einstein's Miraculous Year*, p. 106.
43. Einstein, Albert, Electrodynamics of Moving Bodies; [CPAE2, p. 282–283; CPAE2 ET, p. 146]. This notation differs from that used in Einstein's paper. Einstein referred to the coordinate systems as K and k (rather than K and K'), and to the coordinates as x, y, z (coordinate system K) and as ξ, η, ζ (coordinate system k, or K').
44. Einstein, Albert, Electrodynamics of Moving Bodies; [CPAE2, p. 281; CPAE2 ET, p. 144].
45. Einstein, Albert, Electrodynamics of Moving Bodies; [CPAE2, p. 281; CPAE2 ET, p. 145].
46. Einstein, Albert, Electrodynamics of Moving Bodies; [CPAE2, p. 282; CPAE2 ET, p. 145].
47. Einstein, Albert, Electrodynamics of Moving Bodies; [CPAE2, p. 283; CPAE2 ET, p. 146].
48. Einstein, Albert, Electrodynamics of Moving Bodies; [CPAE2, p. 283, 284; CPAE2 ET, pp. 146, 147].
49. Einstein, Albert, Electrodynamics of Moving Bodies; [CPAE2, p. 286; CPAE2 ET, p. 149].
50. Einstein, Albert, Electrodynamics of Moving Bodies; [CPAE2, p. 288; CPAE2 ET, p. 152].
51. Einstein, Albert, Electrodynamics of Moving Bodies; [CPAE2, p. 289; CPAE2 ET, p. 153].
52. Einstein, Albert, Electrodynamics of Moving Bodies; [CPAE2, p. 292; CPAE2 ET, p. 156].
53. Miller, Arthur, *Einstein's Special Theory*, p. 24.
54. Einstein, Albert, Electrodynamics of Moving Bodies; [CPAE2, pp. 292–293; CPAE2 ET, pp. 156–157].
55. Einstein, Albert, Electrodynamics of Moving Bodies; [CPAE2, p. 293; CPAE2 ET, p. 157].
56. Einstein, Albert, Electrodynamics of Moving Bodies; [CPAE2, p. 294; CPAE2 ET, p. 158].
57. Einstein, Albert, Electrodynamics of Moving Bodies; [CPAE2, p. 294; CPAE2 ET, p. 158].
58. Einstein, Albert, Electrodynamics of Moving Bodies; [CPAE2, p. 295; CPAE2 ET, p. 160].
59. Einstein, Albert, Electrodynamics of Moving Bodies; [CPAE2, p. 297; CPAE2 ET, p. 162]. This expression is stated in the 1905 paper, but is derived in the 1907 paper, On the Relativity Principle and the Conclusions Drawn from It, *Jahrbuch der Radioaktivität und Elektronik* 4 (1907); [CPAE2, pp. 452–453; CPAE2 ET, pp. 273–274].
60. Einstein, Albert, Electrodynamics of Moving Bodies; [CPAE2, p. 298; CPAE2 ET, p. 162].
61. Einstein, Albert, Electrodynamics of Moving Bodies; [CPAE2, pp. 298–299; CPAE2 ET, p. 163].

62. Einstein, Albert, Electrodynamics of Moving Bodies; CPAE2, pp. 298–299; CPAE2 ET, p. 163].

63. Einstein, Albert, On a Heuristic Point of View Concerning the Production and Transformation of Light; [CPAE2, p. 161; CPAE2 ET, p. 97].

64. Einstein, Albert, Inertia and Energy Content; [CPAE2, pp. 311–314; CPAE2 ET, pp. 172–174].

65. Einstein, Albert, Electrodynamics of Moving Bodies; [CPAE2, p. 302; CPAE2 ET, pp. 166–167].

66. Einstein, Albert, Electrodynamics of Moving Bodies; [CPAE2, p. 297; CPAE2 ET, p. 162].This expression is stated in the 1905 paper, but is derived in the 1907 paper, On the Relativity Principle and the Conclusions Drawn from It, document 47; [CPAE2, p. 452; CPAE2 ET, p. 272].

67. Einstein, Albert, Electrodynamics of Moving Bodies; [CPAE2, pp. 303–304; CPAE2 ET, p. 168].

68. Einstein, Albert, Inertia and Energy Content; [CPAE2, pp. 311–314; CPAE2 ET, pp. 172–174].

69. See Section 5.2.8 of this chapter.

70. Einstein, Albert, Electrodynamics of Moving Bodies; [CPAE2, p. 298; CPAE2 ET, p. 163].

71. Einstein, Albert, Inertia and Energy Content; [CPAE2, p. 313; CPAE2 ET, p. 173].

72. Einstein, Albert, Inertia and Energy Content; [CPAE2, p. 314; CPAE2 ET, p. 174].

73. Einstein, Albert, Inertia and Energy Content; [CPAE2, p. 314; CPAE2 ET, p. 174].

74. Stachel, John, ed., *Einstein's Miraculous Year*, pp. 106–107.

75. Cushing, James T., *Philosophical Concepts*, p. 208.

76. Cushing, James T., *Philosophical Concepts*, p. 215. Translation by Cushing from Kaufmann article, 1905, p. 956. Kaufmann, W. (1905), Über die Konstitution des Elektrons, *Sitzungsberichte der Königlich Preussischen Akademie der Wissenschaften* 45, 949–56.

77. Einstein, Albert, Relativity Principle and Conclusions; [CPAE2, p. 461; CPAE2 ET, p. 284].

78. An excellent re-creation of Planck's analysis can be found in James T. Cushing's article, Electromagnetic Mass, Relativity, and the Kaufmann Experiments, *Am. J. Phys.*, 49 (1981), pp. 1133–1149.

79. Lorentz, H. A., *The Theory of Electrons*, pp. 57–58.

80. Rynasiewicz, Robert, *Lorentz's Local Time*, pp. 68–69.

81. Cushing, James, T., *Philosophical Concepts*, pp. 199–202.

82. Lorentz, H. A., *The Theory of Electrons*, pp. 57–58, p. 198.

83. Einstein, Albert, Electrodynamics of Moving Bodies; [CPAE2, p. 283; CPAE2 ET, p. 147].

84. Einstein, Albert, Electrodynamics of Moving Bodies; [CPAE2, pp. 284–285; CPAE2 ET, p. 148].

85. Einstein, Albert, Electrodynamics of Moving Bodies; [CPAE2, pp. 286–287; CPAE2 ET, p. 150].

86. Einstein, Albert, Electrodynamics of Moving Bodies; [CPAE2, p. 287; CPAE2 ET, p. 151].

87. Einstein, Albert, Electrodynamics of Moving Bodies; [CPAE2, p. 293; CPAE2 ET, p. 157].

88. Einstein, Albert, Electrodynamics of Moving Bodies; [CPAE2, p. 293; CPAE2 ET, p. 158].

89. Einstein, Albert, On the Relativity Principle and the Conclusions Drawn from It, *Jahrbuch der Radioaktivität und Elektronik* 4 (1907); [CPAE2, pp. 452–453; CPAE2 ET, pp. 273–274].

90. Einstein, Albert, Electrodynamics of Moving Bodies; [CPAE2, pp. 302–304; CPAE2 ET, pp. 167–169].

91. Einstein, Albert, Relativity Principle and Conclusions; [CPAE2, p. 454; CPAE2 ET, pp. 275–276].

4.7 Bibliography

Boorse, Henry A., and Motz, Lloyd, editors, *The World of the Atom*, Volume 1, Basic Books, Inc., New York, 1966.

Buchwald, Diana Kormos, Rosenkranz, Zéev, Sauer, Tilman, Illy, József, and Holmes, Virginia Iris, editors, *The Collected Papers of Albert Einstein*, Volume 12 [CPAE12], Princeton University Press, Princeton, NJ, 2009.

Cushing, James T., Electromagnetic Mass, Relativity and the Kaufmann Experiments, *Am. J. Phys.* 49 (1981), pp. 1133–1149.

Cushing, James T., *Philosophical Concepts in Physics*, Cambridge University Press, Cambridge, 1998.

Dijksterhuis, E. J., *The Mechanization of the World Picture*, Princeton University Press, Princeton, NJ, 1986.

Einstein, Albert, On the Electrodynamics of Moving Bodies, *Annalen der Physik* 322 (10), (1905) [CPAE2, CPAE2 ET].

Einstein, Albert, Does the Inertia of a Body Depend Upon Its Energy Content? *Annalen der Physik* 323 (13), (1905) [CPAE2, CPAE2 ET].

Einstein, Albert, On the Relativity Principle and the Conclusions Drawn from It, *Jahrbuch der Radioaktivität und Elektronik* 4 (1907) [CPAE2, pp. 432–489; CPAE2 ET, pp. 252–311].

Einstein, Albert, On a Heuristic Point of View Concerning the Production and Transformation of Light, *Annalen der Physik* 322 (6), (1905). [CPAE2, CPAE2 ET].

Hoffmann, Banesh, *Relativity and Its Roots*, Scientific American Books, W. H. Freeman and Company, New York, 1983.

Holton, Gerald, *Thematic Origins of Scientific Thought*, revised edition, Harvard University Press, Cambridge, MA, 1988.

Howard, Don, and Stachel, John, editors, *Einstein: The Formative Years, 1879–1909*, Birkhäuser, Boston, MA, 2000.

Kaufmann, W. (1905), Über die Konstitution des Elektrons, *Sitzungsberichte der Königlich Preussischen Akademie der Wissenschaften* **45** (1905).

Lorentz, H. A., *The Theory of Electrons*, 2nd edition, Dover Publications, New York, 1952.

Lorentz, Hendrik, *The Principle of Relativity*. (Citation in Holton, Gerald, *Thematic Origins of Scientific Thought*.)

Miller, Arthur, *Albert Einstein's Special Theory of Relativity*, Addison-Wesley Publishing Company, Reading, MA, 1981.

Newton, Isaac, *The Principia*, A new translation by I. Bernard Cohen and Anne Whitman, University of California Press, Berkeley, CA, 1999.

Nienkamp, Paul K., *Lorentz' Contributions to the Theory of Relativity*, M. S. Thesis, Creighton University, Omaha, NE, 1999.

Rynasiewicz, Robert, *Lorentz's Local Time and the Theorem of Corresponding States*, Volume 1, Fine, Arthur, and Leplin, Jarrett, editors, Philosophy of Science Association, East Lansing, MI, 1988.

Stachel, John, editor, *Einstein from 'B' to 'Z'*, Birkhäuser Press, Boston, MA, 2002.

Stachel, John, editor, *The Collected Papers of Albert Einstein*, Volume 2 [CPAE2], Princeton University Press, Princeton, NJ, 1989; English translation by Anna Beck [CPAE2 ET].

Stachel, John, editor, *Einstein's Miraculous Year*, Princeton University Press, Princeton, NJ, 1998.

The General Theory of Relativity

5

"The Foundation of the General Theory of Relativity"[1]
Received on 20 March, 1916, by *Annalen der Physik*

5.1 Historical Background   161

5.2 Albert Einstein's Paper, "The Foundation of the General Theory of Relativity"   171

5.3 Discussion and Comments   213

5.4 Appendices   223

5.5 Notes   255

5.6 Bibliography   262

5.1 Historical Background

Once Albert Einstein had published the paper, "On the Electrodynamics of Moving Bodies," he began to address why the theory of relativity needed to be restricted to uniform motion. His aesthetic sense told him that if the theory of relativity is valid it should be valid for all motion, not be restricted to uniform motion. Two years later, he published his first remarks on generalizing the theory of relativity to the case of non-uniform motion. Subsequent to this the theory of relativity restricted to uniform motion (the 1905 version) became known as the special theory of relativity, and the case generalized to non-uniform motion became known as the general theory of relativity.

5.1.1 Lingering Questions

Beyond the issues addressed by Einstein in his 1905 papers there remained a number of perplexing questions. Those of pertinence to this chapter include:

1. At the surface of the earth, why do all objects, regardless of their composition, experience the same acceleration due to gravity? Why does g not depend on the substance of a body?

2. Why do all objects, regardless of their mass, have the same acceleration due to gravity? Why do more-massive objects have the same gravitational acceleration as less-massive objects? More-massive objects need a larger force to give them the same acceleration as less-massive objects. How is gravity able to "compensate" to give objects of different mass the same acceleration?

3. Inertial mass and gravitational mass. A lingering problem in physics was the apparent equality of inertial mass and gravitational mass, an equality empirically verified.

The equality of inertial mass and gravitational mass had been accepted by scientists for centuries. In the *Principia*, Newton had clarified the distinction between mass and weight. Although Newton did not distinguish between inertial mass and gravitational mass, he recognized that mass not only is a measure of a body's resistance to a change in motion (inertial mass), it also determines a body's reaction to a gravitational field (gravitational mass).[2] In the late 1800s, many, if not most, physicists (including Einstein) were aware of the equality of inertial and gravitational mass. (In 1889, this equality was experimentally verified to one part in 10^9 by Baron Roland von Eötvös.[3] However, because of his limited acquaintance with the literature, in 1907 Einstein was unaware of the Eötvös experiments. It was not until the "Entwurf" paper of 1913 that Einstein first discussed the Eötvös experiments.)[4]

From Newton's Second Law of Motion, $F = ma$, the acceleration of an object is proportional to the force causing the acceleration; doubling the force doubles the acceleration, etc. If two objects with different mass are to have equal accelerations, a larger force needs to be applied to the more-massive object. More-massive objects have more resistance to a change in their motion. It is in this sense that mass (this is called the inertial mass) enters into Newton's Second Law, $F = ma$.

In Newton's law of gravitation, $F_{gravity} = \frac{Gm_1m_2}{R^2}$, the mass ($m_1$ or m_2) of an object is a measure of the strength of the gravitational attraction (force) between the two objects. It is in this sense that mass (this is the called gravitational mass) enters into Newton's law of gravitation.

Consider an analogous situation of two charged particles. The force between the two particles is given by Coulomb's law, $F = k\left(q_1q_2/r^2\right)$, with q_1 and q_2 the charges on the two particles. The acceleration of one of the particles is given by Newton's Second Law, $F_1 = m_1a_1$. The force on the particle is dependent on one property of the particle, the charge, while the acceleration is dependent on another property of the particle, the mass.

In the gravitational case, the gravitational force is $F = G\left(m_1m_2/r^2\right)$, and the acceleration is given by Newton's Second Law, $F_1 = m_1a_1$. In the gravitational case, the force on the particle and the acceleration of the particle are dependent on the same property of the particle, the mass. Why, in the electrical case, are the force and the acceleration dependent on different properties of the particle (charge and mass), while, in the gravitational case, the force and the acceleration are dependent on the same property of the particle (mass)?[5]

Inertial mass and gravitational mass come from two completely different concepts. Why should the gravitational force depend on the same mass that is a resistance to change in motion? Why should inertial mass and gravitational mass be the same? On the other hand, if they are the same, when one sets the gravitational force on an object equal to its mass times acceleration, the inertial mass and the gravitational mass terms cancel out of the equation, leaving one with the same gravitational acceleration for all objects near the surface of the earth.

$$\frac{Gm_{object}m_{earth}}{R_{earth}^2} = m_{object}a_{gravity} = m_{object}g \quad \Rightarrow \quad g = \frac{Gm_{earth}}{R_{earth}^2}$$

Or, reversing the argument, the equality of gravitational acceleration for objects of different mass leads to the conclusion that the inertial mass and the gravitational mass are equal to one another.

4. *Relative Motion*: In 1905, with the special theory of relativity, Einstein did away with the idea of absolute velocity, or velocity relative to absolute space, because it was not observable. However, in Einstein's view, why should the relativity of motion be restricted to uniform motion? Why should accelerated (non-uniform motion) reference frames be distinguished from uniform motion reference frames? His aesthetic sense told him this distinction should not be. But Einstein knew we can sense an acceleration (when a car accelerates, the passenger feels the seat pushing her forward; a roller coaster is ridden specifically to feel these accelerations), whereas we cannot sense absolute velocity. We can feel when we are accelerated. We cannot sense uniform motion but we can sense changes in uniform motion. Uniform velocity is relative. Accelerations are absolute.

5.1.2 Generalizing the Special Theory of Relativity

In 1907, Einstein came to the realization that, "all the natural phenomena could be discussed in terms of special relativity except for the law of gravitation. I felt a deep desire to understand the reason behind this.... It was most unsatisfactory to me that, although the relation between inertia and energy is so beautifully derived [in special relativity], there is no relation between inertia and weight. I suspected that this relationship was inexplicable by means of special relativity."[6]

In 1907, he published a review of the special theory of relativity, entitled "On the Relativity Principle and the Conclusions Drawn from It," in *Jahrbuch der Radioaktivität und Elektronik*.[7] In the last section of the paper, entitled "Principle of Relativity and Gravitation," he writes:

So far we have applied the principle of relativity, i.e., the assumption that the physical laws are independent of the state of the motion of the reference system, only to *nonaccelerated* reference systems. Is it conceivable that the principle of relativity also applies to systems that are accelerated relative to each other?

...

We consider two systems Σ_1 and Σ_2 in motion. Let Σ_1 be accelerated in the direction of its X-axis, and let γ be the (temporally constant) magnitude of that acceleration. Σ_2 shall be at rest, but it shall be located in a homogeneous gravitational field that imparts to all objects an acceleration $-\gamma$ in the direction of the X-axis.

As far as we know, the physical laws with respect to Σ_1 do not differ from those with respect to Σ_2: this is based on the fact that all bodies are equally accelerated in the gravitational field.... [W]e shall therefore assume the complete physical equivalence of a gravitational field and a corresponding acceleration of the reference system.

This assumption extends the principle of relativity to the uniformly accelerated translational motion of the reference system.[8]

The equivalence of a gravitational field and an accelerating reference frame became known as the strong equivalence principle. This one assumption (strong equivalence principle) started Einstein on his way to answering the aforementioned "lingering questions." But it took ten years, from 1905 to 1915, for Einstein to arrive at the completed form of the general theory of relativity. This ten-year span reflected both the difficulty of developing the general theory of relativity and that much of his attention at this time was focused on the quanta of the radiation field (see Chapter 6).

5.1.3   The Equivalence of a Gravitational Field and an Accelerated Reference Frame

In the 1907 review article, Einstein considered the equivalence of gravitational fields and constant linear acceleration. In subsequent treatments he extended his investigations into the effects of accelerations due to rotating systems.

5.1.3.1   The Equivalence of Gravity and Constant Linear Acceleration

On earth, all objects fall toward the earth with same acceleration due to the effects of gravity (neglecting air resistance). Consider, first, a person enclosed within a windowless laboratory at rest on the surface of the earth performing a number of experiments that are contained entirely within the laboratory room. For example, the person notes that a rubber ball, a lead brick, and a textbook have the same gravitational acceleration; that light sent through a prism separates into a spectrum of colors; and that the force between two charged objects is given by Coulomb's law.

Consider now the windowless room (with the person still in it) transported to a location in space far from the earth where the gravitational forces are zero. With no gravitational fields, objects such as the rubber ball, lead brick, and textbook will not fall to the floor of the room as they did when the room was in the gravitational field of the earth – they will remain wherever they are placed, "floating" in mid-air. See Figure 5.1.

The windowless room is now given an upward acceleration equal to g, the gravitational acceleration on earth. The person in the room will feel the room pushing upward on her feet, giving a feeling of weight, the same as the floor of the room pushing upward on her feet when the room was on earth in the gravitational field. As the person accelerates upward, the objects maintain their position relative to space, while the person

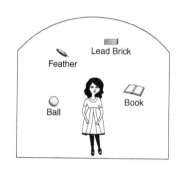

Fig. 5.1 Acceleration is equivalent to gravity.

moves upward past them. To the person accelerating upward, it appears the objects are all accelerating toward the floor of the room with the same acceleration equal to g. This was just what Galileo had found for gravity! All objects experience the same acceleration when falling freely.

The equivalence of gravitational mass and inertial mass is a consequence of the equivalence of gravity and acceleration. As Einstein said, "This law ... of the equality of inertial and gravitational masses was now brought home to me in all its significance. I was in the highest degree amazed at its existence."[9]

However, as Banesh Hoffmann observes, Einstein's aesthetic sense immediately told him this also should be true for all of physics.[10]

But Einstein could not stop here. It would be most inartistic to have so fundamental an equivalence apply only to mechanics and not to all of physics. God would not have made the universe in that way.

So, by a stroke of genius, Einstein broadened the partial equivalence into a total equivalence, saying that every experiment in the [accelerated room], whether mechanical or not, will yield the same results as the corresponding experiment in the earth laboratory. He called this the principle of equivalence.

Einstein returned (in thought) to the space traveler in her laboratory/room in a region of space where there is no gravity. A second laboratory/room is at rest next to the first room, with neither accelerating nor moving. In the first room a beam of light enters from outside and travels across the room in a straight line, creating a spot of light on the wall where it hits. Observer 2, in the second room and looking at the first room, would see the same thing.

As in the previous scenario, the first room is now accelerated upward, the second room remaining stationary, i.e., the second room is not accelerated. Observer 2 again sees the light traveling in a straight line, but sees the first room accelerating upward while the light beam traverses the room. Consequently, the light beam will illuminate a spot on the wall of the first room lower than in the non-accelerated instance. See Figure 5.2.

To the person on the first ship, as the ship accelerates upward the beam of light will appear to bend downward as it traverses the ship. If Einstein's principle of equivalence is correct, the same thing should occur in a gravitational field as occurred in the accelerating room. This bending of light in a gravitational field was anticipated by Einstein in his 1907 review paper.[11]

5.1.3.2 The Rigidly Rotating Disc

From the special theory of relativity, lengths in the direction of relative motion are contracted by a factor of $\sqrt{1 - (\frac{v}{c})^2}$, while lengths perpendicular to the direction of motion remain unaltered (see Section 4.2.4).

Consider a large stationary disc resting on the x–y plane, with its axis of symmetry aligned with the z-axis. A person is sitting on the axis of the disc. The ratio of the circumference of the disc to its diameter is π. If the person measures the diameter of the disc to be 100 m, the

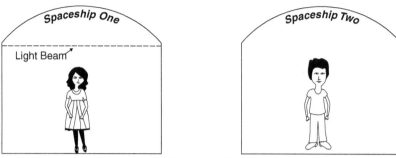

Figure 5.2a. Neither spaceship is accelerating

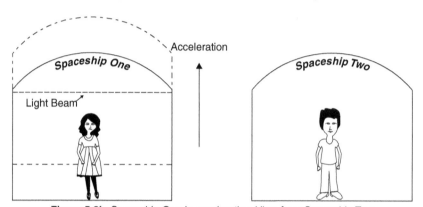

Figure 5.2b. Spaceship One is accelerating. View from Spaceship Two

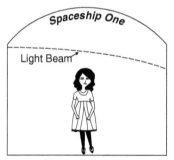

Fig. 5.2 A Beam of light traversing Spaceship One.

Figure 5.2c. Spaceship One is accelerating. View from Spaceship One

circumference of the disc will be measured to be 100π, approximately 314 m. For conceptual convenience, think of 100 "unit rods" fitting along the diameter, and 314 unit rods fitting around the circumference of the disc.

The disc is now set rotating with a large angular velocity, such that points on the circumference are moving at relativistic speeds. The person again is sitting at the axis of the disc, but not rotating with the disc. According to the special theory of relativity, the length of each of the unit rods used to measure the circumference will be contracted, leading to more than 314 "unit rods" fitting around the circumference of the disk. But the length of the unit rods used to measure the diameter will not be contracted since they are perpendicular to the measured motion, leaving the diameter length at 100 unit rods.

In the case of the rotating disc, the ratio of the circumference to the diameter is greater than π $(> \pi)$. Since the ratio of the circumference of a circle to its diameter is $> \pi$ only in curved space, this leads to the thought that an accelerating system corresponds to a curved space and, by the equivalence principle, that gravity is a property of space itself.

5.1.4   The Timeline from 1905 to 1916

The path from 1905 to 1916, from the special theory of relativity to the general theory of relativity, exemplifies the comment of Lorentz that "The way of scientific progress is not a straight one which we can steadfastly pursue. We are continually seeking our course, now trying one path and then another, many times groping in the dark, and sometimes even retracing our steps."[12] We now highlight some of the key points along this path to the completed general theory of relativity of 1916.

1907: In the 1907 paper, Einstein already commented on the possibility of a gravitational redshift (see Section 5.3.1.3 for details) and the bending of light in a gravitational field. However, on his calculation for the bending of light in a gravitational field, he reports, "Unfortunately, the effect of the terrestrial gravitational field is so small according to our theory... that there is no prospect of a comparison of the results of the theory with experience."[13]

1908: Hermann Minkowski showed the equations of the special theory of relativity were simplified when expressed in a four-dimensional space consisting of the three spatial dimensions and one time dimension. Initially, Einstein was unimpressed by Minkowski's elegant treatment but, in 1912, he adopted Minkowski's tensor method (see below) and, in 1916, acknowledged that it facilitated the transition from the special theory of relativity to the general theory of relativity.[14]

1911: The previous several years had been spent establishing his academic career, and directing most of his free time to the perplexing question of the quanta. By 1911, as Einstein was focusing more of his attention on the general theory of relativity, he realized the gravitational field of the sun would be strong enough to allow measurement of the bending of light in its gravitational field (see Section 5.3.1.2 and Appendix 5.4.9 for details). Publishing this insight in a 1911 article, he writes,

... In particular, it turns out that, according to the theory I am going to set forth, rays of light passing near the sun experience a deflection by its gravitational field, so that a fixed star appearing near the sun displays an apparent increase of its angular distance from the latter, which amounts to almost one second of arc....

... *Accordingly, a ray of light traveling past the sun would undergo a deflection amounting to* $4 \times 10^{-6} = 0.83$ *seconds of arc.* This is the amount by which the angular distance of the star from the center of the sun seems to be increased owing to the bending of the ray. Since the fixed stars in the portions of the sky that are adjacent to the sun become visible during the total solar eclipses, it is possible to compare this consequence of the theory with experience.[15]

It also was noted in the same paper that the velocity of light would be affected by the gravitational potential, Φ, where $c = c_0(1 + \frac{\Phi}{c^2})$, with c_0 the velocity of light at the coordinate origin and c the velocity of light in the gravitational potential Φ. Inversion of this relation allows the description of the strength of the gravitational potential by the variable speed of light.[16] But this means the postulate of the constant speed of light of the special theory of relativity no longer applies in the same way in the general theory of relativity. The variation in the speed of light flows "from the fact that clocks run at different rates when located at different gravitational field points, even though with respect to local proper time c still remains a universal constant.... [L]ocally the speed of light remains the same universal value, but the curved path brings about different elapsed times."[17] Also in this paper, Einstein obtained by a second method the expression for the gravitational redshift.[18]

1912: An expedition to check the deflection of starlight passing near to the edge of the sun was unsuccessful in Brazil (because of rain).[19]

Einstein adapted Minkowski's four-dimensional space-time, treating it as a curved space-time rather than flat. Since the geometry of the Minkowski space represents the gravitational potentials, which are determined by the distribution of mass and which, in turn, determine the distribution of mass, the geometry of the space changes with the changing gravitational potentials. In the fall, Einstein returned to the ETH in Zürich to work with Marcel Grossmann on the general theory of relativity, Grossmann introducing Einstein to Riemannian geometry and tensor calculus, and to the work of Christoffel, Ricci, and Levi-Civita.[20] Gravity now would be represented not by one function, the gravitational potential Φ of Newton, but by the ten elements of the metric tensor of the four-dimensional Minkowski space.[21]

In the special theory of relativity, in all inertial reference frames the system of equations describing the physical world was the same. In the general theory of relativity, Einstein was searching for a system of equations whose form was the same in all reference frames, what is called "generally covariant."[22]

1913: In the article, "Outline of a Generalized Theory of Relativity and of a Theory of Gravitation," Einstein and Grossmann published the results of their collaboration.[23] (This paper is sometimes referred to as the "Entwurf" paper from the first word of its title in German, *Entwurf einer verallgemeinerten Relativitätstheorie und einer Theorie der Gravitation*.) In "Einstein's Odyssey," Stachel describes these developments:

...there remained the question of the correct field equations to describe just how the sources generated the gravitational field. There was a tensor, formed from the metric tensor, which described certain aspects of the curvature of space-time. Called the Ricci tensor, it seemed to be just about (but not quite – as later proved to be important) the only possible candidate to be used for generally covariant gravitational field equations. But Einstein and Grossmann convinced themselves that these equations could not be correct since they did not seem to give Newton's theory of gravity as a limiting case. . . . Einstein then reluctantly abandoned the search for generally covariant equations, and set up non-generally covariant equations for the gravitational field. . . . Trying to make a virtue of necessity, Einstein constructed a more general argument purporting to show that the gravitational field equations could not be generally covariant.[24]

1914: An expedition of German scientists traveled to Russia to test the bending of starlight. When Germany declared war on Russia at the start of World War One, the German scientists were taken prisoner. They subsequently were released, but returned home without any photographs or other data.[25]

1915: Looking again at the general theory of relativity, Einstein realized, among other things, the proof against general covariance contained an error. With the removal of this restriction, the pieces fell quickly into place. The simplicity and beauty of the theory convinced Einstein of its correctness. During November, 1915, the full, complete, and correct, theory was assembled.[26] He was now nearing the end of the journey along the path described by Lorentz as " . . . now trying one path and then another, many times groping in the dark, and sometimes even retracing our steps."[27] In four reports to the Prussian Academy of Sciences, delivered over four weeks in the month of November he completed his journey to the general theory of relativity.

November 4, 1915: Einstein returned to the general covariance of the field equations.

In this pursuit I arrived at the demand of general covariance, a demand from which I parted, though with a heavy heart, three years ago when I worked with my friend Grossmann. . . .

Just as the special theory of relativity is based upon the postulate that all equations have to be covariant relative to linear orthogonal transformations, so the theory developed here rests upon the postulate of the *covariance of all systems of equations relative to transformations with the substitution determinant 1.*

Nobody who really grasped it can escape from its charm, because it signifies a real triumph of the differential calculus as founded by Gauss, Riemann, Christoffel, Ricci, and Levi-Civita.

. . .

It can now also be easily shown that the principle of the conservation of energy and momentum is satisfied.[28]

November 11, 1915: Einstein introduces the hypothesis that $\sum T_\mu^\mu = 0$, where T_μ^λ is the energy tensor for matter. With this assumption, he is able to write the gravitational field equations in generally covariant form.[29]

November 18, 1915: Einstein reports on a new value for the bending of starlight passing near to the sun and on his calculation for the value of the precession of the perihelion of the planet Mercury (see Section 5.3.1.1 for details). α is a constant determined by the mass of the sun (the expression for α is obtained in Section 5.2.22, Eq. (5.70a)).

In the present work I find an important confirmation of this most fundamental theory of relativity, showing that it explains qualitatively and quantitatively the secular rotation of the orbit of Mercury (in the sense of the orbital motion itself), which was discovered by Leverrier and which amounts to 45 sec of arc per century. Furthermore, I show that the theory has as a consequence a curvature of light rays due to gravitational fields twice as strong as was indicated in my earlier investigation.

. . .

Upon the application of Huygen's principle, we find . . . after a simple calculation, that a light ray passing at a distance Δ suffers an angular deflection of magnitude $2\alpha/\Delta$, while the earlier calculation, which was not based upon the hypothesis $\sum T^\mu_\mu = 0$, had produced the value α/Δ. A light ray grazing the surface of the sun should experience a deflection of 1.7 sec of arc instead of 0.85 sec of arc. In contrast to this difference, the result concerning the shift of the spectral lines by the gravitational potential, which was confirmed by Mr. Freundlich on the fixed stars (in order of magnitude), remains unaffected, because this result depends only on g_{44}.

. . .

. . . The calculation yields, for the planet Mercury, a perihelion advance of $43''$ per century, while the astronomers assign $45'' \pm 5''$ per century as the unexplained difference between observations and the Newtonian theory. This theory therefore agrees completely with the observations.[30]

November 25, 1915: Einstein added the "finishing touches to the field equations and presented the theory in full tensorial beauty: The crystallization was complete."[31]

In two recently published papers I have shown how to obtain field equations of gravitation that comply with the postulate of general relativity, i.e., which in their general formulation are covariant under arbitrary substitutions of space-time variables.

Historically they evolved in the following sequence. First, I found equations that contain the Newtonian theory as an approximation and are also covariant under arbitrary substitutions of determinant 1. Then I found that these equations are equivalent to generally-covariant ones if the scalar of the energy tensor of "matter" vanishes. . . . [T]his requires the introduction of the hypothesis that the scalar of the energy tensor of matter vanishes.

I now quite recently found that one can get away without this hypothesis about the energy tensor of matter . . . The field equations for vacuum, onto which I based the explanation of the Mercury perihelion, remain unaffected by this modification. . . .

. . .

With this, we have finally completed the general theory of relativity as a logical structure. The postulate of relativity in its most general formulation (which makes space-time coordinates into physically meaningless parameters) leads with compelling necessity to a very specific theory of gravitation that also explains the movement of the perihelion of Mercury.[32]

In December, Einstein wrote to Ehrenfest of the false starts and wrong turns, "Einstein [referring to himself] has it easy. Every year he retracts what he wrote in the preceding year; now the sorry business falls to me of justifying my latest retraction."[33] And to Lorentz he wrote, "My series of gravitation papers are a chain of wrong tracks, which nevertheless did gradually lead closer to the objective."[34]

In March, 1916, a comprehensive review of the completed general theory of relativity was published in the *Annalen der Physik*.

5.2 Albert Einstein's Paper, "The Foundation of the General Theory of Relativity"[35]

This paper is the formal presentation of Einstein's general theory of relativity, built mainly upon his 1914 review of the general theory of relativity, "The Formal Foundation of the General Theory of Relativity,"[36] and the four November reports to the Prussian Academy of Sciences. Included in the paper are necessary background and supporting arguments, and the development of the necessary mathematical tools. Much of the mathematical analysis is similar to the work of Grossmann in the 1913 paper,[37] but with notation corresponding to that of Ricci and Levi-Civita.[38] The paper is divided into four parts:

A. "Fundamental Considerations on the Postulate of Relativity": sections one through four of the paper (pp. 147–156).[39] In these sections, Einstein gives an overview of the special theory of relativity, some problems inherent in it, and the need to express the general laws of nature "by equations which hold good for all systems of coordinates, that is, are co-variant with respect to any substitutions whatever (generally co-variant)."[40]

B. "Mathematical Aids to the Formulation of Generally Covariant Equations": sections five through twelve of the paper (pp. 156–178).[41] In these sections, Einstein develops the mathematical tools necessary to develop the field equations for gravitation.

C. "Theory of the Gravitational Field": sections 13 through 18 (pp. 178–187).[42] This is the heart of the paper. In these sections, Einstein develops the field equations for gravitation.

D. "Material Phenomena": sections 19 through 22 (pp. 187–200).[43] In these sections, Einstein applies the field equations to a number of examples, showing how one can obtain Maxwell's equations, Newton's law of gravity, the bending of light rays in a gravitational field, and the precession of the perihelion of Mercury.

Part A: "Fundamental Considerations on the Postulate of Relativity"

5.2.1 Observations on the Special Theory of Relativity

The paper opens with a very quick overview of the special theory of relativity.

Consider a reference frame K, chosen so that the physical laws hold in their simplest form in K. Consider a second reference frame K′ moving at uniform velocity relative to reference frame K.

The special theory of relativity says that the same laws that are true in reference frame K also are true in reference frame K′. But the theory is restricted to K and K′ moving at constant velocity relative to one another – it does not hold true if the relative motion between K and K′ is non-uniform, i.e., it does not hold true for accelerated motion.

This is consistent with the relativity of Newton. However, the special theory of relativity differs from Newton in a second postulate, that the value of the speed of light in a vacuum will be measured to be the same by an observer at rest in reference frame K and by an observer at rest in reference frame K′. This second postulate leads to the Lorentz transformations, time dilation and length contraction.

In spite of its far-reaching modifications of space and time (length contraction and time dilation), the laws of geometry are "interpreted directly as laws relating to the possible relative positions of solid bodies at rest."[44] Consider a solid body, and two points, A and B, located in (or on) the body. The distance from A to B is a fixed value no matter how the body is oriented, nor where it is located. Similar comments are made regarding an interval of time as measured on a clock being of definite duration, independent of position and time. However, as we shall see, in the general theory of relativity we do not have the same simple physical interpretation of space and time.

5.2.2 The Need for an Extension of the Postulate of Relativity

Einstein, in this section of the paper, gives two justifications for extending the special theory of relativity, the first from an epistemological perspective, the second from a mechanical perspective. Great care is exercised in pointing out the distinction between an observable fact of experience and the postulated causes of the observable facts of experience.

Epistemological perspective: Consider two large masses, call them S_1 and S_2, hovering far out in space. The two masses are so far from each other that they do not interact gravitationally with one another, although each has gravitational interactions within itself. The two masses are so far out in space that there are no gravitational fields from other masses.

Envision a line between the two masses, with the line extending through the center of each mass. The two masses are rotating relative to one another about this line as an axis of rotation. To an observer at rest on S_1, the mass S_2 is seen to be rotating at a constant angular velocity about the line adjoining them. To an observer at rest on S_2, the mass S_1 is seen to be rotating at a constant angular velocity about the line adjoining them. See Figure 5.3.

MASS 1, at rest in space R$_1$ MASS 2, at rest in space R$_2$

Fig. 5.3 The two masses S$_1$ and S$_2$, with the adjoining line, as seen from S$_1$.

A person at rest on each of the masses measures the surface of the mass, the person on S$_1$ determining the surface of S$_1$ is a sphere, while the person on S$_2$ determines the surface of S$_2$ is an ellipsoid of revolution. The question arises as to the reason for this difference. Einstein writes,[45]

...No answer can be admitted as epistemologically satisfactory, unless the reason given is an *observable fact of experience*. The law of causality has not the significance of a statement as to the world of experience, except when *observable facts* ultimately appear as causes and effects.

Newtonian mechanics does not give a satisfactory answer to this question. It pronounces as follows:– The laws of mechanics apply to the space R$_1$, in respect to which the body S$_1$ is at rest, but not to the space R$_2$, in respect to which the body S$_2$ is at rest. But the privileged space R$_1$ of Galileo, thus introduced, is a merely *factitious* cause, and not a thing that can be observed. It is therefore clear that Newton's mechanics does not really satisfy the requirements of causality in the case under consideration, but only apparently does so, since it makes the factitious cause R$_1$ responsible for the observable difference in the bodies S$_1$ and S$_2$.

A factitious cause is defined as a contrived, or artificial, cause as opposed to a genuine, or real, cause: a cause produced by humans rather than by natural forces. Einstein concludes that there are no privileged reference frames, or privileged spaces, R$_1$ or R$_2$, apparently taking an anti-absolute space, or Machian, position. This is stated as,

The laws of physics must be of such a nature that they apply to systems of reference in any kind of motion.[46]

Physical perspective: Consider a reference system K in which Newton's laws of mechanics are valid, and in which a mass M sufficiently far from other masses moves with uniform motion in a straight line, i.e., with constant velocity. Consider a second reference frame, K$'$, moving with constant linear acceleration relative to the reference system K. Relative to K$'$, the mass M would have an accelerated motion, "independent of the material composition and physical state of the mass."[47]

Does the above accelerated motion of the mass M let an observer in K$'$ conclude that the reference system K$'$ is accelerated? The answer is no, because the observed motion may be interpreted equally well as K$'$ being at rest, but in a gravitational field which causes the observed motion of the mass M. The mechanical behavior of the mass M as described from the reference system K$'$ "is the same as presents itself to experience in the case of systems which we are wont to regard as 'stationary' or as 'privileged.' Therefore, from the physical standpoint, the assumption readily suggests itself that the systems K and K′ may both with equal right be looked upon as 'stationary,' that is to say, they have an equal title as systems of reference for the physical description of phenomena."[48]

Einstein concludes this section with two thoughts:

1. In pursuing the general theory of relativity, one is led naturally to a theory of gravitation.
2. The "principle of the constancy of light *in vacuo* must be modified, since we easily recognize that the path of a ray of light with respect to K′ must in general be curvilinear, if with respect to K light is propagated in a straight line with a definite constant velocity."[49]

5.2.3   The Space-Time Continuum. Requirement of General Covariance for the Equations Expressing General Laws of Nature

In classical mechanics, the coordinates of space and time have direct physical meaning. To say an object has a position $x = 4$ means the position is measured to be four unit rigid rods away from the origin. To say a clock has a time $t = 7$ means that a clock stationary relative to the system of coordinates and (nearly) coincident with the event will have measured 7 unit time periods at the occurrence of the event. The same physical interpretation of the coordinates holds true in the special theory of relativity. Einstein then proceeds to show such physical meaning can no longer apply to the coordinates in the general theory of relativity.

The first example is the relativistically rotating rigid disk. Consider a Galilean system of reference K, and a second system K′ in uniform rotation relative to K. The origins of both systems coincide, as well as their z-axes. In Section 5.1.3.2 we showed the ratio of the circumference to the diameter of the relativistically rotating rigid disk was greater than π and, thus, Euclidean geometry broke down in K′. In a similar manner, again using the special theory of relativity, Einstein shows that a clock on the circumference of the rotating disk, as judged from K, goes more slowly than a clock at rest at the origin. An observer at the origin of the coordinates "will interpret his observations as showing the clock at the circumference 'really' goes more slowly than the clock at the origin. So he will be obliged to define time in such a way that the rate of a clock depends upon where the clock may be."[50]

This result is summed up as:

In the general theory of relativity, space and time cannot be defined in such a way that differences of the spatial co-ordinates can be directly measured by the unit measuring-rod, or differences in the time co-ordinate by a standard clock.[51]

Since the old understanding of coordinates and measurement have broken down, and there appears to be no simple way of defining coordinates that would lead to a particularly simple formulation of the laws of nature, Einstein concludes there is nothing left to do "but to regard all imaginable systems of co-ordinates, on principle, as equally suitable for the description of nature."[52] This conclusion is stated as:

The general laws of nature are to be expressed by equations which hold good for all systems of co-ordinates, that is, they are co-variant with respect to any substitutions whatever (generally co-variant).[53]

1. Any physical theory that satisfies this postulate will satisfy the postulate of the general theory of relativity.
2. Covariance for *all* substitutions is overkill, but it would include all those substitutions that correspond to the relative motion of three-dimensional systems of coordinates.
3. The requirement of general covariance removes from space and time the last remnant of physical objectivity.
4. All space-time verifications are but the determination of space-time coincidences.
5. A system of reference only facilitates the description of the totality of such coincidences.
6. To a space-time point event is a corresponding set of values of the four space-time variables x_1, x_2, x_3, x_4. A second event coincident with the first event will have the same set of four space-time variables x_1, x_2, x_3, x_4, i.e., coincidence means the two events occur at the same location at the same time.
7. If we introduce a new system of coordinates x'_1, x'_2, x'_3, x'_4, where the new coordinates are functions of the old coordinates x_1, x_2, x_3, x_4, the equality of all four coordinates in the new system expresses the space-time coincidence of the two point-events.
8. We see there is no obvious reason for preferring one system of coordinates over another, i.e., we arrive at the requirement of general covariance.

5.2.4 The Relation of the Four Coordinates to Measurement in Space and Time

At the outset, Einstein states the general theory of relativity is not as simple as the special theory of relativity, and that his aim is to develop the general theory of relativity in a manner that is natural and based on well-founded assumptions.

In general, the gravitational field is not uniform. Consider the gravitational field from the sun – it extends out radially from the sun. See Figure 5.4.

For an infinitesimally small four-dimensional region, the gravitational field can be considered uniform, both in magnitude and in direction. In this infinitely small region we can select a reference system of coordinates that is accelerating so that no gravitational field occurs. Within this carefully selected system of coordinates the special theory of relativity is valid. See Figure 5.5.

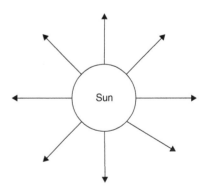

Fig. 5.4 The gravitational field in the vicinity of the sun, extending radially outward.

Fig. 5.5 Infinitesimally small four-dimensional volume in a gravitational field.

Within this carefully selected system of coordinates X_1, X_2, X_3 are the space coordinates and X_4 is the time coordinate (the unit of time is chosen so that $c = 1$ in the local system of coordinates). From the special theory of relativity, the space-time interval

$$ds^2 = -dX_1^2 - dX_2^2 - dX_3^2 + dX_4^2 \qquad (5.1)$$

has a value that is independent of the system of coordinates, and whose value is determined by measurements of space and time. If the interval ds^2 is positive, it is called time-like; if it is negative, it is called space-like. The magnitude of the linear element between two infinitesimally close points in the four-dimensional space is designated ds.

For an arbitrary system of coordinates x_1, x_2, x_3, x_4 (not our carefully selected system), for two infinitely close points A and B in four-dimensional space we can form the differentials $dx_1 = x_{B1} - x_{A1}$, $dx_2 = x_{B2} - x_{A2}$, $dx_3 = x_{B3} - x_{A3}$, $dx_4 = x_{B4} - x_{A4}$. If the "linear elements" dx_1, dx_2, dx_3, dx_4, as well as the local system dX_1, dX_2, dX_3, dX_4, are given for the region under consideration the dX_ν can be represented by linear homogeneous expressions of the dx_σ.

$$dX_\nu = \sum_{\sigma=1}^{4} a_{\nu\sigma} dx_\sigma \qquad (5.2)$$

Inserting these expressions into Eq. (5.1),

$$ds^2 = -dX_1^2 - dX_2^2 - dX_3^2 + dX_4^2$$

$$= -\sum_{\substack{j=1 \\ \sigma=1 \\ \tau=1}}^{\substack{j=3 \\ \sigma=4 \\ \tau=4}} a_{j\sigma} dx_\sigma a_{j\tau} dx_\tau + \sum_{\substack{\sigma=1 \\ \tau=1}}^{\substack{\sigma=4 \\ \tau=4}} a_{4\sigma} dx_\sigma a_{4\tau} dx_\tau$$

$$ds^2 = \sum_{\substack{\sigma=1 \\ \tau=1}}^{\substack{\sigma=4 \\ \tau=4}} g_{\sigma\tau} dx_\sigma dx_\tau \qquad (5.3)$$

with $g_{\sigma\tau} = -\sum_{j=1}^{j=3} a_{j\sigma} a_{j\tau} + a_{4\sigma} a_{4\tau}$

By definition $g_{\sigma\tau} = -\sum_{j=1}^{j=3} a_{j\sigma} a_{j\tau} + a_{4\sigma} a_{4\tau} = -\sum_{j=1}^{j=3} a_{j\tau} a_{j\sigma} + a_{4\tau} a_{4\sigma} = g_{\tau\sigma} \Rightarrow g_{\sigma\tau}$ is symmetric. The $g_{\sigma\tau}$ are functions of the x_σ, and are symmetric, i.e., $g_{\sigma\tau} = g_{\tau\sigma}$.

If it is possible to choose the local set of coordinates so that the $g_{\sigma\tau}$ take on the values to transform Eq. (5.3) into Eq. (5.1), then this is the set of coordinates in which the special theory of relativity is valid.

$$g_{\sigma\tau} = \begin{vmatrix} -1 & 0 & 0 & 0 \\ 0 & -1 & 0 & 0 \\ 0 & 0 & -1 & 0 \\ 0 & 0 & 0 & +1 \end{vmatrix} \qquad (5.4)$$

The set of values for the $g_{\sigma\tau}$ forms the base for much of the subsequent development of the general theory of relativity, in particular referring back to the particular set of values contained in Eq. (5.4), the values of the $g_{\sigma\tau}$ for the special theory of relativity. Einstein makes a number of observations regarding the $g_{\sigma\tau}$.[54]

1. From the discussion in the two preceding sections of the paper, sections 2 and 3, from the physical standpoint we would expect the $g_{\sigma\tau}$ to describe the gravitational field relative to the chosen set of coordinates.
2. If the special theory of relativity applies to the coordinates chosen for the infinitesimal four-volume (four-dimensional volume) we are investigating, then the $g_{\sigma\tau}$ are given by the values shown in Eq. (5.4). This is the coordinate system we denoted by the coordinates X_1, X_2, X_3, X_4.
3. If the region of space has the $g_{\sigma\tau}$ given by Eq. (5.4), the motion of a particle in this space will be uniform and in a straight line.
4. Transforming to another set of space-time variables x_1, x_2, x_3, x_4, the $g_{\sigma\tau}$ no longer will be constants but will be functions of the space and time variables x_1, x_2, x_3, x_4.
5. In the new set of coordinates x_1, x_2, x_3, x_4, the motion of the free particle will be a curvilinear path of non-uniform motion, rather than the straight-line uniform motion in the special theory of relativity coordinate system X_1, X_2, X_3, X_4.
6. The law(s) of motion of the particle in coordinate system x_1, x_2, x_3, x_4 will be independent of the properties of the moving particle (since it was independent of the properties in the coordinate system X_1, X_2, X_3, X_4).
7. Given the above, it is natural to interpret the motion of a particle in the new coordinate system x_1, x_2, x_3, x_4 as the motion of a particle under the influence of a gravitational field.
8. This leads to the conclusion that the description of a gravitational field is connected with the space-time variability of the $g_{\sigma\tau}$.
9. These conclusions are based on the existence of a coordinate system X_1, X_2, X_3, X_4 in which the special theory of relativity is valid, and then a transformation from that system to a second one which is accelerating relative to the initial system. These conclusions are based on the description of the motion in the second coordinate system x_1, x_2, x_3, x_4. Einstein then extends his conclusion to the claim that the $g_{\sigma\tau}$ describe the gravitational field, even in those cases where we are unable to find a coordinate system X_1, X_2, X_3, X_4 in which the special theory of relativity is valid. In other words, starting with the special theory of relativity coordinate system, we

arrived at the conclusion that the $g_{\sigma\tau}$ describe the gravitational field, and then Einstein claims the conclusion is correct even in those cases where we are unable to start with the special theory of relativity coordinate system.

10. This leads to the conclusion that gravitation is different in a fundamental way from the other forces, in particular, the electromagnetic force. For the gravitational force, the ten independent quantities representing the gravitational field, $g_{\sigma\tau}$, at the same time define the metrical properties of the space-time of the four-dimensional space. (Subsequent to the general theory of relativity were attempts to include the electromagnetic field in a manner similar to the gravitational field, what was known as the unified field theory. See Section 5.3.2.2 for a discussion of the unified field theory.)

Part B: "Mathematical Aids to the Formulation of Generally Covariant Equations"

Part B of this paper is a further development of the work by Marcel Grossman in the 1913 "Entwurf" paper.[55] This was a necessary portion of the paper as most physicists of the day were not familiar with the mathematics of tensors. The reader familiar with tensors could scan quickly over this section (or omit it entirely) and move directly to Part C: "The Theory of the Gravitational Field" (see Section 5.2.13 and subsequent sections). For the reader unfamiliar with tensors, a modern text on tensors would be a better source to learn the material, except the development by Einstein has everything in the notation he will be using in the subsequent portions of the paper.

The mathematics needed to develop the general theory of relativity is called the theory of covariants. A fundamental role will be played by the invariant ds where, from Eq. (5.3),

$$ds^2 = \sum_{\substack{\sigma=1 \\ \tau=1}}^{4} g_{\sigma\tau} dx_\sigma dx_\tau.$$

The theory deals with quantities called tensors. Tensors have a number of quantities, called the components, which are functions of the coordinates of the systems. For concreteness, traditional vectors and matrices are tensors, vectors being tensors of rank one and matrices being tensors of rank two.

1. If the components of a tensor are known for an original system of coordinates, the components can be calculated in a new system of coordinates if the transformations connecting the two systems of coordinates are known.
2. The equations of transformation relating the components of a tensor in the original and new coordinate systems are homogeneous.
3. By property 2, above, if all of the components are zero in the original system, all of the components also will be zero in the new system.

4. The components of a tensor are denoted by indices.

    a. A single index, such as μ, is used to represent the components of a vector. Latin indices, such as i, j, or k, are used for three-dimensional space, i.e., they take on the values 1, 2, and 3. Greek indices, such as λ, μ, or ν, are used for four-dimensional space, i.e., they take on the values 1, 2, 3, and 4.

    b. Double indices, such as $\mu\nu$, are used to represent the components of a matrix. The typical tensor used in the development of the general theory of relativity can be represented as a four-by-four matrix.

    c. The "Einstein convention" is introduced in section 5 of the paper: If an index is repeated in an expression, that index is to be summed over, but the summation sign is omitted, i.e.,

$$A_i B_i = \sum_{i=1}^{3} A_i B_i \text{ and } C_\mu D_\mu = \sum_{\mu=1}^{4} C_\mu D_\mu.$$

There are two main results from Part B, "Mathematical Aids to the Formulation of Generally Covariant Equations":

1. The equation of the geodetic line, Eq. (5.22). This is the equation of the path of a particle in the given reference frame.
2. The Riemann–Christoffel tensor, Eqs. (5.43) and (5.44). From the Riemann–Christoffel tensor will be obtained the equations for the gravitational field (Eqs. (5.47) and (5.53) of Part C).

5.2.5 Contravariant and Covariant Four-Vectors

Tensors are classified as contravariant or covariant, depending on their transformation properties. The initial discussion is of four-vectors.

    Contravariant four-vectors: Consider a linear element in four-dimensional space, with components dx_ν, with $dx_\nu = dx_1$, dx_2, dx_3, and dx_4, since the subscript $\nu = 1, 2, 3$, and 4. Transforming to a new system of coordinates, the new components of the linear element are

$$dx'_\sigma = \sum_{\nu=1}^{4} \frac{\partial x'_\sigma}{\partial x_\nu} dx_\nu = \frac{\partial x'_\sigma}{\partial x_\nu} dx_\nu \tag{5.5}$$

The middle and final expressions of Eq. (5.5) are the same, the final expression using the Einstein convention indicating summation over the repeated index ν. The dx'_σ are linear and homogeneous functions of the dx_ν and, thus, these coordinate differentials can be considered the components of a tensor. At this point the notation changes. The components of a contravariant four-vector A are denoted by superscripts (not subscripts), A^ν (for consistency, the elements dx_ν in Eq. (5.5) should have been written as dx^ν. In the 1913 "Entwurf" paper, contravariant tensors were represented by Greek symbols). Using the superscript notation, Eq. (5.5) is rewritten as

$$dx'^{\sigma} = \sum_{\nu=1}^{4} \frac{\partial x'^{\sigma}}{\partial x^{\nu}} dx^{\nu} = \frac{\partial x'^{\sigma}}{\partial x^{\nu}} dx^{\nu} \tag{5.5a}$$

Any quantity that transforms in this manner is called a contravariant four-vector, or a contravariant tensor. The components of the contravariant four-vector A^{ν} transform as:

$$A'^{\sigma} = \sum_{\nu=1}^{4} \frac{\partial x'^{\sigma}}{\partial x^{\nu}} A^{\nu} = \frac{\partial x'^{\sigma}}{\partial x^{\nu}} A^{\nu}$$

The converse transformation is obtained by exchanging the roles of the primed and unprimed coordinates,

$$B^{\nu} = \frac{\partial x^{\nu}}{\partial x'^{\sigma}} B'^{\sigma} \tag{5.5b}$$

If A^{ν} and B^{ν} are components of contravariant four-vectors, the sums $C^{\nu} = A^{\nu} \pm B^{\nu}$ are also the components of a contravariant four-vector.

Covariant four-vectors: Let B^{ν} be the components of any arbitrary contravariant four-vector. The components A_{ν} are called the components of a covariant four-vector if

$$\sum_{\nu=1}^{4} A_{\nu} B^{\nu} = A_{\nu} B^{\nu} = \text{Invariant} \tag{5.6}$$

(i.e., it has the same value after the mathematical operation as before.)

To determine the transformation properties of the covariant four-vector A, Einstein begins with Eq. (5.6). Equation (5.6) is valid for the original system of coordinates. Transforming to a new system of coordinates, Eq. (5.6) must again be valid for the transformed four-vectors,

$$A_{\nu} B^{\nu} = A'_{\sigma} B'^{\sigma}$$

Working with the LHS of the equation, and using Eq. (5.5b),

$$A_{\nu} B^{\nu} = A_{\nu} \sum_{\sigma=1}^{4} \frac{\partial x^{\nu}}{\partial x'^{\sigma}} B'^{\sigma} = A_{\nu} \frac{\partial x^{\nu}}{\partial x'^{\sigma}} B'^{\sigma}$$

$$= \left(\frac{\partial x^{\nu}}{\partial x'^{\sigma}} A_{\nu} \right) B'^{\sigma}$$

But since, from above, this must be equal to $A'_{\sigma} B'^{\sigma}$, A'_{σ} is the expression in the parentheses,

$$\Rightarrow A'_{\sigma} = \frac{\partial x^{\nu}}{\partial x'^{\sigma}} A_{\nu} \tag{5.7}$$

Each form of a tensor, contravariant and covariant, is fully a tensor as each satisfies the definition of a tensor. The difference is only in the law of transformation, Eqs. (5.5a) and (5.7). The contravariant tensor is denoted by placing the index above (superscript), the covariant by placing the subscript below (subscript).

5.2.6   Tensors of the Second and Higher Ranks

Contravariant tensors of rank two: Consider two contravariant four-vectors, A^μ and B^ν. Multiplying the components together gives sixteen possible products ($4 \times 4 = 16$). These form a contravariant tensor of rank two (two indices, μ and ν):

$$A^{\mu\nu} = A^\mu B^\nu \tag{5.8}$$

The components $A^{\mu\nu}$ can be viewed as a 4×4 matrix

$$\begin{pmatrix} A^{11} & A^{12} & A^{13} & A^{14} \\ A^{21} & A^{22} & A^{23} & A^{24} \\ A^{31} & A^{32} & A^{33} & A^{34} \\ A^{41} & A^{42} & A^{43} & A^{44} \end{pmatrix}.$$

$A^{\mu\nu}$ transforms as

$$A'^{\sigma\tau} = A'^\sigma B'^\tau = \left(\frac{\partial x'^\sigma}{\partial x^\mu} A^\mu \right) \left(\frac{\partial x'^\tau}{\partial x^\nu} B^\nu \right)$$

$$= \frac{\partial x'^\sigma}{\partial x^\mu} \frac{\partial x'^\tau}{\partial x^\nu} A^\mu B^\nu = \frac{\partial x'^\sigma}{\partial x^\mu} \frac{\partial x'^\tau}{\partial x^\nu} A^{\mu\nu} \tag{5.9}$$

Not every contravariant tensor of rank two can be formed as the product of two contravariant vectors (tensors of rank one). However, it is always possible to form a contravariant tensor of rank two as the sum of four appropriately selected pairs of four-vectors.

In a similar manner contravariant tensors of rank three, four, or higher can be formed.

Covariant tensors of rank two: In a similar manner, covariant tensors of rank two are defined by

$$A_{\mu\nu} = A_\mu B_\nu \tag{5.10}$$

with the law of transformation,

$$A'_{\sigma\tau} = \frac{\partial x^\mu}{\partial x'^\sigma} \frac{\partial x^\nu}{\partial x'^\tau} A_{\mu\nu} \tag{5.11}$$

In a similar manner, covariant tensors of rank three, four, or higher can be formed.

Mixed tensors of rank two: In a similar manner, mixed tensors of rank two are defined by

$$A_\mu^\nu = A_\mu B^\nu \tag{5.12}$$

with

$$A'^\tau_\sigma = \frac{\partial x'^\tau}{\partial x^\nu} \frac{\partial x^\mu}{\partial x'^\sigma} A_\mu^\nu \tag{5.13}$$

In a similar manner, mixed tensors of rank three, four, or higher can be formed.

Symmetric and antisymmetric tensors: Contravariant or covariant tensors of rank two, $A^{\mu\nu}$ or $A_{\mu\nu}$ (or higher), are symmetric if the two components obtained one from the other by the interchange of two indices are equal,

$$A^{\mu\nu} = A^{\nu\mu} \qquad (5.14)$$

$$A_{\mu\nu} = A_{\nu\mu} \qquad (5.14a)$$

Of the sixteen components of a symmetric tensor of rank two, six are determined by the symmetry condition, $A^{\mu\nu} = A^{\nu\mu}, \mu \neq \nu$, leaving ten independent components.

Contravariant or covariant tensors of rank two, $A^{\mu\nu}$ or $A_{\mu\nu}$ (or higher), are antisymmetric if the two components obtained one from the other by the interchange of two indices are the negative of one another,

$$A^{\mu\nu} = -A^{\nu\mu} \qquad (5.15)$$

$$A_{\mu\nu} = -A_{\nu\mu} \qquad (5.15a)$$

Of the sixteen components of an antisymmetric tensor of rank two, six are determined by the antisymmetry condition and the four with equal indices are equal to zero ($A^{11} = -A^{11} \Rightarrow A^{11} = 0$, etc.), leaving six independent components. An antisymmetric tensor with six independent components is referred to as a six-vector. (N.B. Typically a vector is thought of as a single row or a single column matrix. These six independent components of the antisymmetric matrix, although not a row or column matrix, are referred to as a six-vector.)

A tensor of rank one, A^μ, has four components (4^1), a tensor of rank two, $A^{\mu\nu}$, has sixteen components ($4 \times 4 = 4^2$), a tensor of rank three has 4^3 components, etc. A vector is a tensor of rank one, while a scalar is a tensor of rank zero.

5.2.7   Multiplication of Tensors

Outer multiplication: Straightforward multiplication of the components of a tensor of rank m and a tensor of rank n gives a tensor of rank $m + n$. (See Appendix 5.4.1.1 for an example of the outer multiplication of two vectors.)

$$T_{\mu\nu\sigma} = A_{\mu\nu}B_\sigma$$

$$T^{\mu\nu\sigma\tau} = A^{\mu\nu}B^{\sigma\tau}$$

$$T_{\mu\nu}^{\sigma\tau} = A_{\mu\nu}B^{\sigma\tau}$$

Contraction of a mixed tensor: In a mixed tensor, by equating an index of contravariant character (upper index) with one of covariant character (lower index) and summing over the repeated index (contraction), the result is a tensor of rank less by two,

$$A_\nu^\tau = A_{\mu\nu}^{\mu\tau}$$

Inner product of tensors: From the outer multiplication of a covariant tensor and a contravariant tensor, a mixed tensor is formed. For example, consider the covariant tensor $A_{\mu\nu}$ and the contravariant tensor B^σ, $D_{\mu\nu}^\sigma = A_{\mu\nu}B^\sigma$. Contraction of $D_{\mu\nu}^\sigma$ with respect to one covariant and one contravariant index, for example, the indices ν and σ, is called the inner product of the tensors $A_{\mu\nu}$ and B^σ.

It is pointed out that if $A_{\mu\nu}$ and $B^{\mu\nu}$ are tensors, the product $A_{\mu\nu}B^{\mu\nu}$ is a scalar. But the converse is also true. "If $A_{\mu\nu}B^{\mu\nu}$ is a scalar *for any choice of the tensor $B^{\mu\nu}$*, then $A_{\mu\nu}$ has tensor character."[56] (See Appendix 5.4.1.2 for the details of this statement.)

The following is a summary of the properties of $A_{\mu\nu}$:

1. If $A_{\mu\nu}$ is a covariant tensor and $B^{\mu\nu}$ is a contravariant tensor, the product $A_{\mu\nu}B^{\mu\nu}$ is a scalar.
2. If $B^{\mu\nu}$ is a contravariant tensor and $A_{\mu\nu}B^{\mu\nu}$ is a scalar, then $A_{\mu\nu}$ is a covariant tensor.
3. If B^μ and C^ν are contravariant four-vectors and $A_{\mu\nu}B^\mu C^\nu$ is a scalar, then $A_{\mu\nu}$ is a covariant tensor.
4. Let B^μ be a contravariant four-vector. If $A_{\mu\nu}B^\mu B^\nu$ is a scalar, and if $A_{\mu\nu}$ is symmetric, then $A_{\mu\nu}$ is a covariant tensor.
5. Let B^μ be a contravariant four-vector. If $A_{\mu\nu}B^\nu$ is a tensor of first rank, then $A_{\mu\nu}$ is a tensor of second rank.

Similar results are true for a contravariant tensor $A^{\mu\nu}$.

5.2.8   Some Aspects of the Fundamental Tensor $g_{\mu\nu}$

5.2.8.1   The Covariant Fundamental Tensor

The invariant expression for the square of the linear element is

$$ds^2 = g_{\mu\nu}\,dx_\mu\,dx_\nu$$

The dx_μ are the components of a contravariant tensor (since the dx_μ transform as Eq. (5.5). To be correct, the dx_μ should be written with a superscript, dx^μ, i.e., $ds^2 = g_{\mu\nu}\,dx^\mu\,dx^\nu$), and the $g_{\mu\nu}$ is a covariant tensor of rank two (by statement 4 of the summary in the preceding section). The tensor $g_{\mu\nu}$ has the following properties:

1. $g_{\mu\nu}$ is symmetric.
2. $g_{\mu\nu}$ is a covariant tensor.
3. $g_{\mu\nu}$ is a tensor of rank two.
4. $g_{\mu\nu}$ is called the "fundamental tensor."
5. $g_{\mu\nu}$ plays a special role in the theory of gravitation.

5.2.8.2   The Contravariant Fundamental Tensor

Section 8 is devoted to developing the properties of the fundamental tensor, $g_{\mu\nu}$. Related to the covariant fundamental tensor $g_{\mu\nu}$ is the inverse of the fundamental tensor, $g^{\mu\nu}$, defined such that

$$g_{\mu\sigma}g^{\nu\sigma} = \delta_\mu^\nu \tag{5.16}$$

δ_μ^ν is the Kronecker delta function, with $\delta_\mu^\nu = 1$ when $\mu = \nu$, and $\delta_\mu^\nu = 0$ when $\mu \neq \nu$. Einstein proceeds to show that $g^{\mu\nu}$ is a contravariant tensor, beginning with the expression for ds^2:

$$ds^2 = g_{\mu\nu}\,dx^\mu\,dx^\nu$$
$$= g_{\mu\sigma}\delta^\sigma_\nu\,dx^\mu dx^\nu$$
$$= g_{\mu\sigma}\left(g_{\nu\tau}g^{\sigma\tau}\right)dx^\mu\,dx^\nu$$
$$= \left(g_{\mu\sigma}\,dx^\mu\right)g^{\sigma\tau}\left(g_{\nu\tau}\,dx^\nu\right)$$
$$= \left(d\xi_\sigma\right)g^{\sigma\tau}\left(d\xi_\tau\right)$$
$$ds^2 = g^{\sigma\tau}d\xi_\sigma d\xi_\tau$$

The $d\xi_\sigma = g_{\mu\sigma}\,dx^\mu$, being the contraction (summation) over μ of the covariant tensor $g_{\mu\sigma}$ and the contravariant tensor dx^μ, is a covariant tensor. Since ds^2 is a scalar, and since $g^{\sigma\tau}$ is symmetric, by result four of the preceding section, $g^{\sigma\tau}$ is a contravariant tensor.

5.2.8.3   The Determinant of the Fundamental Tensor

The determinant of $g = |g_{\mu\nu}|$. Then, $-g$ is always finite and positive and, by the appropriate choice of coordinates, can be made equal to one, i.e., if g is calculated in a system with no gravitational fields (special theory of relativity, see Eq. (5.4)), $-g = +1$. In the general case,

$$|g_{\mu\nu}| \times |g^{\mu\nu}| = +1 \tag{5.17}$$

See Appendix 5.4.2.1 for details.

5.2.8.4   The Volume Scalar

Einstein also shows that the four-dimensional volume elements in the two reference frames, $d\tau$ and $d\tau'$, are related as:

$$\sqrt{-g'}d\tau' = \sqrt{-g}d\tau \tag{5.18}$$

See Appendix 5.4.2.2 for details.

5.2.8.5   Note on the Character of the Space-Time Continuum

It is assumed that within an infinitely small volume $d\tau$, the special theory of relativity can be applied, meaning that ds^2 can be expressed as $ds^2 = -dX_1^2 - dX_2^2 - dX_3^2 + dX_4^2$. The volume element expressed in these coordinates is denoted $d\tau_0$ and is called the "natural" volume element. In these coordinates, $-g = +1$. Thus, $\sqrt{-g_0}d\tau_0 = \sqrt{-g}d\tau \Rightarrow$

$$d\tau_0 = \sqrt{-g}d\tau \tag{5.18a}$$

Einstein proceeds to make a number of observations:

1. If $\sqrt{-g}$ were to vanish at some point in the four-dimensional continuum, "it would mean that at this point an infinitely small 'natural' volume would correspond to a finite volume in the coordinates."[57] In $d\tau_0 = \sqrt{-g}d\tau$, even if $d\tau_0$ is infinitely small, if $g = 0$ then $d\tau$ would not necessarily need to be infinitely small.
2. It will be assumed that never does $g = 0$.

3. If g is never equal to zero, then g can never change sign from negative to positive, or vice versa.
4. Since in the special theory of relativity g has a finite negative value, it will be assumed that g always has a finite negative value (but not necessarily equal to negative one).
5. The value of g always being finite and negative is, in consequence, a hypothesis about the four- dimensional continuum.
6. The value of g always being finite and negative is a convention as to the choice of coordinates.

    a. If g is always finite and negative, it is convenient to select a choice of coordinates such that $-g$ always is equal to positive unity, i.e., $-g = +1$. Such a choice would simplify many of the laws of nature, i.e., simplify the equations describing the laws of nature.
    b. For the case of $g = -1$, Eq. (5.18a) becomes $d\tau' = d\tau$.
    c. For $d\tau' = d\tau$, i.e., for $g = -1$, the Jacobian of the transformation is equal to one. See Appendix 5.4.2.2.

    $$\left| \frac{\partial x'^\sigma}{\partial x^\mu} \right| = +1 \qquad (5.19)$$

    d. The condition $-g = +1$ restricts the choice of possible coordinate systems to those for which the Jacobian of transformation is equal to unity, i.e., $\left| \frac{\partial x'^\sigma}{\partial x^\mu} \right| = +1$.

7. Einstein then addresses the question of how this restriction of $-g = +1$ might restrict the search for the generally covariant laws of nature. The order is first to search for the generally covariant laws of nature. After the generally covariant laws of nature have been found, the second task is to "simplify their expression by a particular choice of the system of reference."[58]

5.2.8.6 The Formation of New Tensors by Means of the Fundamental Tensor

To expedite the development of the theory in later sections, examples of multiplication of various tensors by the fundamental tensor, $g_{\mu\nu}$ or $g^{\mu\sigma}$, are given:

$$A^\mu = g^{\mu\sigma} A_\sigma$$

$$A = g_{\mu\nu} A^{\mu\nu}$$

The "complements" of the covariant and contravariant tensors, $A_{\alpha\beta}$ and $A^{\alpha\beta}$, are defined as

$$A^{\mu\nu} = g^{\mu\alpha} g^{\nu\beta} A_{\alpha\beta}$$

$$A_{\mu\nu} = g_{\mu\alpha} g_{\nu\beta} A^{\alpha\beta}$$

By definition, the complement of $g_{\alpha\beta}$ is

$$g^{\mu\alpha} g^{\nu\beta} g_{\alpha\beta} = g^{\mu\alpha} \left(g^{\nu\beta} g_{\alpha\beta} \right) = g^{\mu\alpha} \delta_\alpha^\nu = g^{\mu\nu}$$

From the covariant tensor $A_{\alpha\beta}$, a reduced tensor associated with $A_{\alpha\beta}$, $B_{\mu\nu}$ is defined as

$$B_{\mu\nu} = g_{\mu\nu}g^{\alpha\beta}A_{\alpha\beta}$$

Similarly, for the contravariant tensor $A^{\alpha\beta}$ a reduced tensor $B^{\mu\nu}$ is defined as

$$B^{\mu\nu} = g^{\mu\nu}g_{\alpha\beta}A^{\alpha\beta}$$

5.2.9   The Equation of the Geodetic Line. The Motion of a Particle

Equation (5.22) of this section is at the center of the description of motion under the influence of the gravitational field in the remainder of the paper. Because of its importance its derivation is given in detail in Appendix 5.4.3.

$ds^2 =$ invariant, meaning the linear element ds is defined independently of the system of coordinates. A geodetic line is the shortest line that can be drawn between two points in that space. For example, in three dimensions, the geodetic line on the surface of a sphere drawn between any two points is the arc of a great circle. The line drawn between two points P and P' in four dimensional space is a geodetic line if

$$\delta \int_{P}^{P'} ds = 0 \tag{5.20}$$

The geodetic line is the physical path followed. Einstein derives the equation of the geodetic line by "[c]arrying out the variation in the usual way."[59] (See Appendices 5.4.3.1 and 5.4.3.2 for details of the calculation.)

$$0 = g_{\mu\sigma}\frac{d^2x^\mu}{ds^2} + \frac{\partial g_{\mu\sigma}}{\partial x^\nu}\frac{dx^\nu}{ds}\frac{dx^\mu}{ds} - \frac{1}{2}\frac{\partial g_{\mu\nu}}{\partial x^\sigma}\frac{dx^\mu}{ds}\frac{dx^\nu}{ds} \tag{5.20e}$$

This equation is modified by using the notation of Christoffel, where Christoffel defines

$$[\mu\nu, \sigma] = \frac{1}{2}\left(\frac{\partial g_{\mu\sigma}}{\partial x^\nu} + \frac{\partial g_{\nu\sigma}}{\partial x^\mu} - \frac{\partial g_{\mu\nu}}{\partial x^\sigma}\right) \tag{5.21}$$

Using Christoffel's notation, the equation of the geodetic line becomes

$$0 = g_{\mu\sigma}\frac{d^2x^\mu}{ds^2} + [\mu\nu, \sigma]\frac{dx^\mu}{ds}\frac{dx^\nu}{ds} \tag{5.20d}$$

Since, in the second term, the summation over the μ and ν indices for the first two terms give the same sum, they have been combined into one common term:

$$[\mu\nu,\sigma]\frac{dx^\mu}{ds}\frac{dx^\nu}{ds} = \left[\frac{1}{2}\frac{\partial g_{\mu\sigma}}{\partial x^\nu} + \frac{1}{2}\frac{\partial g_{\nu\sigma}}{\partial x^\mu} - \frac{1}{2}\frac{\partial g_{\mu\nu}}{\partial x^\sigma}\right]\frac{dx^\mu}{ds}\frac{dx^\nu}{ds}$$

$$[\mu\nu,\sigma]\frac{dx^\mu}{ds}\frac{dx^\nu}{ds} = \left[\frac{\partial g_{\mu\sigma}}{\partial x^\nu} - \frac{1}{2}\frac{\partial g_{\mu\nu}}{\partial x^\sigma}\right]\frac{dx^\mu}{ds}\frac{dx^\nu}{ds}$$

Einstein then introduces a second change of notation "where, following Christoffel,"[60] he sets

$$\{\mu\nu,\tau\} = g^{\tau\alpha}[\mu\nu,\alpha] \tag{5.23}$$

Multiplying Eq. (5.20d) by $g^{\sigma\tau}$, one obtains

$$0 = \frac{d^2x^\tau}{ds^2} + \{\mu\nu,\tau\}\frac{dx^\mu}{ds}\frac{dx^\nu}{ds} \tag{5.22}$$

Equation (5.22) is the form of the geodetic equation Einstein uses in the remainder of the paper, and is the foundation on which much of the subsequent material is based.

5.2.10 The Formation of Tensors by Differentiation

From a tensor, by appropriate definitions of differentiation, other tensors can be obtained. Using these appropriate definitions of differentiation is the first step to obtaining the generally covariant differential equations.

The derivative of a function is defined first along an arbitrary curve, but soon becomes restricted to the curve being the geodetic. Let ϕ be an invariant function of space. (The following distinction between an invariant and a constant should be noted: An invariant has the same value no matter which reference frame is used to measure it, while a constant has the same value at all points in a given space.) ds is the distance along some arbitrary curve (soon to become the geodetic), where $s = s(x^1, x^2, x^3, x^4) = s(x^\mu)$. The derivative of ϕ with respect to s is

$$\frac{d\phi}{ds} = \frac{\partial\phi}{\partial x^\mu}\frac{dx^\mu}{ds} = \psi$$

Since both ϕ and s are invariants, the derivative (the differential quotient) is also an invariant, i.e., ψ is an invariant and, Einstein shows, $A_\mu = \frac{\partial\phi}{\partial x^\mu}$ is a covariant four-vector, the gradient of ϕ (see Appendix 5.4.4).

Since ψ is an invariant on the curve, we can repeat the derivative operation,

$$\chi = \frac{d\psi}{ds} = \frac{d}{ds}\left(\frac{d\phi}{ds}\right) = \frac{d}{ds}\left(\frac{\partial\phi}{\partial x^\mu}\frac{dx^\mu}{ds}\right) = \left(\frac{\partial^2\phi}{\partial x^\mu\partial x^\nu}\frac{dx^\nu}{ds}\right)\frac{dx^\mu}{ds} + \frac{\partial\phi}{\partial x^\mu}\frac{d^2x^\mu}{ds^2}$$

Restricting the curve to the geodetic, and substituting the expression for $\frac{d^2 x^\mu}{ds^2}$ from Eq. (5.22) (see Appendix 5.4.4 for the derivation),

$$\chi = \left(\frac{\partial^2 \phi}{\partial x^\mu \partial x^\nu} - \{\mu\nu, \tau\} \frac{\partial \phi}{\partial x^\tau} \right) \frac{dx^\mu}{ds} \frac{dx^\nu}{ds} = A_{\mu\nu} \frac{dx^\mu}{ds} \frac{dx^\nu}{ds}$$

Since χ is a scalar, and since $\frac{dx_\mu}{ds}$ and $\frac{dx_\nu}{ds}$ are four-vectors with arbitrary components, Einstein shows that the quantity in the parentheses is a covariant vector of rank two. Denoting it as $A_{\mu\nu}$,

$$A_{\mu\nu} = \frac{\partial^2 \phi}{\partial x^\mu \partial x^\nu} - \{\mu\nu, \tau\} \frac{\partial \phi}{\partial x^\tau} \qquad (5.25)$$

From the invariant function ϕ, one can form the invariant tensor of rank one (a four-vector),

$$A_\mu = \frac{\partial \phi}{\partial x^\mu}$$

From this, one can form by differentiation a covariant tensor of rank two,

$$A_{\mu\nu} = \frac{\partial^2 \phi}{\partial x^\mu \partial x^\nu} - \{\mu\nu, \tau\} \frac{\partial \phi}{\partial x^\tau}$$

$$A_{\mu\nu} = \frac{\partial A_\mu}{\partial x^\nu} - \{\mu\nu, \tau\} A_\tau \qquad (5.26)$$

Formed in this manner, $A_{\mu\nu}$ is not simply the derivative of A_μ but is modified to be a covariant tensor. The modified derivative of A_μ is called the "extension" of A_μ, or the covariant derivative of A_μ. These results remain true even if A_μ is not the gradient of ϕ.

Similarly, one can form the extension of a rank-two tensor to give a rank-three tensor:

$$A_{\mu\nu\sigma} = \frac{\partial}{\partial x^\sigma} (A_{\mu\nu}) - \{\sigma\mu, \tau\} A_{\tau\nu} - \{\sigma\nu, \tau\} A_{\mu\tau} \qquad (5.27)$$

In summary, the covariant derivatives, or extensions, are

$$A_\mu = \frac{\partial \phi}{\partial x^\mu} \qquad (5.24)$$

$$A_{\mu\nu} = \frac{\partial A_\mu}{\partial x^\nu} - \{\mu\nu, \tau\} A_\tau \qquad (5.26)$$

$$A_{\mu\nu\sigma} = \frac{\partial (A_{\mu\nu})}{\partial x^\sigma} - \{\sigma\mu, \tau\} A_{\tau\nu} - \{\sigma\nu, \tau\} A_{\mu\tau} \qquad (5.27)$$

5.2.11   Some Cases of Special Importance

In section 11 of the paper, Einstein obtains a number of relations that will be of use later.

The fundamental tensor $g_{\mu\nu}$:

From Eq. (5.16), $g_{\mu\sigma}g^{\nu\sigma} = \delta^{\nu}_{\mu}$ which, for the case $\mu = \nu$ becomes

$$g_{\mu\sigma}g^{\mu\sigma} = \delta^{\mu}_{\mu} = \delta^1_1 + \delta^2_2 + \delta^3_3 + \delta^4_4 = 4$$
$$\Rightarrow d\left(g_{\mu\nu}g^{\mu\nu}\right) = \left(dg_{\mu\nu}\right)g^{\mu\nu} + g_{\mu\nu}\left(dg^{\mu\nu}\right) = d\left(4\right) = 0$$
$$\Rightarrow g^{\mu\nu}\left(dg_{\mu\nu}\right) = -g_{\mu\nu}\left(dg^{\mu\nu}\right)$$

From the definition of the differentiation of a determinant (see Appendix 5.4.5.1),

$$dg = g^{\mu\nu}g\,dg_{\mu\nu} = -g_{\mu\nu}g\,dg^{\mu\nu} \tag{5.28}$$

Consider the quantity $\frac{1}{\sqrt{-g}}\frac{\partial\sqrt{-g}}{\partial x^{\sigma}}$. For future convenience, a number of forms of this expression are obtained (see Appendix 5.4.5.1):

$$\frac{1}{\sqrt{-g}}\frac{\partial\sqrt{-g}}{\partial x^{\sigma}} = \frac{1}{2}\frac{\partial\log\left(-g\right)}{\partial x^{\sigma}} = \frac{1}{2}g^{\mu\nu}\frac{\partial g_{\mu\nu}}{\partial x^{\sigma}} = -\frac{1}{2}g_{\mu\nu}\frac{\partial g^{\mu\nu}}{\partial x^{\sigma}} \tag{5.29}$$

From $g_{\mu\sigma}g^{\nu\sigma} = \delta^{\nu}_{\mu}$, upon differentiation one obtains

$$g_{\mu\sigma}g^{\nu\sigma} = \delta^{\nu}_{\mu} \Rightarrow g_{\mu\sigma}dg^{\nu\sigma} = -g^{\nu\sigma}dg_{\mu\sigma} \tag{5.30a}$$

$$\Rightarrow g_{\mu\sigma}\frac{\partial g^{\nu\sigma}}{\partial x^{\lambda}} = -g^{\nu\sigma}\frac{\partial g_{\mu\sigma}}{\partial x^{\lambda}} \tag{5.30b}$$

Multiplying (5.30a) by $g^{\mu\tau}$, and relabeling the indices, $\nu \to \mu, \tau \to \nu, \sigma \to \alpha, \mu \to \beta$,

$$dg^{\mu\nu} = -g^{\mu\alpha}g^{\nu\beta}dg_{\alpha\beta} \tag{5.31a}$$

$$\frac{\partial g^{\mu\nu}}{\partial x^{\sigma}} = -g^{\beta\nu}g^{\mu\alpha}\frac{\partial g_{\beta\alpha}}{\partial x^{\sigma}} \tag{5.31b}$$

In a similar manner, multiplying Eq. (5.30b) by $g_{\nu\tau}$, and relabeling the indices as $\{\mu \to \nu, \tau \to \mu, \nu \to \alpha, \sigma \to \beta, \lambda \to \sigma\}$, one obtains

$$dg_{\nu\mu} = -g_{\nu\beta}g_{\alpha\mu}dg^{\alpha\beta} \tag{5.32a}$$

$$\frac{\partial g_{\nu\mu}}{\partial x^{\sigma}} = -g_{\nu\beta}g_{\alpha\mu}\frac{\partial g^{\alpha\beta}}{\partial x^{\sigma}} \tag{5.32b}$$

Einstein proceeds to put Eqs. (5.31) and (5.32) into an alternative form that will prove more useful later in the paper. From Eq. (5.21), the definition of the Christoffel symbol,

$$[\mu\nu, \sigma] = \frac{1}{2}\left(\frac{\partial g_{\mu\sigma}}{\partial x^{\nu}} + \frac{\partial g_{\nu\sigma}}{\partial x^{\mu}} - \frac{\partial g_{\mu\nu}}{\partial x^{\sigma}}\right)$$

Adding together the following two terms,

$$[\alpha\sigma, \beta] + [\beta\sigma, \alpha] = \frac{1}{2}\left(\frac{\partial g_{\alpha\beta}}{\partial x^{\sigma}} + \frac{\partial g_{\sigma\beta}}{\partial x^{\alpha}} - \frac{\partial g_{\alpha\sigma}}{\partial x^{\beta}}\right) + \frac{1}{2}\left(\frac{\partial g_{\beta\alpha}}{\partial x^{\sigma}} + \frac{\partial g_{\sigma\alpha}}{\partial x^{\beta}} - \frac{\partial g_{\beta\sigma}}{\partial x^{\alpha}}\right)$$

$$= \frac{\partial g_{\alpha\beta}}{\partial x^{\sigma}} \tag{5.33}$$

Substituting the expression $\frac{\partial g_{\alpha\beta}}{\partial x^\sigma} = [\alpha\sigma, \beta] + [\beta\sigma, \alpha]$ into Eq. (5.31b),

$$\frac{\partial g^{\mu\nu}}{\partial x^\sigma} = -g^{\mu\alpha}g^{\nu\beta}\left([\alpha\sigma, \beta] + [\beta\sigma, \alpha]\right) = -g^{\mu\alpha}\left(g^{\nu\beta}[\alpha\sigma, \beta]\right) - g^{\nu\beta}\left(g^{\mu\alpha}[\beta\sigma, \alpha]\right)$$

$$\frac{\partial g^{\mu\nu}}{\partial x^\sigma} = -g^{\mu\alpha}\{\alpha\sigma, \nu\} - g^{\nu\beta}\{\beta\sigma, \mu\}$$

Since the indices α and β are dummy variables over which one sums, they can each be replaced by another dummy variable, τ, giving

$$\frac{\partial g^{\mu\nu}}{\partial x^\sigma} = -g^{\mu\tau}\{\tau\sigma, \nu\} - g^{\nu\tau}\{\tau\sigma, \mu\} \tag{5.34}$$

Substituting Eq. (5.34) into Eq. (5.29), one obtains

$$\frac{1}{\sqrt{-g}}\frac{\partial\sqrt{-g}}{\partial x^\sigma} = \{\mu\sigma, \mu\} \tag{5.29a}$$

The "Divergence" of a contravariant vector:

As the divergence of a contravariant vector is included for completeness and is not used in the subsequent development of the material, the result is simply stated (see Appendix 5.4.5.2). The scalar Φ is the divergence of the contravariant vector A^ν:

$$\Phi = g^{\mu\nu}A_{\mu\nu} = \frac{1}{\sqrt{-g}}\frac{\partial}{\partial x^\nu}\left(\sqrt{-g}A^\nu\right) \tag{5.35}$$

The "Curl" of a covariant vector:

From Eq. (5.26), $A_{\mu\nu} = \frac{\partial A_\mu}{\partial x^\nu} - \{\mu\nu, \tau\}A_\tau$. Since $\{\mu\nu, \tau\}$ is symmetrical in the indices μ and ν, the difference of $A_{\mu\nu}$ and $A_{\nu\mu}$ is an antisymmetric tensor (see Appendix 5.4.5.3):

$$B_{\mu\nu} = A_{\mu\nu} - A_{\nu\mu} = \frac{\partial A_\mu}{\partial x^\nu} - \frac{\partial A_\nu}{\partial x^\mu} \tag{5.36}$$

Antisymmetrical extension of a six-vector:

An antisymmetrical tensor of rank two has six independent components and is referred to as a six-vector. Applying Eq. (5.27) to an antisymmetrical tensor of rank two, $A_{\mu\nu}$, yields a tensor of rank three, $A_{\mu\nu\sigma}$ (see Appendix 5.4.5.4). Adding to $A_{\mu\nu\sigma}$ its cyclic permutations gives

$$B_{\mu\nu\sigma} = A_{\mu\nu\sigma} + A_{\nu\sigma\mu} + A_{\sigma\mu\nu} = \frac{\partial A_{\mu\nu}}{\partial x^\sigma} + \frac{\partial A_{\nu\sigma}}{\partial x^\mu} + \frac{\partial A_{\sigma\mu}}{\partial x^\nu} \tag{5.37}$$

The divergence of a six-vector:

Multiplying Eq. (5.27) by $g^{\mu\alpha}g^{\nu\beta}$, and after substantial manipulation of the equations, the divergence of a contravariant six-vector is obtained (see Appendix 5.4.5.5.):

$$A^\alpha = \frac{\partial A^{\alpha\beta}}{\partial x^\beta} + A^{\alpha\beta}\frac{1}{\sqrt{-g}}\frac{\partial\sqrt{-g}}{\partial x^\beta} = \frac{1}{\sqrt{-g}}\frac{\partial}{\partial x^\beta}\left(A^{\alpha\beta}\sqrt{-g}\right) \tag{5.40}$$

The divergence of a mixed tensor of second rank:

The divergence of a mixed tensor of rank two, A_τ^σ, is shown to be (see Appendix 5.4.5.6):

$$\sqrt{-g}A_\mu = \frac{\partial}{\partial x^\sigma}\left(\sqrt{-g}A_\mu^\sigma\right) - \{\sigma\mu, \tau\}\sqrt{-g}A_\tau^\sigma \qquad (5.41)$$

The second term on the right-hand side can be rewritten as

$$-\{\sigma\mu, \tau\}\sqrt{-g}A_\tau^\sigma = -[\sigma\mu, \rho]g^{\rho\tau}\sqrt{-g}A_\tau^\sigma = -[\sigma\mu, \rho]\sqrt{-g}A^{\rho\sigma}$$

If $A^{\sigma\rho}$ is symmetrical, Eq. (5.41) reduces to

$$\sqrt{-g}A_\mu = \frac{\partial}{\partial x^\sigma}\left(\sqrt{-g}A_\mu^\sigma\right) - \frac{1}{2}\sqrt{-g}\frac{\partial g_{\sigma\rho}}{\partial x^\mu}A^{\sigma\rho} \qquad (5.41a)$$

In a similar manner, using Eq. (5.32) to replace the term $\frac{\partial g_{\rho\sigma}}{\partial x^\mu}$, Eq. (5.41) becomes

$$\sqrt{-g}A_\mu = \frac{\partial}{\partial x^\sigma}\left(\sqrt{-g}A_\mu^\sigma\right) + \frac{1}{2}\frac{\partial g^{\rho\sigma}}{\partial x^\mu}\sqrt{-g}A_{\rho\sigma} \qquad (5.41b)$$

5.2.12 The Riemann–Christoffel Tensor

In this section of the paper, Einstein completes his search for a "tensor that can be obtained from the fundamental tensor *alone*, by differentiation."[61] He first tries placing the fundamental tensor $g_{\mu\nu}$ into Eq. (5.27) to obtain the extension of $g_{\mu\nu}$. However, because of the symmetry of $g_{\mu\nu}$, the extension vanishes identically (see Appendix 5.4.6).

As a second approach, into Eq. (5.27) place the extension of the four-vector A_μ. From Eq. (5.26), the extension of A_μ is $A_{\mu\nu} = \frac{\partial A_\mu}{\partial x^\nu} - \{\mu\nu, \rho\}A_\rho$. Placing this into Eq. (5.27), after considerable manipulation one obtains (see Appendix 5.4.6 for details of this derivation):

$$A_{\mu\sigma\tau} = \frac{\partial^2 A_\mu}{\partial x^\sigma \partial x^\tau} - \{\mu\sigma, \rho\}\frac{\partial A_\rho}{\partial x^\tau} - \{\mu\tau, \rho\}\frac{\partial A_\rho}{\partial x^\sigma} - \{\tau\sigma, \alpha\}\frac{\partial A_\rho}{\partial x^\alpha}$$
$$+ \left[-\frac{\partial\{\mu\sigma, \rho\}}{\partial x^\tau} + \{\mu\tau, \alpha\}\{\alpha\sigma, \rho\} + \{\sigma\tau, \alpha\}\{\alpha\mu, \rho\}\right]A_\rho$$

Because of the symmetry in a number of the terms, it is convenient to form the difference of this expression with the same expression, but with the σ and τ indices reversed:

$$A_{\mu\sigma\tau} - A_{\mu\tau\sigma} = R_{\mu\sigma\tau}^\rho A_\rho \qquad (5.42)$$

where $R_{\mu\sigma\tau}^\rho$ is identified as the Riemann–Christoffel tensor:

$$R_{\mu\sigma\tau}^\rho = -\frac{\partial\{\mu\sigma, \rho\}}{\partial x^\tau} + \frac{\partial\{\mu\tau, \rho\}}{\partial x^\sigma} - \{\mu\sigma, \alpha\}\{\alpha\tau, \rho\} + \{\mu\tau, \alpha\}\{\alpha\sigma, \rho\}$$
$$(5.43)$$

The tensor $R_{\mu\sigma\tau}^\rho$ is composed of a number of terms that are derivatives of the $g_{\mu\nu}$.

1. If there is a coordinate system in which all of the $g_{\mu\nu}$ are constant, all of the components of the Riemann–Christoffel tensor vanish, $R_{\mu\sigma\tau}^\rho = 0$.

2. In a new coordinate system, the $g_{\mu\nu}$ may not be constants.
3. But, since $R^\rho_{\mu\sigma\tau}$ is a tensor, the transformed components in the new coordinate system must still be zero.
4. Conversely, the vanishing of the Riemann tensor is a necessary condition that, by an appropriate choice of the coordinate system, the $g_{\mu\nu}$ may be constants.
5. This corresponds physically to the case in which, for a suitable choice of coordinate system, the special theory of relativity holds good for a finite region of the space.

$R^\rho_{\mu\sigma\tau}$ is a tensor of rank four. Contracting with respect to the indices ρ and τ (set $\rho = \tau$ and sum over the repeated index), one obtains a covariant tensor of rank two:

$$G_{\mu\nu} = R^\rho_{\mu\nu\rho} = R_{\mu\nu} + S_{\mu\nu}$$

$$R_{\mu\nu} = -\frac{\partial \{\mu\nu, \alpha\}}{\partial x^\alpha} + \{\mu\alpha, \beta\}\{\nu\beta, \alpha\}$$

with

$$S_{\mu\nu} = \frac{\partial^2 (\log \sqrt{-g})}{\partial x^\mu \partial x^\nu} - \{\mu\nu, \alpha\}\frac{\partial (\log \sqrt{-g})}{\partial x^\alpha} \tag{5.44}$$

It was noted earlier certain conveniences arise if one sets $\sqrt{-g} = 1$. That convenience continues. For $\sqrt{-g} = 1$, $S_{\mu\nu} = 0$. For the rest of the paper the equations obtained will be for the particular choice of coordinates for which $\sqrt{-g} = 1$.

Part C: "Theory of the Gravitational Field"

After devoting roughly ten pages to background material and twice as many to developing the mathematics he would be needing, in Part C it takes Einstein only about ten pages to obtain the theory of (and equations representing) the gravitational field. The remainder of the paper is then devoted to applications of the results, showing they contain Maxwell's equations and Newton's theory of gravitation, and that they predict the bending of starlight and the precession of the perihelion of the planet Mercury.

Einstein introduces the notation $\Gamma^\tau_{\mu\nu} = -\{\mu\nu, \tau\}$ and identifies the $\Gamma^\tau_{\mu\nu}$ as the components of the gravitational field. Obtaining first the equation of the fields in matter-free space, Einstein then generalizes this result to obtain the general field equations of gravitation.

5.2.13   Equations of Motion of a Material Point in the Gravitational Field. Expression for the Field-Components of Gravitation

Consider a material point not subject to any external forces. This is described by the special theory of relativity, the object moving in a straight line with constant speed. Designate this reference frame as K_0. In K_0, the $g_{\mu\nu}$ have the constant values $0, +1, -1$ as given in Eq. (5.4).

Consider a second reference frame K_1. Reference frame K_1 is accelerating relative to reference frame K_0. From K_1, the particle will be seen as moving in a gravitational field. In K_0, the motion is a four-dimensional straight line, i.e., a geodetic line. From Eq. (5.22), the equation of the geodetic line is

$$\frac{d^2 x^\tau}{ds^2} = -\{\mu\nu, \tau\} \frac{dx^\mu}{ds} \frac{dx^\nu}{ds}$$

Since the geodetic line is defined independently of the reference system, the equation of motion of the material point in K_1 will be the same. Defining

$$\Gamma^\tau_{\mu\nu} = -\{\mu\nu, \tau\} \tag{5.45}$$

$$\frac{d^2 x^\tau}{ds^2} = \Gamma^\tau_{\mu\nu} \frac{dx^\mu}{ds} \frac{dx^\nu}{ds} \tag{5.46}$$

This result is true, based on the existence of the geodetic in a reference frame K_0 where the special theory of relativity holds in a finite region. Einstein then assumes the above system of equations holds also in the case in which there is no K_0.

If the $\Gamma^\tau_{\mu\nu}$ vanish ($=0$) the motion is uniform in a straight line, i.e., there are no effects of gravity. The $\Gamma^\tau_{\mu\nu}$ thus give the deviations from uniform motion in a straight line. As such, the $\Gamma^\tau_{\mu\nu}$ are identified as the components of the gravitational field.

5.2.14 The Field Equations of Gravitation in the Absence of Matter

The concept of matter in this section is very generalized. Einstein turns his attention to the equations for the gravitational field, and then defines all things not gravitational field to be "matter," i.e., "matter" is not only things we normally consider to be matter but also things such as the electromagnetic field. In this section, the field equations are derived for the simpler case of no matter present. (The equations are derived for a region where there is no matter present, although gravitational fields may be present because of matter outside the region. In Section 5.2.16 the results obtained in this section are generalized to the case with matter present.)

Consider first the reference frame K_0, in which the special theory is satisfied. In K_0, all of the $g_{\mu\nu}$ have constant values $(0, +1, -1)$. Since all of the $g_{\mu\nu}$ have constant values, in K_0 the Riemann tensor, being defined in terms of the derivatives of the $g_{\mu\nu}$ (see Eq. (5.43)), $R^\rho_{\mu\sigma\nu} \equiv B^\rho_{\mu\sigma\nu} = 0$. Since the components of the Riemann tensor vanish for the finite space under the reference frame K_0, they vanish also under any other system of coordinates (see introduction to Part B: "the equations of transformation of [the components of tensors] are...homogeneous. Accordingly, all the components in the new system vanish, if they all vanish in the original system."[62]).[63] Thus, if there exists a reference frame K_0, the equations for the matter-free gravitational field must be satisfied if the components of the Riemann tensor vanish ($K_0 \Rightarrow R^\rho_{\mu\sigma\nu} = 0$). However, Einstein notes, a gravitational field generated by a material point has a spherical symmetry that cannot be "transformed away" by moving to a single accelerating reference frame K_1 ($R^\rho_{\mu\sigma\nu} = 0$ does not $\Rightarrow K_0$ exists).

Einstein concludes the condition that the Riemann tensor vanish is too strong. He then requires only that the symmetrical tensor, $G_{\mu\nu} = R^\rho_{\mu\nu\rho}$, $G_{\mu\nu} = 0$. By inspection of Eq. (5.44), it is seen that $G_{\mu\nu}$ is symmetric in the exchange of the indices $\mu \leftrightarrow \nu$. Thus $G_{\mu\nu}$ is a symmetric 4×4 tensor. Being symmetric, only ten of the sixteen components are independent. From Eq. (5.44), $G_{\mu\nu} = R_{\mu\nu} + S_{\mu\nu}$. Setting $\sqrt{-g} = +1$, $S_{\mu\nu} = 0$, and $G_{\mu\nu}$ becomes

$$G_{\mu\nu} = R_{\mu\nu} = 0 = -\frac{\partial}{\partial x^\alpha}\{\mu\nu, \alpha\} + \{\mu\alpha, \beta\}\{\nu\beta, \alpha\}$$

$$0 = +\frac{\partial}{\partial x^\alpha}\Gamma^\alpha_{\mu\nu} + \Gamma^\alpha_{\mu\beta}\Gamma^\beta_{\nu\alpha} \tag{5.47}$$

with $\sqrt{-g} = +1$

In the second equation of Eq. (5.47), the $\Gamma^\alpha_{\mu\nu}$ follows immediately from its definition in Eq. (5.45). In the second term of the second equation, $\Gamma^\alpha_{\mu\beta}\Gamma^\beta_{\nu\alpha}$, the subscripts α and β appear to be reversed from the first equation. This appearance of a reversal is correct, but immaterial, as the α and β indices are summed over in the term and, as such, are "dummy" variables.

Einstein at this point comments, "These equations, which proceed, by the method of pure mathematics, from the requirement of the general theory of relativity, give us, in combination with the equations of motion [Eq. (5.46)] to a first approximation Newton's law of attraction, and to a second approximation the explanation of the motion of the perihelion of the planet Mercury... These facts must, in my opinion, be taken as convincing proof of the correctness of the theory."[64]

5.2.15   The Hamiltonian Function for the Gravitational Field. Laws of Momentum and Energy

In this section, Einstein obtains three alternative expressions of Eq. (5.47).

The first of the three alternative expressions is written in the Hamiltonian form, Eq. (5.47a):

$$\delta \int H d\tau = 0$$

$$H = g^{\mu\nu}\Gamma^\alpha_{\mu\beta}\Gamma^\beta_{\nu\alpha} \tag{5.47a}$$

$$\sqrt{-g} = 1$$

Einstein stipulates that "on the boundary of the finite four-dimensional region of integration which we have in view, the variations vanish."[65] H is a function of the $g^{\mu\nu}$ and their derivatives, $g^{\mu\nu}_\alpha = \frac{\partial g^{\mu\nu}}{\partial x^\alpha}$, $H = H(g^{\mu\nu}, g^{\mu\nu}_\alpha, x^\alpha)$. The variation of the integral $\int H d\tau$ gives the four-dimensional version of Lagrange's equations for the function H. (See Appendix 5.4.7.1 for the details showing the equivalence of Eqs. (5.47) and (5.47a), and for the derivation of Eqs. (5.48) and (5.47b).)

The variation of the second of the Eqs. (5.47a), $H = g^{\mu\nu}\Gamma^\alpha_{\mu\beta}\Gamma^\beta_{\nu\alpha}$, gives

$$\delta H = \delta\left(g^{\mu\nu}\Gamma^\alpha_{\mu\beta}\Gamma^\beta_{\nu\alpha}\right) = \ldots = -\Gamma^\alpha_{\mu\beta}\Gamma^\beta_{\nu\alpha}\delta g^{\mu\nu} + \Gamma^\alpha_{\mu\beta}\delta\left(g^{\mu\beta}_\alpha\right)$$

$$\Rightarrow \quad \begin{aligned} \frac{\partial H}{\partial g^{\mu\nu}} &= -\Gamma^\alpha_{\mu\beta}\Gamma^\beta_{\nu\alpha} \\ \frac{\partial H}{\partial g^{\mu\nu}_\sigma} &= +\Gamma^\sigma_{\mu\nu} \end{aligned} \tag{5.48}$$

Carrying out the variation of Eq. (5.47a), one obtains

$$\frac{\partial}{\partial x^\alpha}\left(\frac{\partial H}{\partial g^{\mu\nu}_\alpha}\right) - \frac{\partial H}{\partial g^{\mu\nu}} = 0 \tag{5.47b}$$

An alternate form of Eq. (5.47b) is then obtained

$$\frac{\partial}{\partial x^\alpha}\left(g^{\mu\nu}_\sigma\frac{\partial H}{\partial g^{\mu\nu}_\alpha} - \delta^\alpha_\sigma H\right) = \frac{\partial}{\partial x^\alpha}\left(t^\alpha_\sigma\right) = 0 \tag{5.49}$$

Equation (5.49a) is the second alternative form of Eq. (5.47). For future convenience, the definition of t^α_σ includes a constant multiplicative constant, -2κ :

$$-2\kappa t^\alpha_\sigma = g^{\mu\nu}_\sigma\frac{\partial H}{\partial g^{\mu\nu}_\alpha} - \delta^\alpha_\sigma H \tag{5.49a}$$

And the expression for t^α_σ is manipulated to yield (see Appendix 5.4.7.2 for details)

$$\kappa t^\alpha_\sigma = \frac{1}{2}\delta^\alpha_\sigma g^{\mu\nu}\Gamma^\lambda_{\mu\beta}\Gamma^\beta_{\nu\lambda} - g^{\mu\nu}\Gamma^\alpha_{\mu\beta}\Gamma^\beta_{\nu\sigma} \tag{5.50}$$

t^α_σ is identified as the energy tensor, with Eq. (5.49a) expressing the law of conservation of momentum and of energy for the gravitational field. Einstein notes that t^α_σ "has tensorial character only under linear transformations." These results are contained in Einstein's November 4, 1915, report to the Prussian Academy of Sciences, "On the General Theory of Relativity."[66] In Eq. (5.20) of this reference, $T^\lambda_\sigma = 0$ since there is no matter present.

To arrive at the third alternative form of Eq. (5.47), Einstein multiplies Eq. (5.47) by $g^{\nu\sigma}$ and manipulates the resulting equations to obtain (see Appendix 5.4.7.3 for details)

$$\frac{\partial}{\partial x^\alpha}\left(g^{\nu\sigma}\Gamma^\alpha_{\mu\nu}\right) = -\kappa\left(t^\sigma_\mu - \frac{1}{2}\delta^\sigma_\mu t\right) \tag{5.51a}$$

$$\sqrt{-g} = 1 \tag{5.51b}$$

with $t = t^\alpha_\alpha$

5.2.16   The General Form of the Field Equations of Gravitation

The primary field equations for the gravitational field in the absence of matter are Eq. (5.47):

$$0 = +\frac{\partial}{\partial x^\alpha}\Gamma^\alpha_{\mu\nu} + \Gamma^\alpha_{\mu\beta}\Gamma^\beta_{\nu\alpha} \tag{5.47}$$

with $\sqrt{-g} = +1$

Three alternative expressions of Eq. (5.47) are obtained as Eqs. (5.47a), (5.49), and (5.51):

$$\left\{ \begin{array}{l} \delta \int H d\tau = 0 \\ H = g^{\mu\nu}\Gamma^\alpha_{\mu\beta}\Gamma^\beta_{\nu\alpha} \\ \sqrt{-g} = 1 \end{array} \right\} \tag{5.47a}$$

$$\left\{ \begin{array}{l} \dfrac{\partial}{\partial x^\alpha}\left(t^\alpha_\sigma\right) = 0 \\ -2\kappa t^\alpha_\sigma = g^{\mu\nu}_\sigma \dfrac{\partial H}{\partial g^{\mu\nu}_\alpha} - \delta^\alpha_\sigma H \\ \kappa t^\alpha_\sigma = \dfrac{1}{2}\delta^\alpha_\sigma g^{\mu\nu}\Gamma^\lambda_{\mu\beta}\Gamma^\beta_{\nu\lambda} - g^{\mu\nu}\Gamma^\alpha_{\mu\beta}\Gamma^\beta_{\nu\sigma} \end{array} \right\} \tag{5.49}$$

$$\left\{ \begin{array}{l} \dfrac{\partial}{\partial x^\alpha}\left(g^{\nu\sigma}\Gamma^\alpha_{\mu\nu}\right) = -\kappa\left(t^\sigma_\mu - \tfrac{1}{2}\delta^\sigma_\mu t\right) \\ \sqrt{-g} = 1 \end{array} \right\} \tag{5.51}$$

Newton's gravitational equation for matter-free space is Laplace's equation:

$$\nabla^2\phi = 0$$

while Newton's gravitational equation with matter present is Poisson's equation:

$$\nabla^2\phi = 4\pi\kappa\rho$$

where ρ denotes the density of matter.

Einstein's field equations of gravitation in matter-free space correspond to the case of Newton's gravitational equation in matter-free space. In this section of the paper, Einstein generalizes the matter-free equations of gravitation (the form given in Eq. (5.51)) to include contributions from matter. The development of the general form proceeds in a "logical discussion" of the factors involved, rather than in a mathematical derivation.

In the special theory of relativity it had been shown that "inert mass is nothing more or less than energy, which finds its complete mathematical expression in a symmetrical tensor of second rank, the energy-tensor."[67] Discussion of this and development of the energy tensor are contained in "*Einstein's 1912 Manuscript on Special Relativity*", sections 20 and 21.[68] In the general theory of relativity, he introduces a corresponding energy tensor of matter, designated T^σ_μ. This energy tensor corresponds to the density of matter ρ in Poisson's equation above.

Equation (5.51) indicates how this energy tensor is to be introduced. In Eq. (5.51), the effect of the gravitational energy is represented by the term t^σ_μ. Einstein then considered a system composed of gravitational fields plus mass, e.g., the solar system. The total gravitating action will depend on the total energy of the system: the ponderable energy, i.e., the mass, along with the gravitational energy. In Eq. (5.51), the term t^σ_μ will be replaced by the total energy-tensor, the sum of the matter and gravitational terms, T^σ_μ plus t^σ_μ. Changing the index $\nu \to \beta$, with matter present Eq. (5.51) becomes (replacing t^σ_μ with $t^\sigma_\mu + T^\sigma_\mu$)

$$\left\{ \begin{array}{l} \frac{\partial}{\partial x^\alpha} \left(g^{\beta\sigma} \Gamma^\alpha_{\mu\beta} \right) = -\kappa \left[\left(t^\sigma_\mu + T^\sigma_\mu \right) - \frac{1}{2} \delta^\sigma_\mu \left(t + T \right) \right] \\ \sqrt{-g} = 1 \end{array} \right\} \tag{5.52}$$

T, in analogy to $t = t^\alpha_\alpha$, is defined as $T = T^\alpha_\alpha$. This is termed the "Laue scalar" and is discussed in the 1913 "Entwurf" paper.[69] Just as Eq. (5.51) was obtained from Eq. (5.47), Eq. (5.52) will need to be obtained from a modified Eq. (5.47). Working backward from Eq. (5.52) to find the modified Eq. (5.47):

Previously, from Eq. (5.47) (matter-free field) we obtained Eq. (5.51),

$$\left\{ \begin{array}{l} \frac{\partial}{\partial x^\alpha} \Gamma^\alpha_{\mu\nu} + \Gamma^\alpha_{\mu\beta} \Gamma^\beta_{\nu\alpha} = 0 \\ \sqrt{-g} = 1 \end{array} \right\} \Rightarrow \frac{\partial}{\partial x^\alpha} \left(g^{\sigma\beta} \Gamma^\alpha_{\mu\beta} \right) = -\kappa \left(t^\sigma_\mu - \frac{1}{2} \delta^\sigma_\mu t \right)$$

$$\tag{5.51}$$

To obtain Eq. (5.52), Eq. (5.47) must be modified to add the terms $-\kappa \left(T^\sigma_\mu - \frac{1}{2} \delta^\sigma_\mu T \right)$ to Eq. (5.51). Noting Eq. (5.51) is a tensor equation of the form $A_{\mu\nu}$, two terms of this form are added to Eq. (5.47), $A_{\mu\nu} + B_{\mu\nu}$, with the anticipation the term $A_{\mu\nu}$ will yield the term T^σ_μ and the term $B_{\mu\nu}$ will yield the term $-\frac{1}{2} \delta^\sigma_\mu T$.

From Eq. (5.47),

$$\frac{\partial}{\partial x^\alpha} \Gamma^\alpha_{\mu\nu} + \Gamma^\alpha_{\mu\beta} \Gamma^\beta_{\nu\alpha} = 0 \to \frac{\partial}{\partial x^\alpha} \Gamma^\alpha_{\mu\nu} + \Gamma^\alpha_{\mu\beta} \Gamma^\beta_{\nu\alpha} = A_{\mu\nu} + B_{\mu\nu}$$

Equation (5.47) was multiplied by $g^{\nu\sigma}$ and manipulated to arrive at Eq. (5.51). Multiplying the above equation by $g^{\nu\sigma}$, the LHS gives the same result (Eq. (5.51)):

$$g^{\nu\sigma} \left(\frac{\partial \Gamma^\alpha_{\mu\nu}}{\partial x^\alpha} + \Gamma^\alpha_{\mu\beta} \Gamma^\beta_{\nu\alpha} \right) = g^{\nu\sigma} A_{\mu\nu} + g^{\nu\sigma} B_{\mu\nu}$$

$$\Rightarrow \frac{\partial}{\partial x^\alpha} \left(g^{\nu\sigma} \Gamma^\alpha_{\mu\nu} \right) + \kappa \left(t^\sigma_\mu - \frac{1}{2} \delta^\sigma_\mu \right) = g^{\nu\sigma} A_{\mu\nu} + g^{\nu\sigma} B_{\mu\nu}$$

To arrive at the desired result, Eq. (5.52), $g^{\nu\sigma} A_{\mu\nu} = -\kappa T^\sigma_\mu$ and $g^{\nu\sigma} B_{\mu\nu} = +\kappa \frac{1}{2} \delta^\sigma_\mu T$. We identify $A_{\mu\nu} = -\kappa T_{\mu\nu}$, with $g^{\nu\sigma} T_{\mu\nu} \equiv T^\sigma_\mu$, and $B_{\mu\nu} = \kappa g_{\mu\nu} \frac{1}{2} T$. Thus, to arrive at Eq. (5.52), Eq. (5.47) is modified to

$$\left\{ \begin{array}{l} \frac{\partial \Gamma^\alpha_{\mu\nu}}{\partial x^\alpha} + \Gamma^\alpha_{\mu\beta} \Gamma^\beta_{\nu\alpha} = -\kappa \left(T_{\mu\nu} - \frac{1}{2} g_{\mu\nu} T \right) \\ \sqrt{-g} = 1 \end{array} \right\} \tag{5.53}$$

Einstein comments that this result is justified by:[70]

1. The "energy of the gravitational field shall act gravitatively in the same way as any other kind of energy."
2. The "strongest reason for the choice of these equations lies in their consequence, that the equations of conservation of momentum and energy, corresponding exactly to equations (49) and (49a), hold good for the components of the total energy." This is shown in the following Section 5.2.17.

5.2.17   The Laws of Conservation in the General Case

From Eq. (5.52), Einstein manipulates the equations, obtaining Eq. (5.56):

$$\frac{\partial \left(t_\mu^\sigma + T_\mu^\sigma \right)}{\partial x^\sigma} = 0 \tag{5.56}$$

Analogous to Eqs. (5.49) and (5.49a), Eq. (5.56) shows that the laws of conservation of momentum and energy are satisfied, except in Eq. (5.56) the energy components are those of the total energy, not just the energy components of the gravitational energy (for details of the derivation of Eq. (5.56), see Appendix 5.4.7.4):

1. From the field equations of gravitation the laws of conservation of momentum and energy are satisfied.
2. Instead of the energy components t_μ^σ of the gravitational field in the absence of matter (Eq. (5.49a)), Eq. (5.56) shows the conservation laws are satisfied for the totality of the energy components of matter plus the gravitational field.

(Today these same results can be obtained more easily using the Bianchi identity.[71])

5.2.18   The Laws of Momentum and Energy for Matter, as a Consequence of the Field Equations

Starting with Eq. (5.53), the general field equations of gravitation, Einstein manipulates the equations to obtain Eq. (5.57) (see Appendix 5.4.7.5 for details of the derivation):

$$\frac{\partial T_\sigma^\alpha}{\partial x^\alpha} + \frac{1}{2}\frac{\partial g^{\mu\nu}}{\partial x^\sigma} T_{\mu\nu} = 0 \tag{5.57}$$

With our choice of $\sqrt{-g} = 1$, in accord with Eq. (5.41b), Eq. (5.57) is the divergence of T_σ^α. Einstein comments that Eq. (5.57) "predicates nothing more or less than the vanishing of divergence of the material energy-tensor."[72] Examining Eq. (5.57) further, the second term,

$\frac{1}{2}\frac{\partial g^{\mu\nu}}{\partial x^\sigma}T_{\mu\nu}$, indicates the laws of conservation of momentum and energy do not apply strictly for matter energy alone. The second term represents the momentum and energy transferred from the gravitational field to matter. The transfer of momentum and energy from the gravitational field is more apparent if Eq. (5.57) is rewritten as (see Appendix 5.4.7.5 for details of the derivation):

$$\frac{\partial T_\sigma^\alpha}{\partial x^\alpha} = -\Gamma_{\alpha\sigma}^\beta T_\beta^\alpha \qquad (5.57a)$$

Part D: "Material Phenomena"

The mathematical structure of the general theory of relativity is a complete and coherent structure. The theory is now applied to the description of some physical phenomena. In Part D, it is shown how hydrodynamics and Maxwell's electromagnetic theory are fitted exactly into the general theory of relativity, while in Part E, Einstein uses approximations to the equations of the general theory of relativity to obtain expressions for Newton's law of gravitation, the gravitational redshift, the bending of a light beam in a gravitational field, and the precession of the perihelion of the orbit of Mercury.

5.2.19 Euler's Equations for a Frictionless Adiabatic Fluid

The contravariant energy-tensor of the fluid is postulated to be

$$T^{\alpha\beta} = -g^{\alpha\beta}p + \rho\frac{dx^\alpha}{ds}\frac{dx^\beta}{ds} \qquad (5.58)$$

where p is the pressure of the fluid and ρ is the density of the fluid. From this is obtained the covariant energy tensor:

$$T_{\mu\nu} = g_{\mu\alpha}g_{\nu\beta}T^{\alpha\beta} = g_{\mu\alpha}g_{\nu\beta}\left(-g^{\alpha\beta}p + \rho\frac{dx^\alpha}{ds}\frac{dx^\beta}{ds}\right)$$

$$= -g_{\mu\alpha}\delta_\nu^\alpha p + g_{\mu\alpha}g_{\nu\beta}\frac{dx^\alpha}{ds}\frac{dx^\beta}{ds}\rho$$

$$T_{\mu\nu} = -g_{\mu\nu}p + g_{\mu\alpha}g_{\nu\beta}\frac{dx^\alpha}{ds}\frac{dx^\beta}{ds}\rho \qquad (5.58a)$$

The mixed energy tensor is

$$T_\sigma^\alpha = g_{\beta\sigma}T^{\alpha\beta} = g_{\beta\sigma}\left(-g^{\alpha\beta}p + \rho\frac{dx^\alpha}{ds}\frac{dx^\beta}{ds}\right) = -\delta_\sigma^\alpha p + g_{\beta\sigma}\rho\frac{dx^\alpha}{ds}\frac{dx^\beta}{ds} \qquad (5.58b)$$

The mixed energy tensor, Eq. (5.58b), was developed in a previous paper, "The Formal Foundation of the General Theory of Relativity,"[73] (Eq. (5.31)), with $\rho = \rho_0\sqrt{-g}(1 + \frac{p}{\rho_0} + P)$ and with $\sqrt{-g} = 1$).

Substituting Eq. (5.58b) into Eq. (5.57a),

$$\frac{\partial T^\alpha_\sigma}{\partial x^\alpha} = -\Gamma^\beta_{\alpha\sigma} T^\alpha_\beta \tag{5.57a}$$

$$\frac{\partial}{\partial x_\alpha}\left(-\delta^\alpha_\sigma p + g_{\beta\sigma}\rho\frac{dx^\alpha}{ds}\frac{dx^\beta}{ds}\right) = -\Gamma^\beta_{\alpha\sigma}\left(-\delta^\alpha_\beta p + g_{\beta\mu}\rho\frac{dx^\alpha}{ds}\frac{dx^\mu}{ds}\right)$$

These are identified as the Eulerian hydrodynamic equations of the general theory of relativity.[74] This is a set of four equations (one each for $\sigma = 1, 2, 3$, and 4). With the relation Eq. (5.58) between p and ρ, and the equation $g_{\alpha\beta}\frac{dx^\alpha}{ds}\frac{dx^\beta}{ds} = 1 (w^2 = g_{\alpha\beta}\frac{dx^\alpha}{ds}\frac{dx^\beta}{ds} = 1$, see Appendices 5.4.3.1 and 5.4.3.2), this gives six constraints on the six variables $p, \rho, \frac{dx^1}{ds}, \frac{dx^2}{ds}, \frac{dx^3}{ds}, \frac{dx^4}{ds}$, assuming all of the $g_{\mu\nu}$ are known.

1. If the $g_{\mu\nu}$ are not known, they are determined from Eq. (5.53). Equation (5.53) provides ten independent equations for the $g_{\mu\nu}$.
2. The condition $\sqrt{-g} = 1$ raises to 11 the number of equations for defining the ten independent $g_{\mu\nu}$.
3. But Eq. (5.57a) was obtained from Eq. (5.53). These four equations reduce to seven the number of independent equations for determining the $g_{\mu\nu}$.
4. These three "missing" equations reflect the freedom in the choice of coordinates.
5. If the condition $\sqrt{-g} = 1$ is removed, there remain four functions to be chosen, corresponding to the choice of the four coordinates.

5.2.20   Maxwell's Electromagnetic Field Equations for Free Space

"In this section Einstein shows that Maxwell's equations can be cast into the Minkowski four-tensor formalism. The calculation is basically a special relativity calculation. Einstein then shows Maxwell's equations can easily fit into the general theory framework when we need to consider electromagnetism in the presence of a gravitational field, i.e. electromagnetism in a curved spacetime."[75]

5.2.20.1   Maxwell's Equations

As in the previous section, Einstein begins by defining an appropriate quantity, this time the covariant vector ϕ_ν composed of the components of the electromagnetic potential, the first three components being the components of the vector potential and the fourth component being the scalar potential. In accordance with Eq. (5.36), one can form the antisymmetric covariant tensor, $F_{\rho\sigma}$,

$$F_{\rho\sigma} = \frac{\partial \phi_\rho}{\partial x^\sigma} - \frac{\partial \phi_\sigma}{\partial x^\rho} \tag{5.59}$$

$F_{\rho\sigma}$ being antisymmetric, the diagonal terms $F_{\rho\rho} = 0$. Being antisymmetric there will be only six independent $F_{\rho\sigma}$. Although these are not in the usual vector format of a row or column matrix, these six components are spoken of as a six-vector. Taking the derivatives of the various $F_{\rho\sigma}$ in a cyclic manner we arrive at

$$\frac{\partial F_{\rho\sigma}}{\partial x^\tau} + \frac{\partial F_{\sigma\tau}}{\partial x^\rho} + \frac{\partial F_{\tau\rho}}{\partial x^\sigma} = \left(\frac{\partial^2 \phi_\rho}{\partial x^\tau \partial x^\sigma} - \frac{\partial^2 \phi_\sigma}{\partial x^\tau \partial x^\rho} \right)$$

$$+ \left(\frac{\partial^2 \phi_\sigma}{\partial x^\rho \partial x^\tau} - \frac{\partial^2 \phi_\tau}{\partial x^\rho \partial x^\sigma} \right) + \left(\frac{\partial^2 \phi_\tau}{\partial x^\sigma \partial x^\rho} - \frac{\partial^2 \phi_\rho}{\partial x^\sigma \partial x^\tau} \right)$$

$$\frac{\partial F_{\rho\sigma}}{\partial x^\tau} + \frac{\partial F_{\sigma\tau}}{\partial x^\rho} + \frac{\partial F_{\tau\rho}}{\partial x^\sigma} = 0 \qquad (5.60)$$

This is an antisymmetric tensor of rank 3, easily verified by the exchange of any two indices, say ρ and τ, and comparing the resulting expression with the original form of Eq. (5.60). If any two of the indices ρ, σ and τ of Eq. (5.60) are equal to one another, the left-hand side of the equation vanishes. Writing out Eq. (5.60) for the four distinct cases of

$$\{\rho, \sigma, \tau\} = \{2, 3, 4\}, \{3, 4, 1\}, \{4, 1, 2\}, \{1, 2, 3\}$$

$$\{2, 3, 4\} \Rightarrow \frac{\partial F_{23}}{\partial x^4} + \frac{\partial F_{34}}{\partial x^2} + \frac{\partial F_{42}}{\partial x^3} = 0$$

$$\{3, 4, 1\} \Rightarrow \frac{\partial F_{34}}{\partial x^1} + \frac{\partial F_{41}}{\partial x^3} + \frac{\partial F_{13}}{\partial x^4} = 0 \qquad (5.60a)$$

$$\{4, 1, 2\} \Rightarrow \frac{\partial F_{41}}{\partial x^2} + \frac{\partial F_{12}}{\partial x^4} + \frac{\partial F_{24}}{\partial x^1} = 0$$

$$\{1, 2, 3\} \Rightarrow \frac{\partial F_{12}}{\partial x^3} + \frac{\partial F_{23}}{\partial x^1} + \frac{\partial F_{31}}{\partial x^2} = 0$$

Relabeling the coordinates x^1, x^2, x^3, x^4 as x, y, z, t, and the six independent functions $F_{\rho\sigma}$ in the following manner:

$$\begin{array}{ll} F_{23} = H_x & F_{14} = E_x \\ F_{31} = H_y & F_{24} = E_y \\ F_{12} = H_z & F_{34} = E_z \end{array} \qquad (5.61)$$

Remembering $F_{\rho\sigma}$ is antisymmetric, Eq. (5.60a) becomes

$$\frac{\partial H_x}{\partial t} + \frac{\partial E_z}{\partial y} - \frac{\partial E_y}{\partial z} = 0 \Rightarrow \frac{\partial E_z}{\partial y} - \frac{\partial E_y}{\partial z} = -\frac{\partial H_x}{\partial t} \Rightarrow \left(\nabla \times \vec{E} \right)_x = -\left(\frac{\partial H}{\partial t} \right)_x$$

$$\frac{\partial E_z}{\partial x} - \frac{\partial E_x}{\partial z} - \frac{\partial H_y}{\partial t} = 0 \Rightarrow \frac{\partial E_x}{\partial z} - \frac{\partial E_z}{\partial x} = -\frac{\partial H_y}{\partial t} \Rightarrow \left(\nabla \times \vec{E} \right)_{y=} = -\left(\frac{\partial H}{\partial t} \right)_y$$

$$\frac{-\partial E_x}{\partial y} + \frac{\partial H_z}{\partial t} + \frac{\partial E_y}{\partial x} = 0 \Rightarrow \frac{\partial E_y}{\partial x} - \frac{\partial E_x}{\partial y} = -\frac{\partial H_z}{\partial t} \Rightarrow \left(\nabla \times \vec{E} \right)_z = -\left(\frac{\partial H}{\partial t} \right)_z$$

$$\frac{\partial H_z}{\partial z} + \frac{\partial H_x}{\partial x} - \frac{\partial H_y}{\partial z} = 0 \Rightarrow \dots\dots\dots\dots\dots \Rightarrow \left(\nabla \cdot \vec{H} \right) = 0$$

Equation (5.60a) is identified as two of Maxwell's equations:

$$\frac{\partial F_{\rho\sigma}}{\partial x_\tau} + \frac{\partial F_{\sigma\tau}}{\partial x_\rho} + \frac{\partial F_{\tau\rho}}{\partial x_\sigma} = 0 \Rightarrow \nabla \times \vec{E} = -\frac{\partial \vec{H}}{\partial t}$$

$$\Rightarrow \nabla \cdot \vec{H} = 0 \qquad (5.60b)$$

To obtain the other two Maxwell equations, Einstein uses the contravariant form of the six-vector (antisymmetric tensor of rank 2) associated with $F_{\alpha\beta}$, and a contravariant four-vector J^μ composed of the three components of the current density (j_x, j_y, j_z) and the charge density ρ:

$$F^{\mu\nu} = g^{\mu\alpha} g^{\nu\beta} F_{\alpha\beta} \tag{5.62}$$

Setting the derivative of $F^{\mu\nu}$ equal to J^μ,

$$\frac{\partial}{\partial x^\nu} F^{\mu\nu} = J^\mu$$

$$J^\mu = \{J^1, J^2, J^3, J^4\} = \{j_x, j_y, j_z, \rho\} \tag{5.63}$$

In like manner to the identification of components of $\vec{H}$ and $\vec{E}$ to components of $F_{\alpha\beta}$ in the covariant treatment above, the components of $F^{\mu\nu}$ are equated to the components of $\vec{H}'$ and $\vec{E}'$. In the reference frame where the special theory of relativity is valid, $\vec{H}' = \vec{H}$ and $\vec{E}' = \vec{E}$,

$$
\begin{array}{ll}
F^{23} = H'_X & F^{14} = -E'_x \\
F^{31} = H'_Y & F^{24} = -E'_y \\
F^{12} = H'_Z & F^{34} = -E'_z
\end{array}
$$

$$\mu = 1 \Rightarrow \frac{\partial F^{11}}{\partial x^1} + \frac{\partial F^{12}}{\partial x^2} + \frac{\partial F^{13}}{\partial x^3} + \frac{\partial F^{14}}{\partial x^4} = J^1 \Rightarrow 0 + \frac{\partial H'_z}{\partial y} - \frac{\partial H'_y}{\partial z} - \frac{\partial E'_x}{\partial t}$$

$$= j_x \Rightarrow \left(\nabla \times \vec{H}' = \vec{j} + \frac{\partial \vec{E}}{\partial t} \right)_x$$

$$\mu = 2 \Rightarrow \frac{\partial F^{21}}{\partial x^1} + \frac{\partial F^{22}}{\partial x^2} + \frac{\partial F^{23}}{\partial x^3} + \frac{\partial F^{24}}{\partial x^4} = J^2 \Rightarrow -\frac{\partial H'_z}{\partial x} + 0 + \frac{\partial H'_x}{\partial z} - \frac{\partial E'_y}{\partial t}$$

$$= j_y \Rightarrow \left(\nabla \times \vec{H}' = \vec{j} + \frac{\partial \vec{E}}{\partial t} \right)_y$$

$$\mu = 3 \Rightarrow \frac{\partial F^{31}}{\partial x^1} + \frac{\partial F^{32}}{\partial x^2} + \frac{\partial F^{33}}{\partial x^3} + \frac{\partial F^{34}}{\partial x^4} = J^3 \Rightarrow \frac{\partial H'_y}{\partial x} - \frac{\partial H'_x}{\partial y} + 0 - \frac{\partial E'_z}{\partial t}$$

$$= j_z \Rightarrow \left(\nabla \times \vec{H}' = \vec{j} + \frac{\partial \vec{E}}{\partial t} \right)_z$$

$$\mu = 4 \Rightarrow \frac{\partial F^{41}}{\partial x^1} + \frac{\partial F^{42}}{\partial x^2} + \frac{\partial F^{43}}{\partial x^3} + \frac{\partial F^{44}}{\partial x^4} = J^4 \Rightarrow \frac{\partial E'_x}{\partial x} + \frac{\partial E'_y}{\partial y} + \frac{\partial E'_z}{\partial z} + 0$$

$$= \rho \Rightarrow \left(\nabla \cdot \vec{E}' = \rho \right)$$

Equation (5.63) is identified as the two additional Maxwell's equations:

$$\frac{\partial F^{\mu\nu}}{\partial x^\nu} = J^\mu \Rightarrow \begin{array}{l} \frac{\partial \vec{E}'}{\partial t} + \vec{j} = \nabla \times \vec{H}' \\ \nabla \cdot \vec{E}' = \rho \end{array} \tag{5.63a}$$

Equations (5.60), (5.62) and (5.63) thus form the generalization of Maxwell's equations in the general theory of relativity.

5.2.20.2   The Energy Components of the Electromagnetic Field

In this section, Einstein searches for an expression for T^ν_μ of the electromagnetic field that satisfies the law of conservation of energy and momentum, Eq. (5.57). T^ν_μ would then correspond to the energy components of the electromagnetic field:

$$\frac{\partial T_\sigma^\alpha}{\partial x^\alpha} + \frac{1}{2}\frac{\partial g^{\mu\nu}}{\partial x^\sigma}T_{\mu\nu} = 0 \qquad (5.57)$$

Start by forming the covariant tensor

$$\kappa_\sigma = F_{\sigma\mu}J^\mu \qquad (5.65)$$

Using Eq. (5.61) and the definition of J^μ given immediately after Eq. (5.63),

$$\kappa_1 = F_{11}J^1 + F_{12}J^2 + F_{13}J^3 + F_{14}J^4 = 0 + H_z j_y - H_y j_z + E_x \rho = \left(\rho\vec{E} + \vec{j}\times\vec{H}\right)_x$$

$$\kappa_2 = F_{21}J^1 + F_{22}J^2 + F_{23}J^3 + F_{24}J^4 = -H_z j_x + 0 + H_x j_z + E_y \rho = \left(\rho\vec{E} + \vec{j}\times\vec{H}\right)_y$$

$$\kappa_3 = F_{31}J^1 + F_{32}J^2 + F_{33}J^3 + F_{34}J^4 = H_y j_x - H_x j_y + 0 + E_z \rho = \left(\rho\vec{E} + \vec{j}\times\vec{H}\right)_z$$

$$\kappa_4 = F_{41}J^1 + F_{42}J^2 + F_{43}J^3 + F_{44}J^4 = -E_x j_x - E_y j_y - E_z j_z + 0 = -\vec{j}\cdot\vec{E}$$

κ_σ is identified as the covariant vector with components equal to the negative momentum or, respectively, energy transferred from the electric masses to the electromagnetic field per unit time per unit volume. Einstein states that, "If the electric masses are free, that is, under the sole influence of the electromagnetic field, the covariant vector κ_σ will vanish."[76] Setting $\kappa_\sigma = 0$ the expression for κ_σ is manipulated to arrive at an expression of the form of Eq. (5.57), from which the appropriate terms are identified as being T_μ^ν:

$$\kappa_\sigma = F_{\sigma\mu}J^\mu = F_{\sigma\mu}\frac{\partial F^{\mu\nu}}{\partial x^\nu} = \frac{\partial}{\partial x^\nu}\left(F_{\sigma\mu}F^{\mu\nu}\right) - F^{\mu\nu}\frac{\partial F_{\sigma\mu}}{\partial x^\nu} \quad \text{by Eq. (5.63)}$$
$$(5.65a)$$

Working first with the second term on the right-hand side,

$$F^{\mu\nu}\frac{\partial F_{\sigma\mu}}{\partial x^\nu} = \frac{1}{2}\left(F^{\mu\nu}\frac{\partial F_{\sigma\mu}}{\partial x^\nu} + F^{\mu\nu}\frac{\partial F_{\sigma\mu}}{\partial x^\nu}\right) = \frac{1}{2}\left(F^{\nu\mu}\frac{\partial F_{\sigma\nu}}{\partial x^\mu} + F^{\mu\nu}\frac{\partial F_{\sigma\mu}}{\partial x^\nu}\right)$$

where the first term in the last parentheses is obtained by exchanging the summation indices μ and ν from the corresponding term in the previous parentheses. Since $F^{\mu\nu}$ and $F_{\sigma\nu}$ are each antisymmetric, exchanging the order of indices on both of them leaves their product unchanged in sign:

$$F^{\mu\nu}\frac{\partial F_{\sigma\mu}}{\partial x^\nu} = \frac{1}{2}\left(F^{\nu\mu}\frac{\partial F_{\sigma\nu}}{\partial x^\mu} + F^{\mu\nu}\frac{\partial F_{\sigma\mu}}{\partial x^\nu}\right) = \frac{1}{2}\left(F^{\mu\nu}\frac{\partial F_{\nu\sigma}}{\partial x^\mu} + F^{\mu\nu}\frac{\partial F_{\sigma\mu}}{\partial x^\nu}\right)$$

$$= \frac{1}{2}F^{\mu\nu}\left(\frac{\partial F_{\nu\sigma}}{\partial x^\mu} + \frac{\partial F_{\sigma\mu}}{\partial x^\nu}\right)$$

But, by Eq. (5.60), $\frac{\partial F_{\nu\sigma}}{\partial x^\mu} + \frac{\partial F_{\sigma\mu}}{\partial x^\nu} = -\frac{\partial F_{\mu\nu}}{\partial x^\sigma}$, and using Eq. (5.62),

$$F^{\mu\nu}\frac{\partial F_{\sigma\mu}}{\partial x^\nu} = -\frac{1}{2}F^{\mu\nu}\frac{\partial F_{\mu\nu}}{\partial x^\sigma} = -\frac{1}{2}g^{\mu\alpha}g^{\nu\beta}F_{\alpha\beta}\frac{\partial F_{\mu\nu}}{\partial x^\sigma}$$

This last term is manipulated in a manner similar to that above. The expression is written as one half of it plus itself, and the summation indices are changed on the second term in the square brackets

$(\mu \leftrightarrow \alpha, \nu \leftrightarrow \beta)$:

$$-\frac{1}{2} g^{\mu\alpha} g^{\nu\beta} F_{\alpha\beta} \frac{\partial F_{\mu\nu}}{\partial x^\sigma} = -\frac{1}{4} \left[g^{\mu\alpha} g^{\nu\beta} F_{\alpha\beta} \frac{\partial F_{\mu\nu}}{\partial x^\sigma} + g^{\mu\alpha} g^{\nu\beta} F_{\alpha\beta} \frac{\partial F_{\mu\nu}}{\partial x^\sigma} \right]$$

$$= -\frac{1}{4} \left[g^{\mu\alpha} g^{\nu\beta} F_{\alpha\beta} \frac{\partial F_{\mu\nu}}{\partial x^\sigma} + g^{\alpha\mu} g^{\beta\nu} F_{\mu\nu} \frac{\partial F_{\alpha\beta}}{\partial x^\sigma} \right]$$

$$= -\frac{1}{4} \left[\left(g^{\mu\alpha} g^{\nu\beta} F_{\alpha\beta} \frac{\partial F_{\mu\nu}}{\partial x^\sigma} \right) + F_{\mu\nu} \left(g^{\alpha\mu} g^{\beta\nu} \frac{\partial F_{\alpha\beta}}{\partial x^\sigma} \right) \right]$$

$$= -\frac{1}{4} \left[\left(\frac{\partial}{\partial x^\sigma} \left(g^{\mu\alpha} g^{\nu\beta} F_{\alpha\beta} F_{\mu\nu} \right) - F_{\mu\nu} \frac{\partial}{\partial x^\sigma} \left(g^{\mu\alpha} g^{\nu\beta} F_{\alpha\beta} \right) \right) \right.$$

$$\left. + F_{\mu\nu} \left(\frac{\partial}{\partial x^\sigma} \left(g^{\mu\alpha} g^{\nu\beta} F_{\alpha\beta} \right) - F_{\alpha\beta} \frac{\partial}{\partial x^\sigma} \left(g^{\mu\alpha} g^{\nu\beta} \right) \right) \right]$$

Noting the second and third terms are the same, their difference is zero,

$$F^{\mu\nu} \frac{\partial F_{\sigma\mu}}{\partial x^\nu} = -\frac{1}{4} \frac{\partial}{\partial x^\sigma} \left(g^{\mu\alpha} g^{\nu\beta} F_{\alpha\beta} F_{\mu\nu} \right) + \frac{1}{4} F_{\mu\nu} F_{\alpha\beta} \frac{\partial}{\partial x^\sigma} \left(g^{\mu\alpha} g^{\nu\beta} \right)$$

$$= -\frac{1}{4} \frac{\partial}{\partial x^\sigma} \left(F^{\mu\nu} F_{\mu\nu} \right) + \frac{1}{4} F_{\mu\nu} F_{\alpha\beta} \left(g^{\mu\alpha} \frac{\partial g^{\nu\beta}}{\partial x^\sigma} + g^{\nu\beta} \frac{\partial g^{\mu\alpha}}{\partial x^\sigma} \right) \quad \text{by Eq. (5.62)}$$

$$= -\frac{1}{4} \frac{\partial}{\partial x^\sigma} \left(F^{\mu\nu} F_{\mu\nu} \right) + \frac{1}{4} F_{\mu\nu} F_{\alpha\beta} g^{\mu\alpha} \frac{\partial g^{\nu\beta}}{\partial x^\sigma} + \frac{1}{4} F_{\mu\nu} F_{\alpha\beta} g^{\nu\beta} \frac{\partial g^{\mu\alpha}}{\partial x^\sigma}$$

Exchanging the summation indices $\mu \leftrightarrow \beta$ and $\nu \leftrightarrow \alpha$ in the final expression on the right-hand side,

$$F^{\mu\nu} \frac{\partial F_{\sigma\mu}}{\partial x^\nu} = -\frac{1}{4} \frac{\partial}{\partial x^\sigma} \left(F^{\mu\nu} F_{\mu\nu} \right) + \frac{1}{4} F_{\mu\nu} F_{\alpha\beta} g^{\mu\alpha} \frac{\partial g^{\nu\beta}}{\partial x^\sigma} + \frac{1}{4} F_{\beta\alpha} F_{\nu\mu} g^{\alpha\mu} \frac{\partial g^{\beta\nu}}{\partial x^\sigma}$$

$$= -\frac{1}{4} \frac{\partial}{\partial x^\sigma} \left(F^{\mu\nu} F_{\mu\nu} \right) + \frac{1}{4} F_{\mu\nu} F_{\alpha\beta} g^{\mu\alpha} \frac{\partial g^{\nu\beta}}{\partial x^\sigma} + \frac{1}{4} F_{\alpha\beta} F_{\mu\nu} g^{\mu\alpha} \frac{\partial g^{\nu\beta}}{\partial x^\sigma}$$

$$= -\frac{1}{4} \frac{\partial}{\partial x^\sigma} \left(F^{\mu\nu} F_{\alpha\nu} \right) + \frac{1}{2} F_{\mu\nu} F_{\alpha\beta} g^{\mu\alpha} \frac{\partial g^{\nu\beta}}{\partial x^\sigma}$$

$$= -\frac{1}{4} \frac{\partial}{\partial x^\sigma} \left(F^{\mu\nu} F_{\mu\nu} \right) + \frac{1}{2} F_{\mu\nu} F_{\alpha\beta} g^{\mu\alpha} \left(-g^{\nu\rho} g^{\beta\tau} \frac{\partial g_{\rho\tau}}{\partial x^\sigma} \right) \quad \text{by Eq. (5.31)}$$

$$= -\frac{1}{4} \frac{\partial}{\partial x^\sigma} \left(F^{\mu\nu} F_{\mu\nu} \right) - \frac{1}{2} F_{\mu\nu} \left(g^{\mu\alpha} g^{\beta\tau} F_{\alpha\beta} \right) g^{\nu\rho} \frac{\partial g_{\rho\tau}}{\partial x^\sigma}$$

$$= -\frac{1}{4} \frac{\partial}{\partial x^\sigma} \left(F^{\mu\nu} F_{\mu\nu} \right) - \frac{1}{2} F_{\mu\nu} F^{\mu\tau} g^{\nu\rho} \frac{\partial g_{\rho\tau}}{\partial x^\sigma} \quad \text{by Eq. (5.62)}$$

Substituting this expression into the expression for κ_σ, Eq. (5.65a),

$$\kappa_\sigma = \frac{\partial}{\partial x^\nu} \left(F_{\sigma\mu} F^{\mu\nu} \right) - F^{\mu\nu} \frac{\partial F_{\sigma\mu}}{\partial x^\nu}$$

$$\kappa_\sigma = \frac{\partial}{\partial x^\nu} \left(F_{\sigma\mu} F^{\mu\nu} \right) + \frac{1}{4} \frac{\partial}{\partial x^\sigma} \left(F^{\mu\nu} F_{\mu\nu} \right) + \frac{1}{2} F^{\mu\tau} F_{\mu\nu} g^{\nu\rho} \frac{\partial g_{\rho\tau}}{\partial x^\sigma}$$

$$\kappa_\sigma = \frac{\partial}{\partial x^\nu} \left(-F_{\sigma\mu} F^{\nu\mu} + \frac{1}{4} \delta^\nu_\sigma F^{\alpha\beta} F_{\alpha\beta} \right) + \frac{1}{2} F^{\mu\tau} F_{\mu\nu} g^{\nu\rho} \frac{\partial g_{\rho\tau}}{\partial x^\sigma} \quad \text{by } F^{\mu\nu} = -F^{\nu\mu}$$

Define T_σ^ν as the term in the first parentheses (with $\mu \to \alpha$ in the first term), $T_\sigma^\nu = -F_{\sigma\alpha}F^{\nu\alpha} + \frac{1}{4}\delta_\sigma^\nu F^{\alpha\beta}F_{\alpha\beta}$. The expression $F^{\mu\tau}F_{\mu\nu}$ in the last term can be expressed as $F^{\mu\tau}F_{\mu\nu} = -T_\nu^\tau + \frac{1}{4}\delta_\nu^\tau F^{\alpha\beta}F_{\alpha\beta}$:

$$\kappa_\sigma = \frac{\partial T_\sigma^\nu}{\partial x^\nu} + \frac{1}{2}g^{\nu\rho}\frac{\partial g_{\rho\tau}}{\partial x^\sigma}\left(-T_\nu^\tau + \frac{1}{4}\delta_\nu^\tau F^{\alpha\beta}F_{\alpha\beta}\right)$$

$$= \frac{\partial T_\sigma^\nu}{\partial x^\nu} - \frac{1}{2}g^{\nu\rho}\frac{\partial g_{\rho\tau}}{\partial x^\sigma}T_\nu^\tau + \frac{1}{8}g^{\nu\rho}\frac{\partial g_{\rho\tau}}{\partial x^\sigma}\delta_\nu^\tau F^{\alpha\beta}F_{\alpha\beta}$$

The third term vanishes since, by Eq. (5.29), $g^{\nu\rho}\frac{\partial g_{\rho\tau}}{\partial x^\sigma}\delta_\nu^\tau = g^{\tau\rho}\frac{\partial g_{\rho\tau}}{\partial x^\sigma}$ vanishes (see Appendix 5.4.7.5). Thus, with the change of indices $\rho \to \mu$, $\nu \leftrightarrow \tau$ in the second term,

$$\kappa_\sigma = \frac{\partial T_\sigma^\nu}{\partial x^\nu} - \frac{1}{2}g^{\tau\mu}\frac{\partial g_{\mu\nu}}{\partial x^\sigma}T_\tau^\nu$$

$$T_\sigma^\nu = -F_{\sigma\alpha}F^{\nu\alpha} + \frac{1}{4}\delta_\sigma^\nu F^{\alpha\beta}F_{\alpha\beta} \tag{5.66}$$

For $\kappa_\sigma = 0$, Eq. (5.66) is the same as Eq. (5.57). To see this, Eq. (5.66) becomes

$$0 = \frac{\partial T_\sigma^\nu}{\partial x^\nu} - \frac{1}{2}g^{\tau\mu}\frac{\partial g_{\mu\nu}}{\partial x^\sigma}T_\tau^\nu = \frac{\partial T_\sigma^\nu}{\partial x^\nu} - \frac{1}{2}g^{\tau\mu}\left(-g_{\mu\alpha}g_{\nu\beta}\frac{\partial g^{\alpha\beta}}{\partial x^\sigma}\right)T_\tau^\nu \quad \text{by Eq.(5.32)}$$

$$= \frac{\partial T_\sigma^\nu}{\partial x^\nu} + \frac{1}{2}\left(g^{\tau\mu}g_{\mu\alpha}\right)g_{\nu\beta}\frac{\partial g^{\alpha\beta}}{\partial x^\sigma}T_\tau^\nu = \frac{\partial T_\sigma^\nu}{\partial x^\nu} + \frac{1}{2}\left(\delta_\alpha^\tau\right)g_{\nu\beta}\frac{\partial g^{\alpha\beta}}{\partial x^\sigma}T_\tau^\nu$$

$$= \frac{\partial T_\sigma^\nu}{\partial x^\nu} + \frac{1}{2}\frac{\partial g^{\alpha\beta}}{\partial x^\sigma}g_{\nu\beta}T_\alpha^\nu = \frac{\partial T_\sigma^\nu}{\partial x^\nu} + \frac{1}{2}\frac{\partial g^{\alpha\beta}}{\partial x^\sigma}T_{\alpha\beta}$$

For comparison, Eq. (5.57) is $\frac{\partial T_\sigma^\alpha}{\partial x^\alpha} + \frac{1}{2}\frac{\partial g^{\mu\nu}}{\partial x^\sigma}T_{\mu\nu} = 0$. Since the T_σ^ν satisfy the energy conservation equation, Eq. (5.57), the T_σ^ν are the energy components of the electromagnetic field.

Part E
5.2.21 Newton's Theory as a First Approximation

Einstein considers two different perspectives in arriving at his first order approximations to the general theory of relativity.

Perspective 1: In the zeroth order, the general theory of relativity is expected to reduce to the special theory of relativity, with the $g_{\mu\nu}$ having the constant values

$$g_{\mu\nu} = \begin{bmatrix} -1 & 0 & 0 & 0 \\ 0 & -1 & 0 & 0 \\ 0 & 0 & -1 & 0 \\ 0 & 0 & 0 & +1 \end{bmatrix} \tag{5.4}$$

It is assumed,

a. in the first approximation, the $g_{\mu\nu}$ differ from the above constant values (Eq. (5.4)) by an amount that is small in comparison to 1,
b. that all terms of second order and higher in these small amounts can be neglected,

c. that, with a suitable choice of coordinates, at spatial infinity, far from the matter generating the gravitational fields, the $g_{\mu\nu}$ approach the constant values given in Eq. (5.4), i.e., the gravitational fields are generated exclusively by the matter in the finite region.

Einstein then focuses his attention on the equations describing the motion of a material point in the gravitational field, Eq. (5.46). Consider the quantities dx^1, dx^2, dx^3, dx^4, with $dx^4 = dt$.

d. Dividing by ds, the expressions $\frac{dx^1}{ds}, \frac{dx^2}{ds}, \frac{dx^3}{ds}$, in the case of the special theory of relativity, can take on any values.

e. The velocity of an object is given as

$$v = \sqrt{\left(\frac{dx^1}{dx^4}\right)^2 + \left(\frac{dx^2}{dx^4}\right)^2 + \left(\frac{dx^3}{dx^4}\right)^2}$$

Perspective 2:

f. $v <$ speed of light in a vacuum.
g. Our experience is almost exclusively that v is very small compared to the speed of light.
h. The components $\frac{dx^1}{ds}, \frac{dx^2}{ds}, \frac{dx^3}{ds}$ are small compared to $\frac{dx^4}{ds}$, which is equal to one, at least to second order in the small terms.

From Perspective 1, the components of the gravitational field, $\Gamma^\tau_{\mu\nu}$, will be small. From Perspective 2, the equation of the motion of a point, $\frac{d^2x^\tau}{ds^2} = \Gamma^\tau_{\mu\nu}\frac{dx^\mu}{ds}\frac{dx^\nu}{ds}$, Eq. (5.46), will have significant contributions only for $\mu = 4$ and $\nu = 4$ ($\frac{dx^\mu}{ds} \ll 1, \mu = 1, 2, 3$).

$$\frac{d^2x^\tau}{ds^2} = \Gamma^\tau_{\mu\nu}\frac{dx^\mu}{ds}\frac{dx^\nu}{ds} \approx \Gamma^\tau_{44}\frac{dx^4}{ds}\frac{dx^4}{ds} \Rightarrow \frac{d^2x^\tau}{dt^2} = \Gamma^\tau_{44} \quad \text{with } ds = dx^4 = dt$$

$$\frac{d^2x^\tau}{dt^2} = \Gamma^\tau_{44} = -\{44, \tau\} = -g^{\tau\alpha}[44, \alpha]$$

From Perspective 1, $g^{\tau\alpha} \approx 0$ for $\tau \neq \alpha$; $g^{\tau\alpha} \approx -1$ for $\tau = \alpha = 1, 2$, or 3; and $g^{\tau\alpha} \approx +1$ for $\tau = \alpha = 4$.

For τ and $\alpha = 1, 2$, or 3, $g^{\tau\alpha} = 0$ unless $\alpha = \tau$. For $\alpha = \tau$, $g^{\tau\tau} = -1(g^{11} = g^{22} = g^{33} = -1)$.

$$\frac{d^2x^\tau}{dt^2} = +[44, \tau], \qquad \tau = 1, 2, 3$$

For τ and $\alpha = 4$, $g^{\tau\beta} = g^{4\beta} = 0$ unless $\beta = 4$. For $\beta = 4$, $g^{44} = +1$,

$$\frac{d^2x^4}{dt^2} = -[44, 4], \qquad \tau = 4$$

If the field is quasi-static, the time derivatives on the right-hand side of these equations may be neglected in relation to the space derivatives. For $\tau = 1, 2$, or 3,

$$\frac{d^2 x^\tau}{dt^2} = +[44, \tau] = \frac{1}{2} \left(\frac{\partial g_{\tau 4}}{\partial x^4} + \frac{\partial g_{\tau 4}}{\partial x^4} - \frac{\partial g_{44}}{\partial x^\tau} \right)$$

$$= \frac{1}{2} \left(\frac{\partial g_{\tau 4}}{\partial t} + \frac{\partial g_{\tau 4}}{\partial t} - \frac{\partial g_{44}}{\partial x^\tau} \right) \approx -\frac{1}{2} \frac{\partial g_{44}}{\partial x^\tau}$$

This is the form of Newton's second law, $F = ma$, for a unit mass moving in a gravitational potential, $\phi = \frac{1}{2} g_{44}$.

Comment 1: Of the 16 components of $g_{\mu\nu}$, 10 are independent, yet, to first order, only one of these, g_{44}, is needed to determine the motion of a material point.

Comment 2: Since $g_{44} = -c^2$, this shows the connection Einstein was pursuing for a while of considering a variable speed of light, c, to represent the gravitational potential.

Working now with the field equations, Eqs. (5.53),

$$\frac{\partial}{\partial x^\alpha} \Gamma^\alpha_{\mu\nu} + \Gamma^\alpha_{\mu\beta} \Gamma^\beta_{\nu\alpha} = -\kappa \left(T_{\mu\nu} - \frac{1}{2} g_{\mu\nu} T \right)$$

$$\sqrt{-g} = 1$$

(5.53)

Modification a: From Perspective 1, the components of the gravitational field, $\Gamma^\tau_{\mu\nu}$, will be small and terms of the form $(\Gamma^\tau_{\mu\nu})^2$ will be of second order and can be neglected. Thus the second term on the left-hand side of the above equation can be neglected.

Modification b: The expression for $T_{\mu\nu}$, Eq. (5.58a), is

$$T_{\mu\nu} = -g_{\mu\nu} p + g_{\mu\alpha} g_{\nu\beta} \frac{dx^\alpha}{ds} \frac{dx^\beta}{ds} \rho$$

Since the energy tensor is expected to be dominated by the matter present, the mass density term, ρ, will be much greater than the pressure, p, term. From Perspective 2, the terms $\frac{dx^\alpha}{ds}$ for $\alpha = 1, 2$, or 3 are small compared to the term for $\alpha = 4$, $\frac{dx^4}{ds}$. The expression for $T_{\mu\nu}$ becomes

$$T_{\mu\nu} = g_{\mu 4} g_{\nu 4} \frac{dx^4}{ds} \frac{dx^4}{ds} \rho$$

But $g_{\mu\nu} \approx 0$ except for $\mu = \nu$ Thus both μ and ν must each be equal to 4 and, since $g_{44} = 1$, $T_{\mu\nu}$ becomes $T_{44} = \rho$, and all other $T_{\mu\nu} = 0$. $T = \sum_{\mu,\nu=1}^4 T_{\mu\nu} = \rho$.

Modification c: Turning now to the left-hand side of the field equations, Eqs. (5.53) above,

$$\frac{\partial}{\partial x^\alpha} \Gamma^\alpha_{\mu\nu} = \frac{-\partial}{\partial x^\alpha} \{\mu\nu, \alpha\} = \frac{-\partial}{\partial x^\alpha} (g^{\tau\alpha} [\mu\nu, \tau])$$

But $g^{\tau\alpha} = 0$ for $\tau \neq \alpha$; $= -1$ for $\tau = \alpha = 1, 2$, or 3; and $= +1$ for $\tau = \alpha = 4$. The above equation becomes

$$\frac{\partial}{\partial x^\alpha}\Gamma^\alpha_{\mu\nu} = \frac{-\partial}{\partial x^\alpha}\left(g^{\alpha\alpha}\left[\mu\nu,\alpha\right]\right)$$

$$= -\left(\frac{-\partial}{\partial x^1}\left[\mu\nu,1\right] - \frac{\partial}{\partial x^2}\left[\mu\nu,2\right] - \frac{\partial}{\partial x^3}\left[\mu\nu,3\right] + \frac{\partial}{\partial x^4}\left[\mu\nu,4\right]\right)$$

Looking at the term $\mu = 4$ and $\nu = 4$, and using again the quasi-static approximation that time variations are so slow that time derivatives can be neglected:

$$\frac{\partial}{\partial x^\alpha}\Gamma^\alpha_{44} = -\left(\frac{-\partial}{\partial x^1}\left[44,1\right] - \frac{\partial}{\partial x^2}\left[44,2\right] - \frac{\partial}{\partial x^3}\left[44,3\right] + \frac{\partial}{\partial x^4}\left[44,4\right]\right)$$

On the right-hand side, the first term becomes

$$\frac{+\partial}{\partial x^1}\left[44,1\right] = \frac{1}{2}\frac{\partial}{\partial x^1}\left(\frac{\partial g_{14}}{\partial x^4} + \frac{\partial g_{14}}{\partial x^4} - \frac{\partial g_{44}}{\partial x^1}\right) \approx -\frac{1}{2}\frac{\partial^2 g_{44}}{\partial\left(x^1\right)^2}$$

In a similar manner, the second and third terms become $-\frac{1}{2}\frac{\partial^2 g_{44}}{\partial(x^2)^2}$ and $-\frac{1}{2}\frac{\partial^2 g_{44}}{\partial(x^3)^2}$. The fourth term vanishes as each term is a time derivative.

$$\frac{\partial}{\partial x^\alpha}\Gamma^\alpha_{44} = -\frac{1}{2}\frac{\partial^2 g_{44}}{\left(\partial x^1\right)^2} - \frac{1}{2}\frac{\partial^2 g_{44}}{\left(\partial x^2\right)^2} - \frac{1}{2}\frac{\partial^2 g_{44}}{\left(\partial x^3\right)^2} = -\frac{1}{2}\nabla^2 g_{44}$$

Remembering we are now reduced to the $\mu, \nu = 4,4$ term. From Comment 2, Modification a, Eq. (5.53) is

$$\frac{\partial}{\partial x^\alpha}\Gamma^\alpha_{\mu\nu} = -\kappa\left(T_{\mu\nu} - \frac{1}{2}g_{\mu\nu}T\right) - \frac{1}{2}\nabla^2 g_{44} = -\kappa\left(T_{44} - \frac{1}{2}g_{44}T\right)$$

$$= -\kappa\left(\rho - \frac{1}{2}\rho\right) = -\kappa\frac{\rho}{2} \Rightarrow \nabla^2 g_{44} = \kappa\rho \qquad (5.68)$$

The solution to this equation is Newton's gravitational potential, ϕ, where $\phi = \frac{1}{2}g_{44}$ and is given by

$$\phi(r) = \frac{1}{2}g_{44} = -\frac{\kappa}{8\pi}\int\frac{\rho d\tau}{r} \qquad (5.68a)$$

From Newton's theory, $\phi(r) = -\frac{G}{c^2}\int\frac{\rho d\tau}{r}$. By comparison to Newton's expression for the gravitational potential,

$$\kappa = \frac{8\pi G}{c^2} = \frac{8\pi(6.7\times 10^{-8})}{(3\times 10^{10})} = 1.87\times 10^{-27}\text{in cgs units.} \qquad (5.69)$$

5.2.22 The Behaviour of Rods and Clocks in the Static Gravitational Field. Bending of Light Rays. Motion of the Perihelion of a Planetary Orbit

In the previous Section 5.2.21, only one component, g_{44}, of the $g_{\mu\nu}$ was needed to arrive at Newton's theory of gravity as a first approximation. But if the determinant of the $g_{\mu\nu}$ is to remain equal to –1, when g_{44}

differs slightly from -1 there must be corresponding modifications to the other $g_{\mu\nu}$.

For a point mass at the origin, the potential given by Eq. (5.68a) is $\phi(r) = -\frac{\kappa}{8\pi} \int \frac{\rho d\tau}{r} = -\frac{\kappa}{8\pi} \frac{M}{r}$, where $r =$ the distance from the origin (location of the point mass) to the location of the measurement. Since g_{44} is equal to 2ϕ plus a constant, and since g_{44} is equal to $+1$ in the absence of a gravitational field, $\phi = 0$, we obtain $g_{44} = 2\phi(r) + 1 = 1 - \frac{\kappa}{4\pi} \frac{M}{r} = 1 - \frac{\alpha}{r}$, with

$$\alpha = \frac{\kappa M}{4\pi}. \tag{5.70a}$$

For a field-producing point mass at the origin, to a first approximation, the radially symmetrical solution is, with $r = \sqrt{(x^1)^2 + (x^2)^2 + (x^3)^2}$,

$$g_{44} = 1 - \frac{\alpha}{r}$$

$$g_{\rho 4} = g_{4\rho} = 0 \dots\dots\dots\dots\dots \rho = 1, 2, 3 \tag{5.70}$$

$$g_{\rho\sigma} = -\delta_\rho^\sigma - \alpha \frac{x^\rho x^\sigma}{r^3} \dots \rho, \sigma = 1, 2, 3$$

These Eqs. (5.70) are taken from the November 18, 1915, report to the Prussian Academy of Sciences. In the report, Einstein states this is an "assumed solution."[77]

To see the effect of the gravitational field from the mass M on the metrical properties of space, consider the quantity $ds^2 = g_{\mu\nu} dx^\mu dx^\nu$.

1. *Measurement Parallel to the Gravitational Field*: At some distance r from the origin a measuring rod of unit length is laid parallel to the x^1 axis: $dx^1 = 1$; $dx^2 = dx^3 = dx^4 = 0$. In K_0, $ds^2 = -(dx^1)^2 - (dx^2)^2 - (dx^3)^2 + (dx^4)^2 = -(1)^2 - 0 - 0 + 0 = -1$. Thus $ds^2 = -1 = g_{11}(dx^1)^2$. See Figure 5.6.

If the point lies on the x^1 axis (a distance r away from the origin), from Eq. (5.70) $g_{11} = -1 - \frac{\alpha r^2}{r^3} = -1 - \frac{\alpha}{r}$. Putting these together,

$$-1 = g_{11}(dx^1)^2 = -\left(1 + \frac{\alpha}{r}\right)(dx^1)^2 \Rightarrow dx^1 = \frac{1}{\sqrt{1 + \frac{\alpha}{r}}} \approx 1 - \frac{1}{2}\frac{\alpha}{r} \tag{5.71}$$

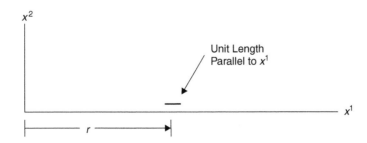

Unit Length
Parallel to x^1

Fig. 5.6 Measurement parallel to the gravitational field.

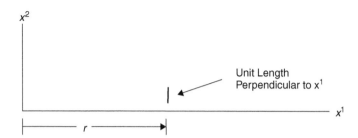

Fig. 5.7 Measurement perpendicular to the gravitational field.

The unit-length measuring rod laid along a radius, i.e., parallel to the field, appears to be shortened by an amount $= \alpha/2r$.

2. *Measurement Perpendicular to the Gravitational Field*: Again move out the x^1 axis some distance r, but this time position the unit measuring rod perpendicular to the field, i.e., parallel to the x^2 axis. See Figure 5.7.

On the x^1 axis, $x^1 = r, x^2 = 0, x^3 = 0$. $dx^2 = 1$, while $dx^1 = dx^3 = dx^4 = 0$, i.e., only $dx^2 \neq 0$. $ds^2 = -1 = g_{22} dx_2^2$. In this case, again using Eq. (5.70), $g_{22} = -1 - \frac{\alpha(x^2=0)^2}{r^3} = -1$, and we have

$$-1 = g_{22}(dx^2)^2 = -(dx^2)^2 \Rightarrow dx^2 = 1$$

Thus, the unit-length measuring rod laid perpendicular to a radius, i.e., perpendicular to the field, is not shortened. The gravitational field of the mass at the origin has no effect on the length of a rod perpendicular to the field.

3. *Non-Euclidean Space*: The previous two sections show the length of a measuring rod will depend on its orientation (parallel to or perpendicular to the gravitational field) and its location r. In this first approximation it is seen Euclidean geometry is no longer valid.

4. *Gravitational Redshift*: Consider a clock at rest in a static gravitational field. $dx^1 = dx^2 = dx^3 = 0$. $ds^2 = +1 = g_{44}(dx^4)^2$. Using Eq. (5.70) and Eq. (5.68a),

$$dx^4 = \frac{1}{\sqrt{g_{44}}} = \frac{1}{\sqrt{1+(g_{44}-1)}} \approx 1 - \frac{1}{2}(g_{44}-1) = 1 + \frac{\kappa}{8\pi} \int \frac{\rho d\tau}{r} = 1 + \frac{\alpha}{2r}$$

$$(5.72)$$

This indicates a clock runs more slowly when in the vicinity of mass, i.e., in a gravitational field. A consequence of this is that the spectral lines coming to us from the vicinity of a more massive object (stars) will appear to be shifted to lower frequencies, what is known as the redshift.

5. *The Bending of Starlight*: Consider a light ray traveling in a direction given by the quantities dx^1, dx^2, dx^3. In the special theory of relativity the speed of light is given by

$$-(dx^1)^2 - (dx^2)^2 - (dx^3)^2 + (dx^4)^2 = 0$$
$$\Rightarrow -(dx^1)^2 - (dx^2)^2 - (dx^3)^2 + (cdt)^2 = 0$$
$$\Rightarrow c^2 = \left(\frac{dx^1}{dt}\right)^2 + \left(\frac{dx^2}{dt}\right)^2 + \left(\frac{dx^3}{dt}\right)^2$$

And, therefore, in the general theory of relativity by

$$ds^2 = g_{\mu\nu}\, dx^\mu\, dx^\nu = 0 \qquad (5.73)$$

Given the quantities dx^1, dx^2, dx^3, one can form the quantities $\frac{dx^1}{dx^4}, \frac{dx^2}{dx^4}, \frac{dx^3}{dx^4}$ and, from these, the speed γ of the light ray:

$$\gamma = \sqrt{\left(\frac{dx^1}{dx^4}\right)^2 + \left(\frac{dx^2}{dx^4}\right)^2 + \left(\frac{dx^3}{dx^4}\right)^2}$$

If the $g_{\mu\nu}$ are not constant, the direction of the light ray will be bent.

Consider again a gravitational field produced by a point mass M at the origin. The gravitational field is a distance Δ from the origin, measured along the x^1 axis. As one moves out the x^1 axis the potential ϕ will change. Consider a ray of light traveling in the x^2 direction, with a small portion of the wave front in the x^1x^3 plane. See Figure 5.8 (the x^3 axis is not shown).

The wave front extends some length dx^1. Since the potential varies with x^1, so also will the speed of light vary with x^1. In time dt, the wave front will move forward in the x^2 direction, the left end moving forward a distance $\gamma_{left}\,dt$, and the right end a distance $\gamma_{right}\,dt$. See Figure 5.9.

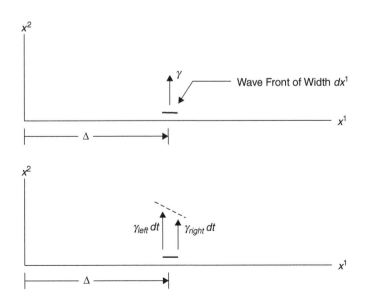

Wave Front of Width dx^1

Fig. 5.8 On the influence of gravitation on the propagation of light.

$\gamma_{left}\,dt$ $\gamma_{right}\,dt$

Fig. 5.9 γ initial.

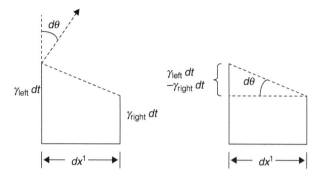

Fig. 5.10 The change in direction $d\theta$.

The change in direction $d\theta$ is given by (see Figure 5.10):

$$d\theta \approx \frac{\gamma_{left}\,dt - \gamma_{right}\,dt}{dx^1} = -\frac{(\gamma_{right} - \gamma_{left})\,dt}{dx^1}$$

$$= -\frac{(\gamma_{right} - \gamma_{left})}{dx^1}\frac{dx^2}{\gamma_{avg}} = -\frac{1}{\gamma}\frac{\partial\gamma}{\partial x^1}\,dx^2$$

The total deflection is the sum (integral) of the $d\theta$s. Remembering that, in these units, $\gamma = 1$, the total deflection B is given as

$$B = \int\limits_{-\infty}^{+\infty} \frac{\partial\gamma}{\partial x^1}\,dx^2$$

For light traveling in the x^2 direction, $dx^1 = 0$, $dx^2 \neq 0$, $dx^3 = 0$, $dx^4 = dt \neq 0$. From Eq. (5.73), for a beam of light,

$$ds^2 = 0 = g_{22}(dx^2)^2 + g_{44}\,dt^2 + g_{24}\,dx^2\,dx^4 + g_{42}\,dx^4\,dx^2.$$

But, by Eq. (5.70), $g_{24} = 0$, $g_{42} = 0$. The speed of light is $\gamma = dx^2/dt = \sqrt{-(g_{44})/g_{22}}$. Using Eqs. (5.70) for g_{22} and g_{44},

$$\gamma = \sqrt{\frac{-g_{44}}{g_{22}}} = \sqrt{\frac{-\left(1 - \frac{\alpha}{r}\right)}{-\left(1 + \frac{\alpha(x^2)^2}{r^3}\right)}} \approx \left(1 - \frac{1}{2}\frac{\alpha}{r}\right)\left(1 - \frac{1}{2}\frac{\alpha\left(x^2\right)^2}{r^3}\right)$$

$$\approx 1 - \frac{1}{2}\frac{\alpha}{r}\left(1 + \frac{\left(x^2\right)^2}{r^2}\right) \tag{5.74}$$

Carrying out the calculation gives (see Appendix 5.4.8 for details)

$$B = \frac{2\alpha}{\Delta} = \frac{\kappa M}{2\pi\Delta}$$

Considering the situation of the sun being a point mass of mass M, Δ being the radius of the sun (the light ray "grazing" the edge of the sun), and κ given by Eq. (5.69), the total deflection of a light ray will be 1.7 seconds of arc.

This result is twice the deflection predicted in his 1911 paper, "On the Influence of Gravitation on the Propagation of Light."[78] The factor of two arose from the fact that the 1911 calculation took into account the variation of time in the gravitational field,[79] whereas the 1915 calculation included the variation in spatial dimensions, in addition to the variation of time in the gravitational field.

6. *The Precession of the Perihelion of Mercury*: As a last comment, Einstein refers to the precession of the perihelion of the planet Mercury, a calculation in his November 18, 1915, report to the Prussian Academy of Science, "Explanation of the Perihelion Motion of Mercury from the General Theory of Relativity."[80] (See Appendix 5.4.9 for details.)

Classical theory could account for the precession of the perihelion of Mercury, but not exactly. After a number of attempts, there remained an unaccounted $45'' \pm 5''$ of precession advance each century. The general theory of relativity predicted a precession advance of $43''$, well within the $45'' \pm 5''$. As Einstein comments, "This theory therefore agrees completely with the observations."[81] This result, coming out of the general theory in a natural way with no special hypotheses, convinced Einstein of the correctness of the theory. See Section 5.3.1.1 for further discussion of the precession of the perihelion of Mercury.

5.3 Discussion and Comments

5.3.1 Verification of the General Theory of Relativity

5.3.1.1 The Precession of the Perihelion of Mercury

One of the great triumphs of Newton's theory of gravity had been the prediction of the existence of the planets Neptune and Uranus to explain the orbits of the other planets (see Section 1.3). However, Newton's theory of gravity was unable to explain exactly the orbit of the planet Mercury.

Due to the Newtonian gravitational force between the sun and the planet, the orbit of each planet is an ellipse. But due to the gravitational interaction between planets, the elliptical orbit does not exactly close in on itself, the difference being very slight. These orbits were described most conveniently as being a perfect elliptical orbit (due to the planet–sun gravitational interaction), but an ellipse which rotated very slowly (due to the planet–planet interactions). This rotation was known as the precession of the perihelion of the orbit, i.e., a rotation of the line drawn from the sun to the perihelion (point of closest approach) of the planet's orbit. See Figure 5.11.

For Mercury, the calculated precession of the perihelion (from interactions with other planets) was about 8.85 minutes of arc per century; the observed precession of the perihelion was about 9.55 minutes of arc per century – the difference being about 43 seconds of arc per century (this is about one extra revolution of Mercury's orbit every three million years). Solutions were proposed, such as the presence of another planet (the planet Vulcan); or Newton's gravitational force was not exactly a

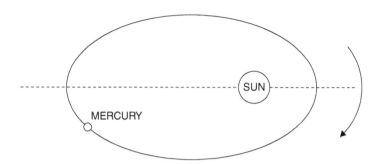

Fig. 5.11 Precession of the perihelion of Mercury.

$1/R^2$ law (perhaps the exponent 2 is not exactly 2 but rather $2 + \varepsilon$, where ε is an extremely small correction); or the non-sphericity of the sun. Each of these, and other, solutions were tried but found wanting, usually because of other effects they predicted that were not observed. Similar results for the precession of the orbits of Venus and the earth were present, but smaller by a factor of five or more.[82]

Einstein's general theory of relativity predicted exactly these missing 43 seconds of rotation each century. More importantly, the result emerged in a most natural way. The discovery of the explanation of the precession of the rotation of the perihelion of Mercury without the need of any special hypotheses was, in the words of Pais, "the strongest emotional experience in Einstein's scientific life... Nature had spoken to him. He had to be right."[83] Einstein writes, in a letter to Ehrenfest, "Imagine my delight at realizing that general covariance was feasible and at finding out that the equations yield Mercury's perihelion motion correctly. I was beside myself with joy and excitement for days."[84] Of all the early tests of the idea of curved spacetime this was the earliest, the most convincing, and the most dramatic.[85]

5.3.1.2   The Bending of Starlight

In his 1915 calculations of the bending of starlight passing near the sun, Einstein arrived at a deflection of 1.7 seconds, twice the value of his previous calculations (0.85 seconds). See Appendix 5.4.9 for a discussion of how the deflection was measured. The failure of the expeditions to measure the bending of starlight prior to 1915 (Brazil in 1912 and Russia in 1914) saved Einstein from the embarrassment of having predicted the wrong value.[86]

Einstein sent a copy of his general theory paper to Willem de Sitter who, in turn, forwarded it to Arthur Eddington, secretary of the Royal Astronomical Society in London. Eddington, a strong mathematician, was captivated by the beauty of the theory. Fortuitously, another solar eclipse, with ideal surrounding factors (bright star fields, full eclipse at certain locations, etc.), would take place in 1919. Eddington set about making the preparations for the May 29, 1919, eclipse.[87]

Two expeditions were dispatched to make eclipse observations, one to Sobral in Brazil (near the equator on the east coast of South America), the other led by Eddington to Principe Island off the coast of Spanish Guinea (now named Equatorial Guinea, on the west coast of Africa). On

June 3, Eddington said, of the sixteen photographs taken on Principe, he could not yet make any announcement, but added, "one plate that I measured gave a result agreeing with Einstein." Later in his life, Eddington referred back to this moment as "the greatest moment of his life." But further confirmation awaited full analysis of the photographs back in England. On September 12, Eddington gave a preliminary report on the results to the British Association for the Advancement of Science. But no one had thought to get the word to Albert Einstein![88] Finally, on September 22, 1919, Lorentz wired to Einstein, "Eddington found stellar shift at solar limb, tentative value between nine-tenths of a second and twice that."[89]

On Thursday, November 6, 1919, Albert Einstein was present at the meeting of the Fellows of the Royal Society and Royal Astronomical Society to hear the official results. These, as announced by Eddington, showed the bending of light as predicted by Einstein's general theory of relativity by an amount of 1.98 ± 0.30 seconds and 1.61 ± 0.30 seconds as observed at Sobral and Principe, respectively. J. J. Thomson, President of the Royal Society, proclaimed it "the most important result obtained in connection with the theory of gravitation since Newton's day.... The result [is] one of the highest achievements of human thought."[90]

When Ilse Rosenthal-Schneider, a student of Einstein's, expressed her joy that the results agreed with his calculations, Einstein replied simply, "But I knew that the theory is correct." Following up, she asked what if there had been no confirmation of the prediction, Einstein countered, "Then I would have been sorry for the Lord – the theory is correct."[91]

At this point, in 1919, Albert Einstein became perhaps the most famous person in the world. He had predicted the impossible – that gravity would bend light and, further, he had predicted the exact amount of this bending.

Einstein was now convinced that gravity was a property of space, and not something reaching out from the sun and pulling on the earth as Newton had said. The effect of the mass of the sun was to distort the space around it. The earth, then moving through this distorted space, followed the path determined by the distortions in space. The sun distorts the space. The earth follows these distortions.

5.3.1.3 The Gravitational Redshift

By the early 1920s, two of the three predictions of the general theory of relativity had been confirmed. The predicted precession of the perihelion of Mercury agreed with long-known observations. The bending of starlight had recently been confirmed (in 1919). The third prediction, the gravitational redshift, was still awaiting certain confirmation.[92] Already, in 1915, Erwin Freundlich had confirmed the redshift, but other astronomers raised questions regarding Freundlich's analysis and were unable to confirm the results.[93]

The first confirmation of substantial accuracy was the Pound–Rebka experiment of 1959. Although difficult in practice, the experiment is simple in concept. An emitter at rest emits a signal of frequency ν_0. A receiver, also at rest, receives the signal at the same frequency ν_0. If there is a constant uniform velocity between the emitter and receiver, the frequency at the receiver will be Doppler shifted by some amount $\Delta\nu_{Doppler}$. A second source of a frequency change is if the system of (emitter plus detector) is accelerating, producing some frequency shift $\Delta\nu_{acceleration}$. From the equivalence principle, the same frequency shift will occur in a gravitational field, $\Delta\nu_{gravity} = \Delta\nu_{acceleration}$. In the Pound–Rebka experiment, ^{57}Fe was used as the emitter, placed at the top of a 22.5 m tower. ^{57}Fe was used also as the receiver, placed at the bottom of the tower. If the gravitational redshift predictions were correct, the frequency of the signal will have been shifted by an amount $\Delta\nu_{gravity}$ when it has reached the receiver. This reduces the absorption efficiency of the detector an amount that was measurable. The emitter then was given a velocity to produce a Doppler shift. The velocity giving the maximum absorption would correspond to the Doppler shift that exactly compensated the gravitational redshift. The ultimate result of the Pound–Rebka experiment was a confirmation of the gravitational redshift prediction to about 1%.[94]

The terrestrial confirmation of the general theory of relativity had been delayed some 45 years because precise enough measurements of the atomic frequencies were not possible until the discovery of the Mössbauer effect in 1958. In 1961, Rudolf Ludwig Mössbauer received the Nobel prize for the Mössbauer effect.

5.3.2   Beyond the General Theory of Relativity: Cosmology and the Unified Field Theory

With the completion of the general theory of relativity in late 1915, and its publication in 1916, Einstein moved on to generalize the theory to include also the electromagnetic fields of Maxwell (the unified field theory), and to use the theory to speculate on the size and structure of the universe (founding the modern theoretical study of cosmology[95]). His first paper on cosmology, "Cosmological Considerations in the General Theory of Relativity,"[96] was published in 1917, and his first papers on the unified field theory were published in 1922 and 1923. The 1922 and 1923 papers were reactions to work done by others. In 1925, he published his own version of a unified field theory.[97]

But, throughout this period, he continued his search for an understanding of quantum mechanics. One of his hopes was that it would emerge out of a properly developed unified field theory. During this period there had arisen a wide-spread belief that Einstein simply did not care anymore about the quantum theory and had moved on. To the contrary, he was pursuing a unified field theory that he was hoping not only would join together gravitational and electromagnetic fields, but also would contain within it the basis for a new interpretation of quantum phenomena.[98]

5.3.2.1   Cosmology

Cosmology, as the philosophical study and explanation of the nature of the universe,[99] dates from the ancient Greeks describing the universe as earth-centered, with the heavens being perfection and composed of quintessence. (See Section 1.2.1.)

The ancient Greek concept of a finite universe with the earth at its center was refuted by the work of Copernicus (earth not at the center) and Newton (a finite material universe would collapse upon itself). Newton proposed an infinite universe populated with an infinitude of stars spread throughout. In the nineteenth century, an "island" universe existing within an infinite and empty space was proposed. Each of these proposed universes was a static, unchanging universe.[100]

In Newton's infinite universe, the infinite number of stars would produce a strong force that would give the stars a high velocity throughout the universe. But observation showed the velocities of stars to be small, contrary to the assumption of an infinitude of stars. The island universe also was ruled out, but for more complex reasons. Einstein's proposed universe was a curved space, which turned back on itself much as the surface of a sphere turns back on itself in three-dimensional space. Let us now begin with the ideas of Isaac Newton.[101]

Isaac Newton saw the universe as playing out against a backdrop of absolute space, and set about to determine our motion relative to this absolute space. When Newton developed his mechanics, he spoke of the laws of mechanics being valid in absolute space. Newton then realized, if the laws were valid in absolute space they would be valid also in any reference system moving with constant linear velocity relative to absolute space, i.e., the reference frame of absolute space could not be distinguished from a reference frame moving through absolute space at a constant linear velocity.

To distinguish absolute motion (motion relative to absolute space) from relative motion, Newton proposed an experiment that has become known as "Newton's Bucket" (see Appendix 5.4.10 for further details in Newton's words). In this experiment, a bucket nearly full of water is sitting at rest on a table. The water is motionless relative to the bucket and the surface of the water remains flat. The bucket is now set rotating and, once the water has begun rotating at the same rate as the bucket, the relative velocity between the water and the bucket is again zero. But now the surface of the spinning water is concave, rising up on the sides of the bucket and being depressed in the center. Since the relative velocity of the water and bucket is the same in both instances, the concave surface of the water in the second instance cannot be due to the relative velocity of the water and bucket. This indicated to Newton the rotation is relative to absolute space.[102]

In the late 1800s, Ernst Mach reinterpreted Newton's bucket experiment, noting, "Newton's experiments with the rotating vessel of water simply informs us that [the curvature of the water's surface] is produced by its relative rotation with respect to the mass of the earth and other celestial bodies."[103]

Thus the question remained unanswered. Is there an absolute space and, if so, are there any effects that would show its existence? Einstein's special theory of relativity had shown velocities are relative, not absolute, while his general theory of relativity had shown accelerations also are relative, not absolute. Einstein believed there were no effects that depended on the concept of absolute space and, thus, in a Machian perspective, believed there was no place for the concept of absolute space in science.

After 1915, Einstein's general theory of relativity became the accepted theory of gravitation and the foundation on which to discuss the large-scale structure of the universe.[104] Building on the general theory, in 1917 he published "Cosmological Considerations in the General Theory of Relativity."[105] This is the beginning of cosmology as the scientific study of the origin and structure of the universe.

In the cosmology paper, Einstein looked first at Newton's theory of gravitation. Newton's gravitational potential, ϕ, satisfies Poisson's equation:

$$\nabla^2 \phi = 4\pi K \rho$$

For ϕ to be a constant at spatial infinity, the density of matter, ρ, must go to zero at infinity more rapidly than $1/r^2$. The ratio of densities at infinity and at the center corresponds to the finite difference in potential between infinity and the center. If the density at infinity is zero, to achieve a finite difference in the potential at infinity and at the center, the density at the center must also be zero.[106] Looking at these, and other concerns, Einstein believed it would not be possible to overcome them within Newton's theory. To overcome them he suggested a modification of the Poisson equation,

...a method which does not in itself claim to be taken seriously; it merely serves as a foil for what is to follow. In place of Poisson's equation we write

$$\nabla^2 \phi - \lambda \phi = 4\pi K \rho$$

where λ denotes a universal constant. If ρ_0 be the uniform density of a distribution of mass, then

$$\phi = -\frac{4\pi K}{\lambda} \rho_0$$

is a solution [to the modified Poisson equation]. This solution would correspond to the case in which the matter of the fixed stars was distributed uniformly through space, if the density ρ_0 is equal to the actual mean density of the matter in the universe. The solution then corresponds to an infinite extension of the central space, filled uniformly with matter.[107]

λ, the constant introduced by Einstein, is called the cosmological constant. It is introduced in an analogous manner into the field equations of gravitation.[108]

$$R_{\mu\nu} - \frac{1}{2} g_{\mu\nu} R = -\kappa T_{\mu\nu}$$

is replaced by

$$R_{\mu\nu} - \frac{1}{2} g_{\mu\nu} R - \lambda g_{\mu\nu} = -\kappa T_{\mu\nu}$$

Calculation shows

$$\lambda = \frac{1}{R^2} = \frac{1}{2}\kappa\rho c^2$$

Thus λ is found to be proportional to ρ, the mean mass density of the universe, and inversely proportional to the square of R, the radius of curvature of the universe.[109] This Einsteinian universe does not encounter the problem of the Newtonian infinite universe because the Einstein universe is a three-dimensional spherically bounded space of radius $R = \sqrt{1/\lambda}$.[110] Einstein concludes the cosmology paper with these words:

Thus the theoretical view of the actual universe, if it is in correspondence with our reasoning, is the following. The curvature of space is variable in time and place, according to the distribution of matter, but we may roughly approximate to it by means of a spherical space. At any rate, this view is logically consistent, and from the standpoint of the general theory of relativity lies nearest at hand; whether from the standpoint of present astronomical knowledge, it is tenable, will not here be discussed. In order to arrive at this consistent view, we admittedly had to introduce an extension of the field equations of gravitation which is not justified by our actual knowledge of gravitation. It is to be emphasized, however, that a positive curvature of space is given by our results, even if the supplementary term is not introduced. That term is necessary only for the purpose of making possible a quasi-static distribution of matter, as required by the fact of the small velocities of the stars.[111]

"The introduction of [the cosmological constant λ] allowed this first quantitative cosmological model to be uniform in space and time, with no evolution taking place.... No redshifts are predicted in this universe model."[112] Einstein believed that "in a consistent relativity theory there cannot be inertia relative to 'space' but only inertia of masses relative to each other."[113] Calculation showed that if there is no matter, i.e., $\rho = 0$, there is no inertia in the Einstein universe. This satisfied Einstein's strong belief at the time of the relativity of inertia.[114]

Einstein's postulate of a uniform density was an audacious move because it was contrary to generally accepted information at the time. As George Ellis states it:

... [A]t the time Einstein proposed his static universe model, not only was there no evidence available that the universe might be spatially homogeneous, but even the natures and distances of galaxies ('nebulas') were unknown. Indeed it was plausibly thought by many that they might all be subsystems of the Milky Way – a manifestly anisotropic and inhomogeneous structure.... Einstein's universe model implied a completely uniform matter distribution despite the observational evidence then available.[115]

In 1917, upon learning of Einstein's cosmological paper, De Sitter produced a second solution of Einstein's field equations, these for a density $\rho = 0$. This second solution, now termed the de Sitter universe, predicted gravitational redshifts. It also allowed the possibility of "inertia relative to space," raising again the possibility of an absolute space independent of matter. Over the next year, Einstein looked for errors in the de Sitter's calculations, but concluded de Sitter's calculations were correct.[116]

The Einstein and de Sitter universes, Einstein's uniformly full and de Sitter's uniformly empty, were static, non-evolving universes. In 1922, the Russian, Alexander Friedmann, proposed a third model of the universe, one that was not static but was evolving. Friedmann did not determine if there were any redshifts.[117] George Gamow, a student of Friedmann, reported that

Friedmann noticed that Einstein had made a mistake in his alleged proof that the universe must necessarily be stable and unchangeable in time. . . .

It is well known to students of high-school algebra that it is permissible to divide both sides of an equation by any quantity, provided that this quantity is not zero. However, in the course of his proof, Einstein had divided both sides of one of his intermediate equations by a complicated expression which, in certain circumstances, could become zero.

In the case, however, when this expression becomes equal to zero, Einstein's proof does not hold, and Friedmann realized that this opened an entire new world of time-dependent universes: expanding, collapsing, and pulsating ones.[118]

In 1927, Georges Lemaître, apparently unaware of Friedmann's work, developed a model of an expanding and evolving universe similar to Friedmann's, but including an expected redshift. He is given credit for being the first person to seriously propose an expanding universe as a model of the real universe. Lemaître's model begins like an Einstein static universe and evolves into a de Sitter static universe; that is, it lies "between" the Einstein and de Sitter universes. Since the accepted wisdom at that time was that the universe was static, i.e., not changing in time, Lemaître's ideas had little influence on cosmological discussions.[119]

The observed universe of course was neither full nor empty, thus neither really an Einsteinian, everywhere full universe, nor a de Sitter, everywhere empty universe. When Hubble announced his red shift findings in 1929, it was apparent that something different was needed; what Eddington called an intermediate solution, a solution between de Sitter's emptiness and Einstein's fullness. Lemaître's model, languishing on the sidelines the past two years, was recognized as the intermediate solution Eddington was looking for. "After Lemaître's work was called to the attention of the astronomical community by Eddington [in 1930], the expanding and evolving universe concept became widely accepted, and an explosion of papers then explored this concept."[120] "Consensus formed around the concept that the universe began in a static Einstein state, suffered an indeterminate period of 'stagnation,' started expanding due to an instability as described by Lemaître, and would end as a de Sitter universe with galaxies spread out so thinly that a virtual emptiness would result.[121] These universes would have no beginning – they had existed forever in the past as an Einstein universe – and they would have no end – they would exist forever into the future, eventually becoming a de Sitter universe.

With the possibility of a non-static universe there no longer was the need for Einstein's cosmological constant. Einstein dropped the cosmological constant term from the field equations, i.e., set λ equal to zero. He felt removing the cosmological constant removed a "blemish on the beauty of the theory."[122] Years later, looking back on his introduction of the cosmological constant, he said, "If Hubble's expansion had been discovered at the time of the creation of the general theory of relativity, the cosmological member never would have been introduced. It seems now so much less justified to introduce such a member into the field equation, since its introduction loses its sole original justification – that of leading to a natural solution of the cosmological problem."[123] More emphatically, in a conversation with Gamow, Einstein said, "the introduction of the cosmological constant was the greatest blunder [I] ever made in [my] life."[124]

R. W. Smith suggests the addition, and subsequent removal, of the cosmological constant also had an important impact on Einstein's subsequent scientific methodology,

the cosmological constant had the status of an *ad hoc* hypothesis, not required by the kinds of simplicity considerations that were such an important driving force in Einstein's work. Realizing that the constant was not needed could well have renewed Einstein's faith in simplicity as a criterion of theory choice, and in this light it is perhaps no accident that this theme emerged so much more prominently than before in Einstein's writings in the early to mid 1930s.[125]

In 1933, Einstein and de Sitter published a new model, known as the Einstein–de Sitter universe, that was the simplest expanding universe model consistent with a zero cosmological constant. But a zero cosmological constant indicates a singular origin of the universe, i.e., it has a beginning (the big bang) and has existed for a finite time. Lemaître then started a search for remnant radiation from the big bang, the radiation discovered in 1965 by Penzias and Wilson:[126]

Returning to the theme of the reality of absolute space, in 1952 Einstein wrote,

On the basis of the general theory of relativity . . . space as opposed to 'what fills space' . . . has no separate existence. . . . If we imagine the gravitational field, i.e. the functions g_{ik}, to be removed, there does not remain a space . . . , but absolutely *nothing*, and also no 'topological space.' For the functions g_{ik} describe not only the field, but at the same time also the topological and metrical structural properties of the manifold. . . . There is no such thing as an empty space, i.e. a space without field. Space-time does not claim existence on its own, but only as a structural quality of the field.[127]

And in 1954, if there were any lingering doubt on his views, Einstein wrote,

It required a severe struggle to arrive at the concept of independent and absolute space, indispensable for the development of theory. It has required no less strenuous exertions subsequently to overcome this concept – a process which is by no means as yet completed.[128]

5.3.2.2   The Unified Field Theory

By the end of 1915, the general theory of relativity was complete, verified by the precession calculations for Mercury, and awaiting confirming measurements for the bending of starlight and for the redshift of radiation in a gravitational field. Just as he had generalized the special theory of relativity to the general theory of relativity, Einstein was now looking to generalize the general theory of relativity to include the electromagnetic field, at that time the only field other than gravity known to exist. As Pais describes the situation:

> ...the very existence of the gravitational field is inalienably woven into the geometry of the physical world....Riemannian geometry does not geometrize electromagnetism. Should not one therefore try to invent a more general geometry in which electromagnetism would be just as fundamental as gravitation? If the special theory had unified electricity and magnetism and if the general theory had geometrized gravitation, should not one try next to unify and geometrize electromagnetism and gravity?...[The purpose of his program for a unified field theory] was neither to incorporate the unexplained nor to resolve any paradox. It was purely a quest for harmony.[129]

The first unified field theory based upon the general theory of relativity was by David Hilbert in 1915.[130] In 1919, Einstein considered a tentative link in his gravitational equations, suggesting electromagnetism constrains gravitation, this being perhaps his first foray into a unified field theory.[131] Although Einstein's interest in a unified field theory dates back to this time, his first proposal for a theory based on the general theory of relativity was not published until 1925.[132]

In 1919, Theodor Kaluza sent to Einstein a paper, requesting his help in having the paper published in the Prussian Academy.[133] In the paper, Kaluza speculated that

> the electromagnetic field tensor might be a truncated Christoffel symbol. Since, in a four-dimensional world, these symbols are saturated by the components of the gravitational field, one is led 'to the extremely odd decision to ask for help from a new, fifth dimension of the world.'...[This] possibility could be implemented by assuming proportionality of the electromagnetic potential to the mixed ($\mu5$) components of the metric.[134]

Thus from the beginning Einstein was interested in a five-dimensional approach.[135]

In the general theory of relativity, the gravitational potential was represented by a symmetric tensor of rank two (a 4×4 symmetric matrix), whereas in the relativistic treatment of Maxwell's theory, the electromagnetic field was represented by an antisymmetric tensor of rank two (a 4×4 antisymmetric matrix).

> ...Einstein discovered that there existed a mathematical literature on a type of Riemannian space including both curvature and torsion. Torsion is represented by an antisymmetric tensor that generalizes a classical notion in differential geometry. In those spaces that possess a linear connection, there are two limiting cases: if the torsion is zero, but not the curvature, the space

is Riemannian; if the curvature equals zero, but not the torsion, it is a space with 'distant parallelism' (*Fernparallelismus*) or 'absolute parallelism.'...

In the 'old' theory of gravitation (general theory of relativity) Riemannian space-time had curvature but no torsion. Now Einstein suggested a space with torsion and no curvature. Einstein was led in that direction because there is in this case an antisymmetric tensor of rank two, which could be taken to be the electromagnetic tensor, and another symmetric tensor of rank two, which could be used to represent gravitation.[136]

In a 1928 paper, Einstein introduced the new geometry, characterized by the property of distant parallelism.[137]

For the last thirty years of his life, the unified field theory dominated Einstein's thinking. Periodically he would return to the five-dimensional approach of Kaluza, interspersed with work in four-dimensional space. His goal was clear: (1) to join gravity with electromagnetism into one unified theory, and (2) to obtain from this unified theory an understanding of the quantum theory. Although this journey never achieved the success of his journey to the general theory of relativity, he remained in pursuit of the unified field theory until his last days.

5.4 Appendices

5.4.1 Multiplication of Tensors

5.4.1.1 An Example of the Outer Multiplication of Two Vectors

Consider two three dimensional vectors, $\vec{A}$ and $\vec{B}$, expressed in Cartesian coordinates, with unit vectors $\hat{i}, \hat{j}, \hat{k}$ aligned with the x, y, and z axes, respectively. $\vec{A} = A_1\hat{i} + A_2\hat{j} + A_3\hat{k}$ and $\vec{B} = B_1\hat{i} + B_2\hat{j} + B_3\hat{k}$. The outer multiplication of these two vectors is

$$(\vec{A} = A_1\hat{i} + A_2\hat{j} + A_3\hat{k})(\vec{B} = B_1\hat{i} + B_2\hat{j} + B_3\hat{k}) =$$

$$\begin{pmatrix} A_1B_1\hat{i}\hat{i} + A_1B_2\hat{i}\hat{j} + A_1B_3\hat{i}\hat{k}+ \\ A_2B_1\hat{j}\hat{i} + A_2B_2\hat{j}\hat{j} + A_2B_3\hat{j}\hat{k}+ \\ A_3B_1\hat{k}\hat{i} + A_3B_2\hat{k}\hat{j} + A_3B_3\hat{k}\hat{k} \end{pmatrix} \tag{5.75}$$

5.4.1.2 "If $A_{\mu\nu}B^{\mu\nu}$ is a scalar for any choice of the tensor $B^{\mu\nu}$, then $A_{\mu\nu}$ has tensor character."[138]

If $A_{\mu\nu}B^{\mu\nu}$ is a scalar, then $A'_{\sigma\tau}B'^{\sigma\tau} = A_{\mu\nu}B^{\mu\nu}$. But, from Eq. (5.9), $B'^{\sigma\tau} = \frac{\partial x'^{\sigma}}{\partial x^{\mu}}\frac{\partial x'^{\tau}}{\partial x^{\nu}}B^{\mu\nu}$. Inverting this, with $\sigma\tau \leftrightarrow \mu\nu$ and primed $\leftrightarrow$ unprimed, we obtain $B^{\mu\nu} = \frac{\partial x^{\mu}}{\partial x'^{\sigma}}\frac{\partial x^{\nu}}{\partial x'^{\tau}}B'^{\sigma\tau}$. We can write,

$$A'_{\sigma\tau}B'^{\sigma\tau} = A_{\mu\nu}B^{\mu\nu} = A_{\mu\nu}\frac{\partial x^{\mu}}{\partial x'^{\sigma}}\frac{\partial x^{\nu}}{\partial x'^{\tau}}B'^{\sigma\tau}$$

$$\left[A'_{\sigma\tau} - \frac{\partial x^{\mu}}{\partial x'^{\sigma}}\frac{\partial x^{\nu}}{\partial x'^{\tau}}A_{\mu\nu} \right]B'^{\sigma\tau} = 0$$

For this to be true for arbitrary $B'^{\sigma\tau}$, it requires that $[\ ] = 0$. Thus,

$$A'_{\sigma\tau} = \frac{\partial x^\mu}{\partial x'^\sigma}\frac{\partial x\nu}{\partial x'^\tau}A_{\mu\nu}$$

By definition, this is the covariant transformation of $A_{\mu\nu}$. Thus $A_{\mu\nu}$ is a covariant tensor.

5.4.2 Some Aspects of the Fundamental Tensor $g_{\mu\nu}$

5.4.2.1 The Determinant of $g_{\mu\nu}$

The determinant of the identity tensor $\delta^\nu_\mu = |\delta^\nu_\mu| = 1$. Taking the determinant of $g_{\mu\alpha}g^{\alpha\nu} = \delta^\nu_\mu$ we have:

$$|g_{\mu\alpha}g^{\alpha\nu}| = |\delta^\nu_\mu|$$
$$|g_{\mu\alpha}|\,|g^{\alpha\nu}| = 1$$

Or, since the two determinants are calculated separately, each set of indices (subscripts and superscripts) can be set to the same indices, $\mu\nu$.

$$|g_{\mu\nu}|\,|g^{\mu\nu}| = 1$$

5.4.2.2 The Volume Scalar

Expressed in terms of contravariant differentials, the four-dimensional volume element is $d\tau = dx^1\,dx^2\,dx^3\,dx^4$. Changing to another set of coordinates, $\{x'_1, x'_2, x'_3, x'_4\}$ the volume element transforms as $d\tau' = Jd\tau$, where J is the Jacobian of the transformation, $J \equiv |\frac{\partial x'^\sigma}{\partial x^\mu}|$.

The fundamental tensor $g_{\mu\nu}$ transforms as $g'_{\sigma\tau} = \frac{\partial x^\mu}{\partial x'^\sigma}\frac{\partial x^\nu}{\partial x'^\tau}g_{\mu\nu}$. Taking the determinant of this equation:

$$g' = |g'_{\sigma\tau}| = \left|\frac{\partial x^\mu}{\partial x'^\sigma}\frac{\partial x^\nu}{\partial x'^\tau}g_{\mu\nu}\right|$$

$$= \left|\frac{\partial x^\mu}{\partial x'^\sigma}\right|\left|\frac{\partial x^\nu}{\partial x'^\tau}\right||g_{\mu\nu}|$$

$$= \left|\frac{\partial x^\mu}{\partial x'^\sigma}\right|^2 g$$

Taking the square root of this equation, $\sqrt{g'} = |\frac{\partial x^\mu}{\partial x'^\sigma}|\sqrt{g}$. Multiplying the volume element $d\tau\prime$ by $\sqrt{g'}$,

$$\sqrt{g'}d\tau' = \left(\left|\frac{\partial x^\mu}{\partial x'_\sigma}\right|\sqrt{g}\right)d\tau'$$

$$= \left(\left|\frac{\partial x^\mu}{\partial x'^\sigma}\right|\sqrt{g}\right)\left(\left|\frac{\partial x'^\sigma}{\partial x^\mu}\right|d\tau\right)$$

$$= \sqrt{g}d\tau$$

Since $|\frac{\partial x^\mu}{\partial x'^\sigma}|$ and $|\frac{\partial x'^\sigma}{\partial x^\mu}|$ are the determinants of the Jacobian of transformation from the unprimed to the primed coordinates, and then from the primed to the unprimed coordinates, the two transformations return the volume element to the original coordinates and their product must be equal to one.

5.4.3   The Equation of the Geodetic Line

5.4.3.1   Derivation of Eq. (5.20c): The Equations of the Geodetic Line

The line drawn between two points P and P$'$ in four dimensional space is a geodetic line if

$$\delta \int_{P}^{P'} ds = 0 \qquad (5.20)$$

The geodetic line is the physical path followed. Einstein derives the equation of the geodetic line by "carrying out the variation in the usual way." [139] See Figure 5.12.

1. Consider a number of other paths in close proximity to the geodetic line that also pass through the points P and P$'$. See Figure 5.13.
2. Let λ be some function of the contravariant coordinates, $\lambda = \lambda(x^1, x^2, x^3, x^4)$, that defines a family of surfaces that intersect the geodetic line as well as the other lines in proximity to it. See Figure 5.14. (In four-dimensional space the surfaces defined by $\lambda =$ constant are three-dimensional surfaces.)
3. The coordinates x^ν on a given line (geodetic or proximate) can be expressed as a function of the parameter λ, $x^\nu = x^\nu(\lambda)$.
4. Designate the value of λ for the surface passing through the point P as $\lambda_1 = \lambda(P) = \lambda(x_P^1, x_P^2, x_P^3, x_P^4)$ and the value of λ

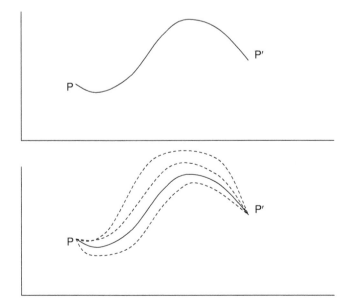

Fig. 5.12 The geodetic line between points P and P$'$.

Fig. 5.13 The geodetic line and some proximate lines.

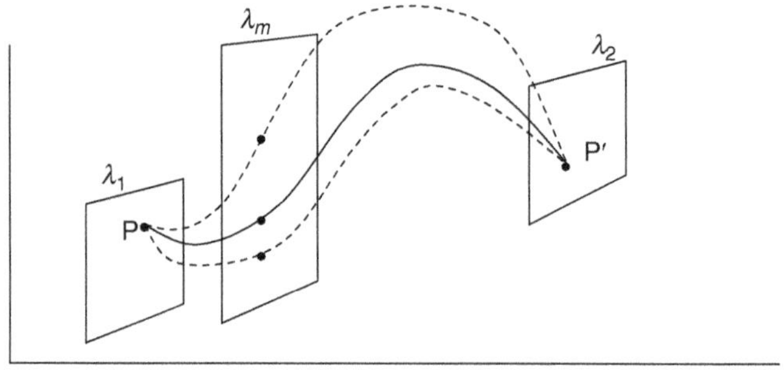

Fig. 5.14 The family of surfaces defined by λ.

for the surface passing through the point P' as $\lambda_2 = \lambda(P') = \lambda(x_{P'}^1, x_{P'}^2, x_{P'}^3, x_{P'}^4)$.

5. For a given λ, δ represents the transition from the point on the geodetic line to the point on the proximate line.

6. The integral $\int_P^{P'} ds$ is from the point P to the point P' or, equivalently, from λ_1 to λ_2.

The expression for ds is manipulated in the following manner:

$$ds^2 = g_{\mu\nu}\, dx^\mu\, dx^\nu$$

$$= g_{\mu\nu}\left(\frac{dx^\mu}{d\lambda}d\lambda\right)\left(\frac{dx^\nu}{d\lambda}d\lambda\right)$$

$$= g_{\mu\nu}\frac{dx^\mu}{d\lambda}\frac{dx^\nu}{d\lambda}(d\lambda)^2$$

$$ds^2 = w^2\,(d\lambda)^2$$

With $w^2 = g_{\mu\nu}\frac{dx^\mu}{d\lambda}\frac{dx^\nu}{d\lambda}$

Thus, Eq. (5.20) can be rewritten as

$$\delta\int_P^{P'} ds = 0 = \delta\int_{\lambda_1}^{\lambda_2} w\,d\lambda = \int_{\lambda_1}^{\lambda_2} \delta w\,d\lambda \qquad (5.20\text{a})$$

The δ and the integral can be interchanged since the integration is over λ and the variation δ is for a given value of λ.

The variation of an arbitrary function f is $\delta f = \frac{\partial f}{\partial x'^\sigma}\delta x'^\sigma$. Written in this form, it is seen that $\delta x'^\sigma$ is a contravariant tensor while, by Eq. (5.7), $\frac{\partial f}{\partial x'^\sigma} = \frac{\partial x^\nu}{\partial x'^\sigma}\frac{\partial f}{\partial x^\nu}$ transforms as a covariant tensor. By Eq. (5.6), the quantity δf is an invariant. (For completeness, it should be noted that we could write $\delta f = \frac{\partial f}{\partial x_\sigma}\delta x_\sigma$. Since δf was just shown to be an invariant and dx_σ is a covariant vector, the derivative $\frac{\partial f}{\partial x_\sigma}$ must have the properties of a contravariant vector.)

It now remains to determine the expression for δw. The variation δ is not a variation in λ (δ is for a fixed value of λ), rather it is a variation in x^ν, i.e., $\delta \Rightarrow \Delta x^\nu$, not $\Delta\lambda$. Working with the expression for w^2,

$$\delta\left(w^2\right) = 2w\delta w = \delta\left(g_{\mu\nu}\frac{dx^{\mu}}{d\lambda}\frac{dx^{\nu}}{d\lambda}\right)$$

$$= \left\{\left(\frac{\partial g_{\mu\nu}}{\partial x^{\sigma}}\delta x^{\sigma}\right)\frac{dx^{\mu}}{d\lambda}\frac{dx^{\nu}}{d\lambda} + g_{\mu\nu}\delta\left(\frac{dx^{\mu}}{d\lambda}\right)\frac{dx^{\nu}}{d\lambda} + g_{\mu\nu}\frac{dx^{\mu}}{d\lambda}\delta\left(\frac{dx^{\nu}}{d\lambda}\right)\right\}$$

(a) Since the indices μ and ν in the second and third terms are each summed over the values 1 to 4, the terms are the same, and can be written as twice either of them.

(b) Since the δ is at constant λ, we can exchange the variation δ and the derivative with respect to λ, i.e., $\delta(\frac{dx^{\mu}}{d\lambda}) = \frac{d}{d\lambda}(\delta x^{\mu})$.

(c) In the second term, replace the index ν by the index σ:

$$\delta w = \frac{1}{2w}\left\{\left(\frac{\partial g_{\mu\nu}}{\partial x^{\sigma}}\delta x^{\sigma}\right)\frac{dx^{\mu}}{d\lambda}\frac{dx^{\nu}}{d\lambda} + 2g_{\mu\sigma}\frac{dx^{\mu}}{d\lambda}\frac{d}{d\lambda}\left(\delta x^{\sigma}\right)\right\}$$

Substituting this expression for δw into Eq. (5.20a):

$$0 = \int_{\lambda_1}^{\lambda_2}\delta w d\lambda = \int_{\lambda_1}^{\lambda_2}\left\{\frac{1}{2w}\frac{\partial g_{\mu\nu}}{\partial x^{\sigma}}\frac{dx^{\mu}}{d\lambda}\frac{dx^{\nu}}{d\lambda}\delta x^{\sigma} + \frac{g_{\mu\sigma}}{w}\frac{dx^{\mu}}{d\lambda}\frac{d}{d\lambda}\left(\delta x^{\sigma}\right)\right\}d\lambda$$

Working with the second term, and using

$$\frac{d}{d\lambda}\left(AB\right) = \left(\frac{dA}{d\lambda}\right)B + A\left(\frac{dB}{d\lambda}\right) \Rightarrow A\left(\frac{dB}{d\lambda}\right) = \frac{d}{d\lambda}\left(AB\right) - \left(\frac{dA}{d\lambda}\right)B$$

We obtain

$$\left(\frac{g_{\mu\sigma}}{w}\frac{dx^{\mu}}{d\lambda}\right)\frac{d}{d\lambda}\left(\delta x^{\sigma}\right) = \frac{d}{d\lambda}\left(\frac{g_{\mu\sigma}}{w}\frac{dx^{\mu}}{d\lambda}\delta x^{\sigma}\right) - \delta x^{\sigma}\frac{d}{d\lambda}\left(\frac{g_{\mu\sigma}}{w}\frac{dx^{\mu}}{d\lambda}\right)$$

The integration over the first term is equal to zero, since $\delta x^{\sigma} = 0$ at λ_1 and at λ_2:

$$\int_{\lambda_1}^{\lambda_2}\frac{d}{d\lambda}\left\{\frac{g_{\mu\sigma}}{w}\frac{dx^{\mu}}{d\lambda}\delta x^{\sigma}\right\}d\lambda = \frac{g_{\mu\sigma}}{w}\frac{dx^{\mu}}{d\lambda}\delta x^{\sigma}\Bigg]_{\lambda_1}^{\lambda_2} = 0$$

Substituting these results into the integral, one obtains

$$0 = \int_{\lambda_1}^{\lambda_2}\delta w d\lambda = \int_{\lambda_1}^{\lambda_2}\left\{\frac{1}{2w}\frac{\partial g_{\mu\nu}}{\partial x^{\sigma}}\frac{dx^{\mu}}{d\lambda}\frac{dx^{\nu}}{d\lambda} - \frac{d}{d\lambda}\left(\frac{g_{\mu\sigma}}{w}\frac{dx^{\mu}}{d\lambda}\right)\right\}\delta x^{\sigma}d\lambda$$

But since the δx^{σ} are arbitrary, for the integral always to be zero, the expression in the brackets {} must be equal to zero. This expression is denoted as $-\kappa_{\sigma}$:

$$0 = -\kappa_{\sigma} = \frac{1}{2w}\frac{\partial g_{\mu\nu}}{\partial x^{\sigma}}\frac{dx^{\mu}}{d\lambda}\frac{dx^{\nu}}{d\lambda} - \frac{d}{d\lambda}\left(\frac{g_{\mu\sigma}}{w}\frac{dx^{\mu}}{d\lambda}\right) \qquad (5.20c)$$

These four equations, one for each of the four values of the index σ, are the equations of the geodetic line.

5.4.3.2   Alternate Form of the Equations of the Geodetic Line

The parameter λ has been specified only in that it designate the surfaces that intersect the geodetic line and the paths (lines) in proximity to the geodetic line. We can choose to define λ as the length along the geodetic line, i.e., $d\lambda = ds$. As long as ds does not vanish, from the relation in Eq. (5.20a), above, we have $ds = wd\lambda$, and $w = 1$. Substituting into Eq. (5.20c) ds for $gd\lambda$ and $w = 1$ (note: the order of the terms has been reversed to agree with the order in Einstein's paper),

$$0 = \frac{d}{ds}\left(g_{\mu\sigma}\frac{dx^{\mu}}{ds}\right) - \frac{1}{2}\frac{\partial g_{\mu\nu}}{\partial x^{\sigma}}\frac{dx^{\mu}}{ds}\frac{dx^{\nu}}{ds}$$

$$= g_{\mu\sigma}\frac{d^2 x^{\mu}}{ds^2} + \frac{\partial g_{\mu\sigma}}{\partial x^{\nu}}\frac{dx^{\nu}}{ds}\frac{dx^{\mu}}{ds} - \frac{1}{2}\frac{\partial g_{\mu\nu}}{\partial x^{\sigma}}\frac{dx^{\mu}}{ds}\frac{dx^{\nu}}{ds} \qquad (5.20e)$$

Following Christoffel, Einstein introduces the following notation:[140]

$$[\mu\nu,\sigma] = \frac{1}{2}\left(\frac{\partial g_{\mu\sigma}}{\partial x^{\nu}} + \frac{\partial g_{\nu\sigma}}{\partial x^{\mu}} - \frac{\partial g_{\mu\nu}}{\partial x^{\sigma}}\right). \qquad (5.21)$$

In the German original,[141] the notation is $\left[\begin{smallmatrix}\mu\nu\\\sigma\end{smallmatrix}\right]$, but represents exactly the same expression as $[\mu\nu,\sigma]$. With this notation, the equations for the geodetic line can be written:

$$0 = g_{\mu\sigma}\frac{d^2 x^{\mu}}{ds^2} + [\mu\nu,\sigma]\frac{dx^{\mu}}{ds}\frac{dx^{\nu}}{ds} \qquad (5.20d)$$

That this is true can be seen by writing out the terms for $[\mu\nu,\sigma]$:

$$0 = g_{\mu\sigma}\frac{d^2 x^{\mu}}{ds^2} + \frac{1}{2}\left(\frac{\partial g_{\mu\sigma}}{\partial x^{\nu}} + \frac{\partial g_{\nu\sigma}}{\partial x^{\mu}} - \frac{\partial g_{\mu\nu}}{\partial x^{\sigma}}\right)\frac{dx^{\mu}}{ds}\frac{dx^{\nu}}{ds}$$

The summation over μ and ν for the first two terms in $()\times \frac{dx^{\mu}}{ds}\frac{dx^{\nu}}{ds}$ gives the same sum. Thus, combining these two terms, one obtains

$$0 = g_{\mu\sigma}\frac{d^2 x^{\mu}}{ds^2} + \left(\frac{\partial g_{\mu\sigma}}{\partial x^{\nu}} - \frac{1}{2}\frac{\partial g_{\mu\nu}}{\partial x^{\sigma}}\right)\frac{dx^{\mu}}{ds}\frac{dx^{\nu}}{ds}$$

Einstein then introduces a second change of notation "where, following Christoffel, we have written"[142]

$$\{\mu\nu,\tau\} = g^{\tau\alpha}[\mu\nu,\alpha] \qquad (5.23)$$

Multiplying Eq. (5.20d) by $g^{\sigma\tau}$, one obtains

$$0 = g^{\sigma\tau}g_{\mu\sigma}\frac{d^2 x^{\mu}}{ds^2} + g^{\sigma\tau}[\mu\nu,\sigma]\frac{dx^{\mu}}{ds}\frac{dx^{\nu}}{ds}$$

$$0 = \delta^{\tau}_{\mu}\frac{d^2 x^{\mu}}{ds^2} + \{\mu\nu,\tau\}\frac{dx^{\mu}}{ds}\frac{dx^{\nu}}{ds}$$

$$0 = \frac{d^2 x^{\tau}}{ds^2} + \{\mu\nu,\tau\}\frac{dx^{\mu}}{ds}\frac{dx^{\nu}}{ds} \qquad (5.22)$$

5.4.4 The Formation of Tensors by Differentiation

Let ϕ be an invariant function of space. ds is the distance along some arbitrary curve (soon to become the geodetic), where $s = s(x^1, x^2, x^3, x^4) = s(x^\mu)$. The derivative of ϕ with respect to s is

$$\frac{d\phi}{ds} = \frac{\partial\phi}{\partial x^\mu}\frac{dx^\mu}{ds} = \psi$$

Since both $d\phi$ and ds are invariants, the derivative also is an invariant, i.e., ψ is an invariant. Since dx^μ is a contravariant four-vector (from section 8, first paragraph), $\frac{dx^\mu}{ds}$ also is a contravariant four-vector. From Eq. (5.6), then,

$$A_\mu = \frac{\partial\phi}{\partial x^\mu} \tag{5.24}$$

the gradient of ϕ, is a covariant four-vector.

Since ψ is an invariant on the curve, we can repeat the derivative operation,

$$\chi = \frac{d\psi}{ds} = \frac{d}{ds}\left(\frac{d\phi}{ds}\right) = \frac{d}{ds}\left(\frac{\partial\phi}{\partial x^\mu}\frac{dx^\mu}{ds}\right) = \left(\frac{\partial^2\phi}{\partial x^\mu \partial x^\nu}\frac{dx^\nu}{ds}\right)\frac{dx^\mu}{ds} + \frac{\partial\phi}{\partial x^\mu}\frac{d^2 x^\mu}{ds^2}$$

At this point the curve is restricted to the geodetic, and the expression for $\frac{d^2 x^\mu}{ds^2}$ is substituted from Eq. (5.22) (note that the index μ in the second term is replaced by the index τ. This has no effect because the index is summed over the values 1, 2, 3, 4.).

$$\chi = \left(\frac{\partial^2\phi}{\partial x^\mu \partial x^\nu}\frac{dx^\nu}{ds}\right)\frac{dx^\mu}{ds} + \frac{\partial\phi}{\partial x^\tau}\frac{d^2 x^\tau}{ds^2}$$

$$\chi = \left(\frac{\partial^2\phi}{\partial x^\mu \partial x^\nu}\frac{dx^\nu}{ds}\right)\frac{dx^\mu}{ds} + \frac{\partial\phi}{\partial x^\tau}\left[-\{\mu\nu,\tau\}\frac{dx^\mu}{ds}\frac{dx^\nu}{ds}\right]$$

$$\chi = \left(\frac{\partial^2\phi}{\partial x^\mu \partial x^\nu} - \{\mu\nu,\tau\}\frac{\partial\phi}{\partial x^\tau}\right)\frac{dx^\mu}{ds}\frac{dx^\nu}{ds} = A_{\mu\nu}\frac{dx^\mu}{ds}\frac{dx^\nu}{ds}$$

$$A_{\mu\nu} = \frac{\partial^2\phi}{\partial x^\mu \partial x^\nu} - \{\mu\nu,\tau\}\frac{\partial\phi}{\partial x^\tau} \tag{5.25}$$

Since

1. χ is a scalar, and
2. $A_{\mu\nu}$ is symmetric in the indices μ and ν, ($\{\mu\nu,\tau\}$ is symmetrical in the exchange of $\mu \leftrightarrow \nu$, and the exchange of $\mu \leftrightarrow \nu$ in $\frac{\partial^2\phi}{\partial x^\mu \partial x^\nu}$ simply changes the order of differentiation.), and
3. $\frac{dx^\mu}{ds}$ and $\frac{dx^\nu}{ds}$ are contravariant four-vectors with arbitrary components,
4. from property three of Section 5.2.7, $A_{\mu\nu}$ is a covariant vector of rank two.

To recap:

From the invariant function ϕ, one can form the invariant tensor of rank one (a four-vector):

$$A_\mu = \frac{\partial \phi}{\partial x^\mu} = \text{gradient of } \phi$$

$$= \text{covariant four-vector}$$

From this, one can form by differentiation, $A_{\mu\nu}$, a covariant tensor of rank two:

$$A_{\mu\nu} = \frac{\partial^2 \phi}{\partial x^\mu \partial x^\nu} - \{\mu\nu, \tau\} \frac{\partial \phi}{\partial x^\tau} \tag{5.25}$$

$$= \frac{\partial A_\mu}{\partial x^\nu} - \{\mu\nu, \tau\} A_\tau \tag{5.26}$$

$$= \text{extension of the tensor} A_\mu$$

$$= \text{covariant derivative of the tensor} A_\mu$$

$$= \text{covariant tensor of rank two}$$

The derivation began with A_μ as the gradient of ϕ. These results also are true even if A_μ is not the gradient of ϕ, i.e., the same operation leads to a tensor $A_{\mu\nu}$ even if A_μ cannot be represented as a gradient. To see this:

1. Let A_μ be an arbitrary covariant four-vector with components A_1, A_2, A_3, A_4, each of which is an arbitrary function of the x^μ, i.e., of x^1, x^2, x^3, x^4.
2. Since $\frac{\partial \phi}{\partial x^\mu}$ is a covariant four-vector, if ψ is a scalar, the product $\psi \frac{\partial \phi}{\partial x^\mu}$ is a covariant four-vector.
3. Consider four scalar functions, $\psi^{(1)}, \psi^{(2)}, \psi^{(3)}, \psi^{(4)}$, and a second set of four scalar functions $\phi^{(1)}, \phi^{(2)}, \phi^{(3)}, \phi^{(4)}$. Form the pair-wise products of them of the form $\psi^{(1)} \frac{\partial \phi^{(1)}}{\partial x^\mu}$, etc., and form the sum of the terms $S_\mu = \psi^{(1)} \frac{\partial \phi^{(1)}}{\partial x^\mu} + \psi^{(2)} \frac{\partial \phi^{(2)}}{\partial x^\mu} + \psi^{(3)} \frac{\partial \phi^{(3)}}{\partial x^\mu} + \psi^{(4)} \frac{\partial \phi^{(4)}}{\partial x^\mu}$.
4. S_μ, being the sum of covariant four-vectors, is a covariant four-vector.
5. But any arbitrary four-vector A_μ can be represented as the sum S_μ. This is easily seen if we set

$$
\begin{array}{lll}
\psi^{(1)} = A_1, & \phi^{(1)} = x^1 & S_1 = A_1 \\
\psi^{(2)} = A_2, & \phi^{(2)} = x^2 & S_2 = A_2 \\
\psi^{(3)} = A_3, & \phi^{(3)} = x^3 \Rightarrow & S_3 = A_3 \\
\psi^{(4)} = A_4, & \phi^{(4)} = x^4 & S_4 = A_4
\end{array}
$$

6. The four-vector $A_\mu = \{A_1, A_2, A_3, A_4\} = \{S_1, S_2, S_3, S_4\}$. Showing Eq. (5.26) holds for an arbitrary component of A_μ, since A_μ is a linear combination of the components, Eq. (5.26) will also hold for the four vector A_μ.

a. Set $A_\mu = \psi \frac{\partial \phi}{\partial x^\mu}$
b. From Eq. (5.25), $A_{\mu\nu}$ is a covariant tensor (this is true when $A_\mu = \frac{\partial \phi}{\partial x^\mu}$). The RHS of Eq. (5.25) is, therefore, also a covariant tensor.

c. If we multiply the RHS of Eq. (5.25) by the scalar ψ, the resultant product remains a covariant tensor: $\psi\frac{\partial^2\phi}{\partial x^\mu\partial x^\nu} - \{\mu\nu,\tau\}\psi\frac{\partial\phi}{\partial x^\tau}$ is a covariant tensor.

d. $B_\mu = \frac{\partial\psi}{\partial x^\mu}$ and $C_\nu = \frac{\partial\phi}{\partial x^\nu}$ both are covariant four-vectors. Their product $B_\mu C_\nu = \frac{\partial\psi}{\partial x^\mu}\frac{\partial\phi}{\partial x^\nu}$ is a covariant tensor of rank two.

e. Adding together the covariant tensors of part c and of part d above, the resulting sum is a covariant tensor: $\psi\frac{\partial}{\partial x^\mu}\left(\frac{\partial\phi}{\partial x^\nu}\right) + \frac{\partial\psi}{\partial x^\mu}\frac{\partial\phi}{\partial x^\nu} - \{\mu\nu,\tau\}\left(\psi\frac{\partial\phi}{\partial x^\tau}\right)$

$$= \frac{\partial}{\partial x^\mu}\left(\psi\frac{\partial\phi}{\partial x^\nu}\right) - \{\mu\nu,\tau\}\left(\psi\frac{\partial\phi}{\partial x^\tau}\right)$$

f. We had shown that any vector A_μ could be represented as $\psi\frac{\partial\phi}{\partial x^\mu}$. Thus, statement e, above, proves that Eq. (5.26) yields a result $A_{\mu\nu}$ that is a covariant tensor for an arbitrary four-vector A_μ.

$A_{\mu\nu}$ is the extension of the four-vector A_μ. This can be generalized to define the extension of a covariant tensor of any rank. The example given is the extension of a covariant tensor of rank two to give a covariant tensor of rank three. From this, one can easily generalize to the extension of tensor of any rank.

(a) Any covariant tensor of rank two can be represented as a sum of tensors of the type $A_\mu B_\nu$. It will be sufficient to deduce the expression for this special type because any arbitrary tensor of rank two can be represented as the sum of a maximum of four such terms.

(b) From Eq. (5.26), the covariant derivative, the extension, of the four-vector A_μ and the four-vector B_ν is

$$A_{\mu\sigma} = \frac{\partial A_\mu}{\partial x^\sigma} - \{\sigma\mu,\tau\}A_\tau$$

$$B_{\nu\sigma} = \frac{\partial B_\nu}{\partial x^\sigma} - \{\sigma\nu,\tau\}B_\tau$$

Each of these expressions is a covariant tensor of rank two.

(c) Outer multiplication of the first of these two equations by B_ν, the second by A_μ, and adding, gives

$$A_{\mu\nu\sigma} = B_\nu\frac{\partial A_\mu}{\partial x^\sigma} - \{\sigma\mu,\tau\}A_\tau B_\nu + A_\mu\frac{\partial B_\nu}{\partial x^\sigma} - \{\sigma\nu,\tau\}B_\tau A_\mu$$

$$= \frac{\partial}{\partial x^\sigma}(A_\mu B_\nu) - \{\sigma\mu,\tau\}A_\tau B_\nu - \{\sigma\nu,\tau\}A_\mu B_\tau$$

$$A_{\mu\nu\sigma} = \frac{\partial}{\partial x^\sigma}(A_{\mu\nu}) - \{\sigma\mu,\tau\}A_{\tau\nu} - \{\sigma\nu,\tau\}A_{\mu\tau} \qquad (5.27)$$

$A_{\mu\nu\sigma}$ is the extension of $A_{\mu\nu}$ and is a covariant tensor of rank three.

Summary of Covariant Derivatives, Extensions:

1. Eq. (5.24): $A_\mu = \frac{\partial\phi}{\partial x^\mu}$

2. Eq. (5.26): $A_{\mu\nu} = \frac{\partial A_\mu}{\partial x^\nu} - \{\mu\nu, \tau\}A_\tau$

3. Eq. (5.27): $A_{\mu\nu\sigma} = \frac{\partial}{\partial x^\sigma}(A_{\mu\nu}) - \{\sigma\mu, \tau\}A_{\tau\nu} - \{\sigma\nu, \tau\}A_{\mu\tau}$

5.4.5 Some Cases of Special Importance

5.4.5.1 The Fundamental Tensor

To show that $dg = g^{\mu\nu} g dg_{\mu\nu}$, with g the determinant of $g_{\mu\nu}$

$$g = \begin{vmatrix} g_{11} & g_{12} & g_{13} & g_{14} \\ g_{21} & g_{22} & g_{23} & g_{24} \\ g_{31} & g_{32} & g_{33} & g_{34} \\ g_{41} & g_{42} & g_{43} & g_{44} \end{vmatrix}$$

The determinant of $g = g = \in_{\alpha\beta\sigma\tau} g_{1\alpha} g_{2\beta} g_{3\sigma} g_{4\tau}$, where $\in_{\alpha\beta\sigma\tau}$ is the Levi-Civita symbol. The variation of g is

$$dg = \in_{\alpha\beta\sigma\tau}(dg_{1\alpha})g_{2\beta}g_{3\sigma}g_{4\tau} + \in_{\alpha\beta\sigma\tau} g_{1\alpha}(dg_{2\beta})g_{3\sigma}g_{4\tau}$$
$$+ \in_{\alpha\beta\sigma\tau} g_{1\alpha}g_{2\beta}(dg_{3\sigma})g_{4\tau} + \in_{\alpha\beta\sigma\tau} g_{1\alpha}g_{2\beta}g_{3\sigma}(dg_{4\tau})$$

Working with the first term,

$$\in_{\alpha\beta\sigma\tau}(dg_{1\alpha})g_{2\beta}g_{3\sigma}g_{4\tau} = \in_{\alpha\beta\sigma\tau}(dg_{1\mu}\delta_\alpha^\mu)g_{2\beta}g_{3\sigma}g_{4\tau}$$
$$= \in_{\alpha\beta\sigma\tau}(dg_{1\mu}g^{\mu\nu}g_{\nu\alpha})g_{2\beta}g_{3\sigma}g_{4\tau}$$
$$= \in_{\alpha\beta\sigma\tau}(g^{\mu\nu}dg_{1\mu})(g_{\nu\alpha})g_{2\beta}g_{3\sigma}g_{4\tau}$$

Writing out explicitly the terms for $\nu = 1, 2, 3, 4$:

$$\in_{\alpha\beta\sigma\tau}(dg_{1\alpha})g_{2\beta}g_{3\sigma}g_{4\tau} = \in_{\alpha\beta\sigma\tau} g^{\mu 1}dg_{1\mu}g_{1\alpha}g_{2\beta}g_{3\sigma}g_{4\tau}$$
$$+ \in_{\alpha\beta\sigma\tau} g^{\mu 2}dg_{1\mu}g_{2\alpha}g_{2\beta}g_{3\sigma}g_{4\tau}$$
$$+ \in_{\alpha\beta\sigma\tau} g^{\mu 3}dg_{1\mu}g_{3\alpha}g_{2\beta}g_{3\sigma}g_{4\tau}$$
$$+ \in_{\alpha\beta\sigma\tau} g^{\mu 4}dg_{1\mu}g_{4\alpha}g_{2\beta}g_{3\sigma}g_{4\tau}$$

The second, third, and fourth terms are each equal to zero. Consider the second term. For a given value of σ and τ, for example if $\alpha = 1, \beta = 2$ and $\alpha = 2, \beta = 1$, we have $g_{2\alpha}g_{2\beta} = g_{21}g_{22} = g_{22}g_{21}$, since g is symmetric. However, the Levi-Civita symbol is antisymmetric in the exchange $\{\alpha, \beta\} = \{1, 2\} \to \{2, 1\}$. Thus the sum of these two terms must be zero. The same is true for each pair of terms $\{\alpha, \beta\}$. In a similar manner the third and fourth terms above are shown to be zero.

Collecting together, the resulting expression for dg is

$$dg = \in_{\alpha\beta\sigma\tau}\left(g^{\mu 1}dg_{1\mu}\right)g_{1\alpha}g_{2\beta}g_{3\sigma}g_{4\tau} + \in_{\alpha\beta\sigma\tau}\left(g^{\mu 2}dg_{2\mu}\right)g_{1\alpha}g_{2\beta}g_{3\sigma}g_{4\tau}$$
$$+ \in_{\alpha\beta\sigma\tau}\left(g^{\mu 3}dg_{3\mu}\right)g_{1\alpha}g_{2\beta}g_{3\sigma}g_{4\tau} + \in_{\alpha\beta\sigma\tau}\left(g^{\mu 4}dg_{4\mu}\right)g_{1\alpha}g_{2\beta}g_{3\sigma}g_{4\tau}$$
$$= \in_{\alpha\beta\sigma\varsigma}\left(g^{\mu\nu}dg_{\nu\mu}\right)g_{1\alpha}g_{2\beta}g_{3\sigma}g_{4\tau} = \left(g^{\mu\nu}dg_{\mu\nu}\right)\left(\in_{\alpha\beta\sigma\varsigma} g_{1\alpha}g_{2\beta}g_{3\sigma}g_{4\tau}\right)$$
$$= g^{\mu\nu}g dg_{\mu\nu}$$
$$dg = g^{\mu\nu}g dg_{\mu\nu}$$

But $\qquad g^{\mu\nu}g_{\mu\nu} = \delta^\mu_\mu = 4 \Rightarrow d\left(g^{\mu\nu}g_{\mu\nu}\right) = d\left(g^{\mu\nu}\right)g_{\mu\nu} + g^{\mu\nu}d\left(g_{\mu\nu}\right) = 0.$
Thus,

$$dg = g^{\mu\nu}g\,dg_{\mu\nu} = -g_{\mu\nu}g\,dg^{\mu\nu} \qquad (5.28)$$

Consider the quantity $\frac{1}{\sqrt{-g}}\frac{\partial\sqrt{-g}}{\partial x^\sigma}$.

$$\frac{1}{\sqrt{-g}}\frac{\partial\sqrt{-g}}{\partial x^\sigma} = \frac{\partial\left(\log\sqrt{-g}\right)}{\partial x^\sigma} = \frac{\partial\left(\log\left(-g\right)^{1/2}\right)}{\partial x^\sigma} = \frac{1}{2}\frac{\partial\log\left(-g\right)}{\partial x^\sigma}$$

This is the second term of Eq. (5.29). Continuing,

$$\frac{1}{2}\frac{\partial\log\left(-g\right)}{\partial x^\sigma} = \frac{1}{2}\frac{1}{\left(-g\right)}\frac{\partial\left(-g\right)}{\partial x^\sigma} = \frac{1}{2}\frac{1}{g}\frac{\partial g}{\partial x^\sigma}$$

But, from Eq. (5.28), $\frac{\partial g}{\partial x^\sigma} = g^{\mu\nu}g\frac{\partial g_{\mu\nu}}{\partial x^\sigma} \Rightarrow= \frac{1}{2}\frac{1}{g}g^{\mu\nu}g\frac{\partial g_{\mu\nu}}{\partial x^\sigma} = \frac{1}{2}g^{\mu\nu}\frac{\partial g_{\mu\nu}}{\partial x^\sigma}.$
This is the third term of Eq. (5.29). Similarly, from Eq. (5.28), $\frac{\partial g}{\partial x^\sigma} = -g_{\mu\nu}g\frac{\partial g^{\mu\nu}}{\partial x^\sigma} \Rightarrow = -\frac{1}{2}g_{\mu\nu}\frac{\partial g^{\mu\nu}}{\partial x^\sigma}.$ This is the fourth term of Eq. (5.29):

$$\frac{1}{\sqrt{-g}}\frac{\partial\sqrt{-g}}{\partial x^\sigma} = \frac{1}{2}\frac{\partial\log(-g)}{\partial x^\sigma} = \frac{1}{2}g^{\mu\nu}\frac{\partial g_{\mu\nu}}{\partial x^\sigma} = -\frac{1}{2}g_{\mu\nu}\frac{\partial g^{\mu\nu}}{\partial x^\sigma} \qquad (5.29)$$

Taking the variation of $g_{\mu\sigma}g^{\nu\sigma} = \delta^\nu_\mu$ we have

$$d\left(g_{\mu\sigma}g^{\nu\sigma}\right) = d\left(\delta^\nu_\mu\right) \Rightarrow d\left(g_{\mu\sigma}\right)g^{\nu\sigma} + g_{\mu\sigma}d\left(g^{\nu\sigma}\right) = 0 \Rightarrow$$

$$g_{\mu\sigma}dg^{\nu\sigma} = -g^{\nu\sigma}dg_{\mu\sigma} \qquad (5.30a)$$

$$g_{\mu\sigma}\frac{\partial g^{\nu\sigma}}{\partial x^\lambda} = -g^{\nu\sigma}\frac{\partial g_{\mu\sigma}}{\partial x^\lambda} \qquad (5.30b)$$

These two equations, Eqs. (5.30), are modified, first by multiplying by $g^{\mu\tau}$ to remove the $g_{\mu\sigma}$ term from the LHS of Eq. (5.30):

$$g^{\mu\tau}g_{\mu\sigma}dg^{\nu\sigma} = -g^{\mu\tau}g^{\nu\sigma}dg_{\mu\sigma}$$

$$\delta^\tau_\sigma dg^{\nu\sigma} = -g^{\mu\tau}g^{\nu\sigma}dg_{\mu\sigma}$$

$$dg^{\nu\tau} = -g^{\mu\tau}g^{\nu\sigma}dg_{\mu\sigma}$$

Relabeling the indices $\{\nu \to \mu, \tau \to \nu, \sigma \to \alpha, \mu \to \beta, \lambda \to \sigma\}$,

$$dg^{\mu\nu} = -g^{\beta\nu}g^{\mu\alpha}dg_{\beta\alpha} \qquad (5.31a)$$

$$\frac{\partial g^{\mu\nu}}{\partial x^\sigma} = -g^{\beta\nu}g^{\mu\alpha}\frac{\partial g_{\beta\alpha}}{\partial x^\sigma} \qquad (5.31b)$$

Repeating the operation on Eqs. (5.30), this time multiplying by $g_{\nu\tau}$ to remove the $g^{\nu\sigma}$ term from the RHS, gives $g_{\mu\sigma}g_{\nu\tau}dg^{\nu\sigma} = -dg_{\mu\tau}$. This time relabel the indices as $\{\mu \to \nu, \tau \to \mu, \nu \to \alpha, \sigma \to \beta, \lambda \to \sigma\}$.

$$dg_{\nu\mu} = -g_{\nu\beta}g_{\alpha\mu}dg^{\alpha\beta} \qquad (5.32a)$$

$$\frac{\partial g_{\nu\mu}}{\partial x^\sigma} = -g_{\nu\beta}g_{\alpha\mu}\frac{\partial g^{\alpha\beta}}{\partial x^\sigma} \qquad (5.32b)$$

Returning to the notation of Christoffel, Eq. (5.21),

$$[\mu\nu, \sigma] = \frac{1}{2}\left(\frac{\partial g_{\mu\sigma}}{\partial x^\nu} + \frac{\partial g_{\nu\sigma}}{\partial x^\mu} - \frac{\partial g_{\mu\nu}}{\partial x^\sigma}\right)$$

$$[\alpha\sigma, \beta]+[\beta\sigma, \alpha] = \frac{1}{2}\left(\frac{\partial g_{\alpha\beta}}{\partial x^\sigma} + \frac{\partial g_{\sigma\beta}}{\partial x^\alpha} - \frac{\partial g_{\alpha\sigma}}{\partial x^\beta}\right) + \frac{1}{2}\left(\frac{\partial g_{\beta\alpha}}{\partial x^\sigma} + \frac{\partial g_{\sigma\alpha}}{\partial x^\beta} - \frac{\partial g_{\beta\sigma}}{\partial x^\alpha}\right)$$

$$= \frac{\partial g_{\alpha\beta}}{\partial x^\sigma} \tag{5.33}$$

Substituting this result into Eq. (5.31b), and using Eq. (5.23),

$$\frac{\partial g^{\mu\nu}}{\partial x^\sigma} = -g^{\mu\alpha}g^{\nu\beta}([\alpha\sigma, \beta] + [\beta\sigma, \alpha])$$

$$= -g^{\mu\alpha}(g^{\nu\beta}[\alpha\sigma, \beta]) - g^{\nu\beta}(g^{\mu\alpha}[\beta\sigma, \alpha])$$

$$= -g^{\mu\alpha}\{\alpha\sigma, \nu\} - g^{\nu\beta}\{\beta\sigma, \mu\}$$

Since in the first term the index α is summed over, and in the second term the index β is summed over, each can be replaced by the same index τ:

$$\frac{\partial g^{\mu\nu}}{\partial x^\sigma} = -g^{\mu\tau}\{\tau\sigma, \nu\} - g^{\nu\tau}\{\tau\sigma, \mu\} \tag{5.34}$$

Substituting Eq. (5.34) into Eq. (5.29) gives

$$\frac{1}{\sqrt{-g}}\frac{\partial\sqrt{-g}}{\partial x^\sigma} = -\frac{1}{2}g_{\mu\nu}\frac{\partial g^{\mu\nu}}{\partial x^\sigma}$$

$$= -\frac{1}{2}g_{\mu\nu}\left(-g^{\mu\tau}\{\tau\sigma, \nu\} - g^{\nu\tau}\{\tau\sigma, \mu\}\right)$$

$$= -\frac{1}{2}\left(g_{\mu\nu}g^{\mu\tau}\{\tau\sigma, \nu\} - g_{\mu\nu}g^{\nu\tau}\{\tau\sigma, \mu\}\right)$$

$$= +\frac{1}{2}\left(\delta_\nu^\tau\{\tau\sigma, \nu\} + \delta_\mu^\tau\{\tau\sigma, \mu\}\right)$$

$$= \frac{1}{2}\left(\{\nu\sigma, \nu\} + \{\mu\sigma, \mu\}\right)$$

$$\frac{1}{\sqrt{-g}}\frac{\partial\sqrt{-g}}{\partial x^\sigma} = \{\mu\sigma, \mu\} \tag{5.29a}$$

since $\{\nu\sigma, \nu\} = \{\mu\sigma, \mu\}$ as both μ and ν are summed over the values 1, 2, 3, and 4.

5.4.5.2   The Divergence of a Contravariant Vector

Taking the inner product of Eq. (5.26) with the contravariant fundamental tensor, $g^{\mu\nu}$, gives

$$g^{\mu\nu}A_{\mu\nu} = g^{\mu\nu}\frac{\partial A_\mu}{\partial x^\nu} - g^{\mu\nu}\{\mu\nu, \tau\}A_\tau \tag{5.35a}$$

The first term on the RHS can be written as

$$g^{\mu\nu}\frac{\partial A_\mu}{\partial x^\nu} = \frac{\partial\left(g^{\mu\nu}A_\mu\right)}{\partial x^\nu} - A_\mu\frac{\partial g^{\mu\nu}}{\partial x^\nu} \qquad (5.35\text{b})$$

Using Eqs. (5.21) and (5.23), the second term on the RHS of Eq. (5.35a) can be expanded as

$$-g^{\mu\nu}\{\mu\nu,\tau\}A_\tau = -\frac{1}{2}g^{\mu\nu}g^{\tau\alpha}[\mu\nu,\alpha]A_\tau$$

$$= -\frac{1}{2}g^{\mu\nu}g^{\tau\alpha}\left(\frac{\partial g_{\mu\alpha}}{\partial x^\nu} + \frac{\partial g_{\nu\alpha}}{\partial x^\mu} - \frac{\partial g_{\mu\nu}}{\partial x^\alpha}\right)A_\tau \quad (5.35\text{c})$$

Using Eq. (5.32b), the first term on the RHS of Eq. (5.35c) becomes

$$-\frac{1}{2}g^{\mu\nu}g^{\tau\alpha}A_\tau\left(\frac{\partial g_{\mu\alpha}}{\partial x^\nu}\right) = -\frac{1}{2}g^{\mu\nu}g^{\tau\alpha}A_\tau\left(-g_{\mu\beta}g_{\gamma\alpha}\frac{\partial g^{\gamma\beta}}{\partial x^\nu}\right)$$

$$= +\frac{1}{2}\left(g^{\mu\nu}g_{\mu\beta}\right)\left(g^{\tau\alpha}g_{\gamma\alpha}\right)A_\tau\frac{\partial g^{\gamma\beta}}{\partial x^\nu}$$

$$= +\frac{1}{2}\delta^\nu_\beta\delta^\tau_\gamma A_\tau\frac{\partial g^{\gamma\beta}}{\partial x^\nu} = +\frac{1}{2}A_\tau\frac{\partial g^{\tau\nu}}{\partial x^\nu}$$

In a similar manner, the second term on the RHS of Eq. (5.35c) gives the same result, $-\frac{1}{2}g^{\mu\nu}g^{\tau\alpha}A_\tau\frac{\partial g_{\nu\alpha}}{\partial x^\mu} = +\frac{1}{2}A_\tau\frac{\partial g^{\tau\mu}}{\partial x^\mu}$. Since the index ν in the first term, and the index μ in the second term, are each summed over the integers 1, 2, 3, and 4, the two expressions can be added together into one expression with the index ν, $\Rightarrow= +A_\tau\frac{\partial g^{\tau\nu}}{\partial x^\nu}$. Substituting this result into Eq. (5.35c), and then substituting the resulting expression plus Eq. (5.35b) into Eq. (5.35a),

$$g^{\mu\nu}A_{\mu\nu} = \left(\frac{\partial(g^{\mu\nu}A_\mu)}{\partial x^\nu} - A_\mu\frac{\partial g^{\mu\nu}}{\partial x^\nu}\right) + \left(A_\tau\frac{\partial g^{\tau\nu}}{\partial x^\nu} + \frac{1}{2}g^{\mu\nu}g^{\tau\alpha}A_\tau\frac{\partial g_{\mu\nu}}{\partial x^\alpha}\right)$$
$$(5.35\text{d})$$

On the RHS, the second term in the first parentheses and the first term in the second parentheses are the negative of one another. Using Eq. (5.29), we obtain

$$g^{\mu\nu}A_{\mu\nu} = \frac{\partial\left(g^{\mu\nu}A_\mu\right)}{\partial x^\nu} + \frac{1}{\sqrt{-g}}\frac{\partial\sqrt{-g}}{\partial x^\alpha}g^{\tau\alpha}A_\tau \qquad (5.35\text{e})$$

Writing $g^{\mu\nu}A_\mu = A^\nu$, Eq. (5.35d) becomes

$$g^{\mu\nu}A_{\mu\nu} = \frac{\partial\left(A^\nu\right)}{\partial x^\nu} + \frac{1}{\sqrt{-g}}\frac{\partial\sqrt{-g}}{\partial x^\alpha}A^\alpha$$

Again, since the indices ν and α are summed over, they can be represented as the same index ν:

$$g^{\mu\nu}A_{\mu\nu} = \frac{\partial(A^\nu)}{\partial x^\nu} + \frac{1}{\sqrt{-g}}\frac{\partial\sqrt{-g}}{\partial x^\nu}A^\nu = \frac{1}{\sqrt{-g}}\frac{\partial}{\partial x^\nu}\left(\sqrt{-g}A^\nu\right)$$

This expression, termed the divergence of the contravariant vector A^ν, is denoted as

$$\Phi = g^{\mu\nu} A_{\mu\nu} = \frac{1}{\sqrt{-g}} \frac{\partial}{\partial x^\nu} \left(\sqrt{-g} A^\nu \right) \qquad (5.35)$$

5.4.5.3   The Curl of a Covariant Vector

From Eq. (5.26), the covariant derivative of the tensor A_μ, i.e., the extension of the tensor A_μ is $A_{\mu\nu} = \frac{\partial A_\mu}{\partial x^\nu} - \{\mu\nu, \tau\} A_\tau$. If we form the quantity $B_{\mu\nu} = A_{\mu\nu} - A_{\nu\mu} = \frac{\partial A_\mu}{\partial x^\nu} - \frac{\partial A_\nu}{\partial x^\mu} - \{\mu\nu, \tau\} + \{\nu\mu, \tau\}$, the third and fourth terms, because of the symmetry in the indices μ and ν, add to zero. The quantity $B_{\mu\nu}$ is an antisymmetric tensor called the curl of the covariant vector A_μ:

$$B_{\mu\nu} = \frac{\partial A_\mu}{\partial x^\nu} - \frac{\partial A_\nu}{\partial x^\mu} \qquad (5.36)$$

5.4.5.4   Antisymmetrical Extension of a Six-Vector

Consider an antisymmetrical tensor of second rank, $A_{\mu\nu}$, where $A_{\mu\nu} = -A_{\nu\mu}$. From Eq. (5.27), the extension of $A_{\mu\nu}$ is

$$A_{\mu\nu\sigma} = \frac{\partial}{\partial x^\sigma} (A_{\mu\nu}) - \{\sigma\mu, \tau\} A_{\tau\nu} - \{\sigma\nu, \tau\} A_{\mu\tau}$$

Form the sum, $B_{\mu\nu\sigma}$, of the above equation plus the two equations formed by a cyclic permutation of the indices μ, ν, and σ.

$$B_{\mu\nu\sigma} = A_{\mu\nu\sigma} + A_{\nu\sigma\mu} + A_{\sigma\mu\nu}$$

$$= \frac{\partial A_{\mu\nu}}{\partial x^\sigma} - \{\sigma\mu, \tau\} A_{\tau\nu} - \{\sigma\nu, \tau\} A_{\mu\tau}$$

$$+ \frac{\partial A_{\nu\sigma}}{\partial x^\mu} - \{\mu\nu, \tau\} A_{\tau\sigma} - \{\mu\sigma, \tau\} A_{\nu\tau}$$

$$+ \frac{\partial A_{\sigma\mu}}{\partial x^\nu} - \{\nu\sigma, \tau\} A_{\tau\mu} - \{\nu\mu, \tau\} A_{\sigma\tau}$$

But, by symmetry, $\{\sigma\mu, \tau\} = \{\mu\sigma, \tau\}$, and, by antisymmetry, $A_{\tau\nu} = -A_{\nu\tau}$. Thus the pair of terms $-\{\sigma\mu, \tau\} A_{\tau\nu} - \{\mu\sigma, \tau\} A_{\nu\tau} = -\{\sigma\mu, \tau\} A_{\tau\nu} + \{\mu\sigma, \tau\} A_{\tau\nu} = 0$. In a similar manner, the two other pairs of comparable terms can be shown to be zero.

$$B_{\mu\nu\sigma} = A_{\mu\nu\sigma} + A_{\nu\sigma\mu} + A_{\sigma\mu\nu} = \frac{\partial A_{\mu\nu}}{\partial x^\sigma} + \frac{\partial A_{\nu\sigma}}{\partial x^\mu} + \frac{\partial A_{\sigma\mu}}{\partial x^\nu} \qquad (5.37)$$

$B_{\mu\nu\sigma}$ is antisymmetric in the exchange of any two of the indices, i.e., $B_{\nu\mu\sigma} = -B_{\mu\nu\sigma}$. Using the antisymmetry property of $A_{\mu\nu}$,

$$B_{\nu\mu\sigma} = \frac{\partial A_{\nu\mu}}{\partial x^\sigma} + \frac{\partial A_{\mu\sigma}}{\partial x^\nu} + \frac{\partial A_{\sigma\nu}}{\partial x^\mu} = \frac{\partial(-A_{\mu\nu})}{\partial x^\sigma} + \frac{\partial(-A_{\sigma\mu})}{\partial x^\nu} + \frac{\partial(-A_{\nu\sigma})}{\partial x^\mu} = -B_{\mu\nu\sigma}$$

5.4.5.5   The Divergence of a Six-Vector

A six-vector is an antisymmetric 4×4 tensor. Multiplying Eq. (5.27) by $g^{\mu\alpha} g^{\nu\beta}$, denoting $A_\sigma^{\alpha\beta} = g^{\mu\alpha} g^{\nu\beta} A_{\mu\nu\sigma}$ and $A^{\alpha\beta} = g^{\mu\alpha} g^{\nu\beta} A_{\mu\nu}$, and using Eq. (5.34),

$$g^{\mu\alpha}g^{\nu\beta}A_{\mu\nu\sigma} = g^{\mu\alpha}g^{\nu\beta}\frac{\partial A_{\mu\nu}}{\partial x^\sigma} - g^{\mu\alpha}g^{\nu\beta}\{\sigma\mu,\gamma\}A_{\gamma\nu} - g^{\mu\alpha}g^{\nu\beta}\{\sigma\nu,\gamma\}A_{\mu\gamma}$$

$$A_\sigma^{\alpha\beta} = \left(\frac{\partial(g^{\mu\alpha}g^{\nu\beta}A_{\mu\nu})}{\partial x^\sigma} - A_{\mu\nu}g^{\nu\beta}\frac{\partial g^{\mu\alpha}}{\partial x^\sigma} - A_{\mu\nu}g^{\mu\alpha}\frac{\partial g^{\nu\beta}}{\partial x^\sigma}\right)$$

$$-g^{\mu\alpha}g^{\nu\beta}\{\sigma\mu,\gamma\}A_{\gamma\nu} - g^{\mu\alpha}g^{\nu\beta}\{\sigma\nu,\gamma\}A_{\mu\gamma}$$

$$A_\sigma^{\alpha\beta} = \frac{\partial A^{\alpha\beta}}{\partial x^\sigma} - A_{\mu\nu}g^{\nu\beta}(-g^{\mu\gamma}\{\gamma\sigma,\alpha\} - g^{\alpha\gamma}\{\gamma\sigma,\mu\})$$

$$-A_{\mu\nu}g^{\mu\alpha}(-g^{\nu\gamma}\{\gamma\sigma,\beta\} - g^{\beta\gamma}\{\gamma\sigma,\nu\})$$

$$-g^{\mu\alpha}g^{\nu\beta}\{\sigma\mu,\gamma\}A_{\gamma\nu} - g^{\mu\alpha}g^{\nu\beta}\{\sigma\nu,\gamma\}A_{\mu\gamma}$$

$$A_\sigma^{\alpha\beta} = \frac{\partial A^{\alpha\beta}}{\partial x^\sigma} + (A_{\mu\nu}g^{\nu\beta}g^{\mu\gamma}\{\gamma\sigma,\alpha\} + A_{\mu\nu}g^{\mu\alpha}g^{\nu\gamma}\{\gamma\sigma,\beta\})$$

$$+(A_{\mu\nu}g^{\nu\beta}g^{\alpha\gamma}\{\gamma\sigma,\mu\} - A_{\gamma\nu}g^{\mu\alpha}g^{\nu\beta}\{\sigma\mu,\gamma\})$$

$$+(A_{\mu\nu}g^{\mu\alpha}g^{\beta\gamma}\{\gamma\sigma,\nu\} - A_{\mu\gamma}g^{\mu\alpha}g^{\nu\beta}\{\sigma\nu,\gamma\})$$

The second parenthesis is equal to zero (this is most easily seen if, in the second term, the indices are changed as follows:$\gamma \leftrightarrow \mu$, using the symmetry of $g^{\gamma\alpha}$ and using the symmetry of $\{\sigma\gamma, u\}$ in the first two indices). By a similar analysis, the third parenthesis also is zero ($\gamma \leftrightarrow \nu$).

$$A_\sigma^{\alpha\beta} = \frac{\partial A^{\alpha\beta}}{\partial x^\sigma} + \{\sigma\gamma,\alpha\}A^{\gamma\beta} + \{\sigma\gamma,\beta\}A^{\alpha\gamma} \tag{5.38}$$

Equation (5.38) is the expression for the extension of the contravariant tensor of second rank, $A^{\alpha\beta}$, just as Eq. (5.27) is the expression for the extension of the covariant tensor of second rank, $A_{\mu\nu}$. Multiplying Eq. (5.27) by $g^{\nu\alpha}$ and proceeding in a similar manner, Einstein obtains for the extension of the mixed tensor A_μ^α,

$$A_{\mu\sigma}^\alpha = \frac{\partial A_\mu^\alpha}{\partial x^\sigma} - \{\sigma\mu,\tau\}A_\tau^\alpha + \{\sigma\tau,\alpha\}A_\mu^\tau \tag{5.39}$$

Contracting Eq. (5.38) with respect to the indices β and σ, i.e., inner multiplication by δ_β^σ, one obtains

$$A^\alpha = \frac{\partial A^{\alpha\beta}}{\partial x^\beta} + \{\beta\gamma,\alpha\}A^{\gamma\beta} + \{\beta\gamma,\beta\}A^{\alpha\gamma}$$

In the interchange of $\beta \leftrightarrow \gamma$, $\{\beta\gamma,\alpha\}$ is symmetric. But, if it is assumed $A^{\gamma\beta}$ is an antisymmetric tensor, the sum of $\{\beta\gamma,\alpha\}A^{\gamma\beta}$ over γ and β will be zero. Using Eq. (5.29a),

$$A^\alpha = \frac{\partial A^{\alpha\beta}}{\partial x^\beta} + \{\beta\gamma,\beta\}A^{\alpha\gamma} = \frac{\partial A^{\alpha\beta}}{\partial x^\beta} + A^{\alpha\gamma}\frac{1}{\sqrt{-g}}\frac{\partial\sqrt{-g}}{\partial x^\gamma}$$

In the second term, changing the index $\gamma \to \beta$,

$$A^\alpha = \frac{\partial A^{\alpha\beta}}{\partial x^\beta} + A^{\alpha\beta}\frac{1}{\sqrt{-g}}\frac{\partial\sqrt{-g}}{\partial x^\beta} = \frac{1}{\sqrt{-g}}\frac{\partial}{\partial x^\beta}(A^{\alpha\beta}\sqrt{-g}) \tag{5.40}$$

Equation (5.40) is the expression for the divergence of a contravariant six-vector $A^{\alpha\beta}$.

5.4.5.6   The Divergence of a Mixed Tensor of Second Rank

Contracting Eq. (5.39) with respect to the indices α and σ, i.e., inner multiplication by δ^σ_α, and using Eq. (5.29a), one obtains

$$A_\mu = \frac{\partial A^\sigma_\mu}{\partial x^\sigma} - \{\sigma\mu,\tau\}A^\sigma_\tau + \{\sigma\tau,\sigma\}A^\tau_\mu$$

$$= \frac{\partial A^\sigma_\mu}{\partial x^\sigma} - \{\sigma\mu,\tau\}A^\sigma_\tau + A^\tau_\mu \frac{1}{\sqrt{-g}} \frac{\partial\sqrt{-g}}{\partial x^\tau}$$

In the final term, changing the index $\tau \to \sigma$,

$$A_\mu = \frac{\partial A^\sigma_\mu}{\partial x^\sigma} - \{\sigma\mu,\tau\}\,A^\sigma_\tau + A^\sigma_\mu \frac{1}{\sqrt{-g}} \frac{\partial\sqrt{-g}}{\partial x^\sigma}$$

$$\sqrt{-g}\,A_\mu = \sqrt{-g}\frac{\partial A^\sigma_\mu}{\partial x^\sigma} - \sqrt{-g}\,\{\sigma\mu,\tau\}\,A^\sigma_\tau + A^\sigma_\mu \frac{\partial\sqrt{-g}}{\partial x^\sigma}$$

$$\sqrt{-g}\,A_\mu = \frac{\partial}{\partial x^\sigma}\left(\sqrt{-g}A^\sigma_\mu\right) - \{\sigma\mu,\tau\}\sqrt{-g}A^\sigma_\tau \qquad (5.41)$$

Using Eq. (5.23) and denoting $A^{\rho\sigma} = g^{\rho\tau} A^\sigma_\tau$, the last term in Eq. (5.41) can be expressed as

$$-\{\sigma\mu,\tau\}\sqrt{-g}A^\sigma_\tau = -[\sigma\mu,\rho]g^{\rho\tau}\sqrt{-g}A^\sigma_\tau = -[\sigma\mu,\rho]\sqrt{-g}A^{\rho\sigma}$$

If $A^{\rho\sigma}$ is symmetrical, this term becomes

$$-[\sigma\mu,\rho]\sqrt{-g}A^{\rho\sigma} = -\sqrt{-g}\frac{1}{2}\left(\frac{\partial g_{\sigma\rho}}{\partial x^\mu} + \frac{\partial g_{\mu\rho}}{\partial x^\sigma} - \frac{\partial g_{\sigma\mu}}{\partial x^\rho}\right)A^{\sigma\rho} = -\sqrt{-g}\frac{1}{2}\frac{\partial g_{\sigma\rho}}{\partial x^\mu}A^{\sigma\rho}$$

Since the last two terms in the parentheses are antisymmetric in the indices σ and ρ (for example, consider the values 1 and 2 for the indices σ and ρ, then the values 2 and 1. Since, by symmetry, $A^{12} = A^{21}$, the sum of the last two terms in the parentheses add up pair-wise to zero):

$$\sqrt{-g}\,A_\mu = \frac{\partial}{\partial x^\sigma}(\sqrt{-g}A^\sigma_\mu) - \frac{1}{2}\sqrt{-g}\frac{\partial g_{\sigma\rho}}{\partial x^\mu}A^{\sigma\rho} \qquad (5.41a)$$

Instead of the contravariant tensor $A^{\sigma\rho}$ in Eq. (5.41a), we can introduce the covariant tensor $A_{\rho\sigma}$. Using Eq. (5.32b),

$$-\frac{1}{2}\sqrt{-g}\frac{\partial g_{\sigma\rho}}{\partial x^\mu}A^{\sigma\rho} = -\frac{1}{2}\sqrt{-g}\left(-g_{\sigma\alpha}g_{\rho\beta}\frac{\partial g^{\alpha\beta}}{\partial x^\mu}\right)A^{\sigma\rho}$$

$$= +\frac{1}{2}\sqrt{-g}\frac{\partial g^{\alpha\beta}}{\partial x^\mu}(g_{\sigma\alpha}g_{\rho\beta}A^{\sigma\rho}) = \frac{1}{2}\sqrt{-g}\frac{\partial g^{\alpha\beta}}{\partial x^\mu}A_{\alpha\beta}$$

Since $g_{\sigma\alpha}g_{\rho\beta}A^{\sigma\rho} = g_{\sigma\alpha}g_{\rho\beta}(g^{\rho\tau}A^\sigma_\tau) = g_{\sigma\alpha}\delta^\tau_\beta A^\sigma_\tau = g_{\sigma\alpha}A^\sigma_\beta = A_{\alpha\beta}$. Changing the indices from α and β to ρ and σ, Eq. (5.41a) becomes

$$\sqrt{-g}\,A_\mu = \frac{\partial}{\partial x^\sigma}\left(\sqrt{-g}A^\sigma_\mu\right) + \frac{1}{2}\frac{\partial g^{\rho\sigma}}{\partial x^\mu}\sqrt{-g}A_{\rho\sigma} \qquad (5.41b)$$

5.4.6   The Riemann–Christoffel Tensor

$g_{\mu\nu}$ is the fundamental tensor. The initial quest is to find a tensor that can be obtained from the fundamental tensor alone, by differentiation. The first thought is to form the extension of the fundamental tensor, i.e., place the fundamental tensor into Eq. (5.27),

$$
\begin{aligned}
A_{\mu\nu\sigma} &= \frac{\partial}{\partial x^\sigma}(g_{\mu\nu}) - \{\sigma\mu, \tau\}g_{\tau\nu} - \{\sigma\nu, \tau\}g_{\mu\tau} \\
&= \frac{\partial}{\partial x^\sigma}(g_{\mu\nu}) - g^{\tau\alpha}[\sigma\mu, \alpha]g_{\tau\nu} - g^{\tau\beta}[\sigma\nu, \beta]g_{\mu\tau} \\
&= \frac{\partial}{\partial x^\sigma}(g_{\mu\nu}) - [\sigma\mu, \alpha]\delta^\alpha_\nu - [\sigma\nu, \beta]\delta^\beta_\mu \\
&= \frac{\partial}{\partial x^\sigma}(g_{\mu\nu}) - [\sigma\mu, \nu] - [\sigma\nu, \mu] \\
&= \frac{\partial}{\partial x^\sigma}(g_{\mu\nu}) - \frac{1}{2}\left(\frac{\partial g_{\nu\sigma}}{\partial x^\mu} + \frac{\partial g_{\nu\mu}}{\partial x^\sigma} - \frac{\partial g_{\sigma\mu}}{\partial x^\nu}\right) \\
&\quad - \frac{1}{2}\left(\frac{\partial g_{\mu\sigma}}{\partial x^\nu} + \frac{\partial g_{\mu\nu}}{\partial x^\sigma} - \frac{\partial g_{\sigma\nu}}{\partial x^\mu}\right) = 0
\end{aligned}
$$

Since the extension of $g_{\mu\nu}$ vanishes identically, it becomes necessary to look for another way. Returning to the general tensor $A_{\mu\nu}$ and using Eq. (5.26) to write it as the extension of the tensor A_μ, $A_{\mu\nu} = \frac{\partial A_\mu}{\partial x^\nu} - \{\mu\nu, \rho\}A_\rho$, this form of $A_{\mu\nu}$ is inserted into the expression for the extension of $A_{\mu\nu}$:

$$
\begin{aligned}
A_{\mu\sigma\tau} &= \frac{\partial}{\partial x^\tau}(A_{\mu\sigma}) - \{\tau\mu, \alpha\}A_{\alpha\sigma} - \{\tau\sigma, \alpha\}A_{\mu\alpha} \\
&= \frac{\partial}{\partial x^\tau}\left(\frac{\partial A_\mu}{\partial x^\sigma} - \{\mu\sigma, \rho\}A_\rho\right) - \{\tau\mu, \alpha\}\left(\frac{\partial A_\alpha}{\partial x^\sigma} - \{\alpha\sigma, \varepsilon\}A_\varepsilon\right) \\
&\quad - \{\tau\sigma, \alpha\}\left(\frac{\partial A_\mu}{\partial x^\alpha} - \{\mu\alpha, \beta\}A_\beta\right) \\
&= \frac{\partial^2 A_\mu}{\partial x^\tau \partial x^\sigma} - A_\rho \frac{\partial\{\mu\sigma, \rho\}}{\partial x^\tau} - \{\mu\sigma, \rho\}\frac{\partial A_\rho}{\partial x^\tau} - \{\tau\mu, \alpha\}\frac{\partial A_\alpha}{\partial x^\sigma} \\
&\quad + \{\tau\mu, \alpha\}\{\alpha\sigma, \varepsilon\}A_\varepsilon - \{\tau\sigma, \alpha\}\frac{\partial A_\mu}{\partial x^\alpha} + \{\tau\sigma, \alpha\}\{\mu\alpha, \beta\}A_\beta
\end{aligned}
$$

Renaming the indices so that $A_\alpha, A_\varepsilon, A_\beta$ each have the subscript ρ,

$$
\begin{aligned}
&= \frac{\partial^2 A_\mu}{\partial x^\tau \partial x^\sigma} - A_\rho \frac{\partial\{\mu\sigma, \rho\}}{\partial x^\tau} - \{\mu\sigma, \rho\}\frac{\partial A_\rho}{\partial x^\tau} - \{\tau\mu, \rho\}\frac{\partial A_\rho}{\partial x^\sigma} \\
&\quad + \{\tau\mu, \alpha\}\{\alpha\sigma, \rho\}A_\rho - \{\tau\sigma, \alpha\}\frac{\partial A_\rho}{\partial x^\alpha} + \{\tau\sigma, \alpha\}\{\mu\alpha, \rho\}A_\rho
\end{aligned}
$$

In this equation, the first and last two terms are symmetric in the indices σ and τ. Using the symmetries present in the indices σ and τ, the tensor

$$
A_{\mu\sigma\tau} - A_{\mu\tau\sigma} = B^\rho_{\mu\sigma\tau}A_\rho \tag{5.42}
$$

is formed. In the subtraction, the corresponding symmetric terms add to zero. The third and fourth terms, when subtracted from the same terms with the indices σ and τ exchanged also add to zero. The remaining terms are

$$B^\rho_{\mu\sigma\tau} A_\rho = A_{\mu\sigma\tau} - A_{\mu\tau\sigma}$$

$$= \left(-\frac{\partial\{\mu\sigma,\rho\}}{\partial x^\tau} + \frac{\partial\{\mu\tau,\rho\}}{\partial x^\sigma} + \{\tau\mu,\alpha\}\{\alpha\sigma,\rho\} - \{\sigma\mu,\alpha\}\{\alpha\tau,\rho\} \right) A_\rho$$

$$B^\rho_{\mu\sigma\tau} = -\frac{\partial\{\mu\sigma,\rho\}}{\partial x^\tau} + \frac{\partial\{\mu\tau,\rho\}}{\partial x^\sigma} - \{\mu\sigma,\alpha\}\{\alpha\tau,\rho\} + \{\mu\tau,\alpha\}\{\alpha\sigma,\rho\} \quad (5.43)$$

1. Equation (5.42), the expression for $A_{\mu\sigma\tau} - A_{\mu\tau\sigma}$, is a function of A_ρ alone, not of its derivatives.
2. $B^\rho_{\mu\sigma\tau}$ is a tensor. From section 10, Eq. (5.27), since $A_{\mu\sigma\tau}$ and $A_{\mu\tau\sigma}$ are tensors their difference $A_{\mu\sigma\tau} - A_{\mu\tau\sigma}$ is a tensor. A_ρ is an arbitrary tensor. By an analysis similar to that of section 7, it follows that $B^\rho_{\mu\sigma\tau}$ is a tensor.
3. Equation (5.43), the equation for $B^\rho_{\mu\sigma\tau}$, is the same as the equation for the Riemann–Christoffel tensor.[143] Thus, $B^\rho_{\mu\sigma\tau}$ is identified as the Riemann–Christoffel tensor $R^\rho_{\mu\sigma\tau}$:

$$B^\rho_{\mu\sigma\tau} = R^\rho_{\mu\sigma\tau}$$

4. $B^\rho_{\mu\sigma\tau}$ is a function of various derivatives of the $g_{\mu\nu}$.
5. If there is a coordinate system in which the $g_{\mu\nu}$ are constants, all of the $B^\rho_{\mu\sigma\tau}$ vanish.
6. In another coordinate system, the $g_{\mu\nu}$ in general will not be constants.
7. Because $B^\rho_{\mu\sigma\tau}$ is a tensor, the components of $B^\rho_{\mu\sigma\tau}$ in the new coordinate system (being linear combinations of the components in the original coordinate system) will vanish (be equal to zero).
8. The converse of the preceding statement (#7) is that the vanishing of the Riemann tensor, $B^\rho_{\mu\sigma\tau}$, is a necessary condition that, by an appropriate choice of the reference system, the $g_{\mu\nu}$ may be constants.
9. The $g_{\mu\nu}$ being constants corresponds physically to the situation of the special theory of relativity holding for a finite region of the continuum.

Contracting the tensor $B^\rho_{\mu\sigma\tau}$ over the indices τ and ρ, one obtains a covariant tensor of rank two. Also changing the index σ to the index ν,

$$B^\rho_{\mu\sigma\rho} = B^\rho_{\mu\nu\rho} = G_{\mu\nu}$$

$$= -\frac{\partial\{\mu\nu,\rho\}}{\partial x^\rho} + \frac{\partial\{\mu\rho,\rho\}}{\partial x^\nu} - \{\mu\nu,\alpha\}\{\alpha\rho,\rho\} + \{\mu\rho,\alpha\}\{\alpha\nu,\rho\}$$

Regrouping the terms, and changing the indices $\alpha \to \beta$ and $\rho \to \alpha$,

$$G_{\mu\nu} = \left(-\frac{\partial\{\mu\nu,\alpha\}}{\partial x^\alpha} + \{\mu\alpha,\beta\}\{\beta\nu,\alpha\} \right) + \left(\frac{\partial\{\mu\alpha,\alpha\}}{\partial x^\nu} - \{\mu\nu,\beta\}\{\beta\alpha,\alpha\} \right)$$

$$= R_{\mu\nu} + S_{\mu\nu}$$

With $R_{\mu\nu} = -\frac{\partial\{\mu\nu,\alpha\}}{\partial x^\alpha} + \{\mu\alpha,\beta\}\{\beta\nu,\alpha\} = -\frac{\partial\{\mu\nu,\alpha\}}{\partial x^\alpha} + \{\mu\alpha,\beta\}\{\nu\beta,\alpha\}$

$S_{\mu\nu} = \frac{\partial\{\mu\alpha,\alpha\}}{\partial x^\nu} - \{\mu\nu,\beta\}\{\beta\alpha,\alpha\} = \frac{\partial\{\alpha\mu,\alpha\}}{\partial x^\nu} - \{\mu\nu,\beta\}\{\alpha\beta,\alpha\}$

By Eq. (5.29a), $\{\mu\sigma,\mu\} = \frac{1}{\sqrt{-g}}\frac{\partial\sqrt{-g}}{\partial x^\sigma} = \frac{\partial}{\partial x^\sigma}(\log\sqrt{-g})$, $S_{\mu\nu}$ becomes

$$S_{\mu\nu} = \frac{\partial}{\partial x^\nu}\frac{\partial(\log\sqrt{-g})}{\partial x^\mu} - \{\mu\nu,\beta\}\frac{\partial(\log\sqrt{-g})}{\partial x^\beta}$$

$$= \frac{\partial^2(\log\sqrt{-g})}{\partial x^\mu\partial x^\nu} - \{\mu\nu,\alpha\}\frac{\partial(\log\sqrt{-g})}{\partial x^\alpha} \qquad (5.44)$$

In the second term, the index β has been changed to the index α.

1. If the choice of coordinates is made such that $\sqrt{-g} = 1$, a number of simplifications to the equations is obtained.
2. For $\sqrt{-g} = 1$, the expression for $S_{\mu\nu}$ becomes zero.
3. For $\sqrt{-g} = 1$, $G_{\mu\nu} = R_{\mu\nu}$.
4. In the remainder of the paper, unless stated specifically otherwise, the choice of coordinates will be made such that $\sqrt{-g} = 1$.

5.4.7 The Hamiltonian Function for the Gravitational Field

5.4.7.1 The Hamiltonian Form of Eq. (5.47)

The variation of the second of the Eqs. (5.47a), $H = g^{\mu\nu}\Gamma^\alpha_{\mu\beta}\Gamma^\beta_{\nu\alpha}$, gives

$$\delta H = \delta(g^{\mu\nu}\Gamma^\alpha_{\mu\beta}\Gamma^\beta_{\nu\alpha}) = (\delta g^{\mu\nu})\Gamma^\alpha_{\mu\beta}\Gamma^\beta_{\nu\alpha} + g^{\mu\nu}[\delta(\Gamma^\alpha_{\mu\beta}\Gamma^\beta_{\nu\alpha})]$$

$$= \Gamma^\alpha_{\mu\beta}\Gamma^\beta_{\nu\alpha}\delta g^{\mu\nu} + g^{\mu\nu}[(\delta\Gamma^\alpha_{\mu\beta})\Gamma^\beta_{\nu\alpha} + \Gamma^\alpha_{\mu\beta}(\delta\Gamma^\beta_{\nu\alpha})]$$

But the two terms in the [ ] are the same if the indices $\alpha \leftrightarrow \beta$ and $\mu \leftrightarrow \nu$ in the first term,

$$\delta H = \Gamma^\alpha_{\mu\beta}\Gamma^\beta_{\nu\alpha}\delta g^{\mu\nu} + g^{\mu\nu}[2\Gamma^\alpha_{\mu\beta}\delta(\Gamma^\beta_{\nu\alpha})]$$

$$= \Gamma^\alpha_{\mu\beta}\Gamma^\beta_{\nu\alpha}\delta g^{\mu\nu} + 2\Gamma^\alpha_{\mu\beta}[g^{\mu\nu}\delta\Gamma^\beta_{\nu\alpha}]$$

$$= \Gamma^\alpha_{\mu\beta}\Gamma^\beta_{\nu\alpha}\delta g^{\mu\nu} + 2\Gamma^\alpha_{\mu\beta}[\delta(g^{\mu\nu}\Gamma^\beta_{\nu\alpha}) - \Gamma^\beta_{\nu\alpha}\delta g^{\mu\nu}]$$

$$= -\Gamma^\alpha_{\mu\beta}\Gamma^\beta_{\nu\alpha}\delta g^{\mu\nu} + 2\Gamma^\alpha_{\mu\beta}\delta(g^{\mu\nu}\Gamma^\beta_{\nu\alpha})$$

In the last term, the expression, $\delta(g^{\mu\nu}\Gamma^\beta_{\nu\alpha})$, can be expanded as

$$\delta(g^{\mu\nu}\Gamma^\beta_{\nu\alpha}) = -\delta(g^{\mu\nu}\{\nu\alpha,\beta\}) \qquad \text{by Eq. (5.45)}$$

$$= -\delta(g^{\mu\nu}g^{\beta\lambda}[\nu\alpha,\lambda]) \qquad \text{by Eq. (5.23)}$$

$$= -\frac{1}{2}\delta\left[g^{\mu\nu}g^{\beta\lambda}\left(\frac{\partial g_{\nu\lambda}}{\partial x^\alpha} + \frac{\partial g_{\alpha\lambda}}{\partial x^\nu} - \frac{\partial g_{\nu\alpha}}{\partial x^\lambda}\right)\right] \qquad \text{by Eq. (5.21)}$$

Substituting this expanded form of $\delta(g^{\mu\nu}\Gamma^\beta_{\nu\alpha})$ into the expression, the last term becomes

$$2\Gamma^{\alpha}_{\mu\beta}\delta(g^{\mu\nu}\Gamma^{\beta}_{\nu\alpha}) = +2\Gamma^{\alpha}_{\mu\beta}\left(-\frac{1}{2}\right)\delta\left[g^{\mu\nu}g^{\beta\lambda}\left(\frac{\partial g_{\nu\lambda}}{\partial x^{\alpha}} + \frac{\partial g_{\alpha\lambda}}{\partial x^{\nu}} - \frac{\partial g_{\alpha\nu}}{\partial x^{\lambda}}\right)\right]$$

Consider the second and third terms in the ( ). If the values of λ and ν are interchanged (say, $\lambda = 1, \nu = 3 \rightarrow \lambda = 3, \nu = 1$), the latter terms are the negative of the initial terms and, over the summation for all values of λ and ν, the sum of these terms will be zero. However, when the values of λ and ν are interchanged in the ( ), the values of λ and ν also must be interchanged in the $g^{\mu\nu}g^{\beta\lambda}$ factor preceding the ( ). Exchanging also the indices μ and β in $g^{\mu\nu}g^{\beta\lambda}$(i.e., $\lambda \leftrightarrow \nu, \mu \leftrightarrow \beta$), and using the symmetry $\Gamma^{\alpha}_{\mu\beta} = \Gamma^{\alpha}_{\beta\mu}$, the second and third terms become

$$\Gamma^{\alpha}_{\mu\beta}\delta\left[g^{\mu\nu}g^{\beta\lambda}\left(\frac{\partial g_{\alpha\lambda}}{\partial x^{\nu}} - \frac{\partial g_{\alpha\nu}}{\partial x^{\lambda}}\right)\right] \rightarrow \Gamma^{\alpha}_{\mu\beta}\delta\left[g^{\beta\lambda}g^{\mu\nu}\left(\frac{\partial g_{\alpha\nu}}{\partial x^{\lambda}} - \frac{\partial g_{\alpha\lambda}}{\partial x^{\nu}}\right)\right]$$

Thus, in the summation over the indices λ, ν, μ, β, the second and third terms pair-wise add up to zero, leaving only the first term.

$$\delta H = -\Gamma^{\alpha}_{\mu\beta}\Gamma^{\beta}_{\nu\alpha}\delta g^{\mu\nu} - \Gamma^{\alpha}_{\mu\beta}\delta\left[g^{\mu\nu}g^{\beta\lambda}\frac{\partial g_{\nu\lambda}}{\partial x^{\alpha}}\right]$$

But, using Eq. (5.31), $g^{\mu\nu}g^{\beta\lambda}\frac{\partial g_{\nu\lambda}}{\partial x^{\alpha}} = -\frac{\partial g^{\mu\beta}}{\partial x^{\alpha}} = -g^{\mu\beta}_{\alpha}$, the last term being an alternative notation for the middle term:

$$\delta H = -\Gamma^{\alpha}_{\mu\beta}\Gamma^{\beta}_{\nu\alpha}\delta g^{\mu\nu} + \Gamma^{\alpha}_{\mu\beta}\delta g^{\mu\beta}_{\alpha}$$

$$\Rightarrow \quad \frac{\partial H}{\partial g^{\mu\nu}} = -\Gamma^{\alpha}_{\mu\beta}\Gamma^{\beta}_{\nu\alpha} \qquad (5.48a)$$

$$\Rightarrow \quad \frac{\partial H}{\partial g^{\mu\nu}_{\sigma}} = \Gamma^{\sigma}_{\mu\nu} \qquad (5.48b)$$

In Eq. (5.48b), the index ν has replaced the index β, and the index σ has replaced the index α.

Returning to Eq. (5.47a), in analogy to the derivation of Lagrange's equations, $\delta \int L(q, \dot{q}, t)dt = 0 \Rightarrow \frac{d}{dt}\left(\frac{\partial L}{\partial \dot{q}}\right) - \frac{\partial L}{\partial q} = 0$, Einstein carries out the variation of Eq. (5.47a). Regarding H as a function of the $g^{\mu\nu}$ and $g^{\mu\nu}_{\alpha} = \frac{\partial g^{\mu\nu}}{\partial x^{\alpha}}$, and "where, on the boundary of the finite four-dimensional region of integration which we have in view, the variations vanish."[144] The $g^{\mu\nu}$ are written as $g^{\mu\nu}(x^{\alpha}, \in) = g^{\mu\nu}(x^{\alpha}, \in= 0)+ \in \eta(x^{\alpha})$, where $g^{\mu\nu}(x^{\alpha}, \in= 0)$ are the solutions to the variation, the $g^{\mu\nu}(x^{\alpha}, \in\neq 0)$ define the proximate paths, $\in$ is a small parameter labeling a set of possible curves, and $\eta(x^{\alpha})$ is an arbitrary but well-behaved function in the interval of integration that vanishes at the boundaries.[145]

$$0 = \delta \int_{1}^{2} H d\tau = \int_{1}^{2} \delta H d\tau = \int_{1}^{2} \delta H(g^{\mu\nu}, g^{\mu\nu}_{\alpha}, x^{\alpha})dx^{1}\,dx^{2}\,dx^{3}\,dx^{4}$$

$$= \int_{1}^{2}\left[\frac{\partial H}{\partial g^{\mu\nu}}\frac{\partial g^{\mu\nu}}{\partial \in}\delta \in + \frac{\partial H}{\partial g^{\mu\nu}_{\alpha}}\frac{\partial g^{\mu\nu}_{\alpha}}{\partial \in}\delta \in\right]dx^{1}\,dx^{2}\,dx^{3}\,dx^{4}$$

$$= \int_1^2 \left[\frac{\partial H}{\partial g^{\mu\nu}} \frac{\partial g^{\mu\nu}}{\partial \epsilon} \delta \epsilon + \frac{\partial H}{\partial g^{\mu\nu}_\alpha} \frac{\partial^2 g^{\mu\nu}}{\partial \epsilon \, \partial x^\alpha} \delta \epsilon \right] dx^1 dx^2 dx^3 dx^4$$

$$= \int_1^2 \left[\frac{\partial H}{\partial g^{\mu\nu}} \frac{\partial g^{\mu\nu}}{\partial \epsilon} \delta \epsilon + \left(\frac{\partial}{\partial x^\alpha} \left(\frac{\partial H}{\partial g^{\mu\nu}_\alpha} \frac{\partial g^{\mu\nu}}{\partial \epsilon} \right) \delta \epsilon \right. \right.$$

$$\left. \left. - \frac{\partial g^{\mu\nu}}{\partial \epsilon} \frac{\partial}{\partial x^\alpha} \left(\frac{\partial H}{\partial g^{\mu\nu}_\alpha} \right) \delta \epsilon \right) \right] dx^1 dx^2 dx^3 dx^4$$

The middle term is equal to zero since all the curves pass through the boundary points 1 and 3, and $\frac{\partial g^{\mu\nu}}{\partial \epsilon} = 0$ at these boundary points:

$$\int_1^2 \frac{\partial}{\partial x^\alpha} \left(\frac{\partial H}{\partial g^{\mu\nu}_\alpha} \frac{\partial g^{\mu\nu}}{\partial \epsilon} \right) dx^1 dx^2 dx^3 dx^4 = \left[\frac{\partial H}{\partial g^{\mu\nu}_\alpha} \frac{\partial g^{\mu\nu}}{\partial \epsilon} \right]_1^2 = 0$$

Returning to the remaining terms,

$$0 = \int_1^2 H d\tau = \int_1^2 \left[\frac{\partial H}{\partial g^{\mu\nu}} - \frac{\partial}{\partial x^\alpha} \frac{\partial H}{\partial g^{\mu\nu}_\alpha} \right] \frac{\partial g^{\mu\nu}}{\partial \epsilon} \delta \epsilon \, dx^1 dx^2 dx^3 dx^4$$

$$0 = \int \left[\frac{\partial H}{\partial g^{\mu\nu}} - \frac{\partial}{\partial x^\alpha} \left(\frac{\partial H}{\partial g^{\mu\nu}_\alpha} \right) \right] \eta(x^\alpha) \delta \epsilon \, dx^1 dx^2 dx^3 dx^4$$

But $\eta(x^\alpha)$ is an arbitrary function, except that it continuous, well behaved, and vanishes at the boundary. For the above integral to vanish for all arbitrary functions $\eta(x^\alpha)$ it is necessary that the expression in the square brackets [ ] vanish. Thus,

$$\delta \int H(g^{\mu\nu}, g^{\mu\nu}_\alpha, x^\alpha) d\tau = 0 \Rightarrow \frac{\partial}{\partial x^\alpha} \left(\frac{\partial H}{\partial g^{\mu\nu}_\alpha} \right) - \frac{\partial H}{\partial g^{\mu\nu}} = 0 \qquad (5.47b)$$

$$\frac{\partial}{\partial x^\alpha} \left(\Gamma^\alpha_{\mu\nu} \right) + \Gamma^\alpha_{\mu\beta} \Gamma^\beta_{\nu\alpha} = 0$$

by Eqs (5.48a) and (5.48b)
This shows that Eq. (5.47a) is the equivalent of Eq. (5.47). Thus Eqs. (5.47) and (5.47a) are equivalent forms of the same equation.

Einstein then manipulates Eq. (5.47b) into an alternative form. Multiplying Eq. (5.47b) by $g^{\mu\nu}_\sigma$, and using $fdg = d(fg) - gdf$, the first term becomes

$$g^{\mu\nu}_\sigma \frac{\partial}{\partial x^\alpha} \left(\frac{\partial H}{\partial g^{\mu\nu}_\alpha} \right) = \frac{\partial}{\partial x^\alpha} \left(g^{\mu\nu}_\sigma \frac{\partial H}{\partial g^{\mu\nu}_\alpha} \right) - \frac{\partial H}{\partial g^{\mu\nu}_\alpha} \frac{\partial g^{\mu\nu}_\sigma}{\partial x^\alpha}$$

but since $\frac{\partial g^{\mu\nu}_\sigma}{\partial x^\alpha} = \frac{\partial}{\partial x^\alpha} \left(\frac{\partial g^{\mu\nu}}{\partial x^\sigma} \right) = \frac{\partial}{\partial x^\sigma} \left(\frac{\partial g^{\mu\nu}}{\partial x^\alpha} \right) = \frac{\partial g^{\mu\nu}_\alpha}{\partial x^\sigma}$ this term becomes

$g^{\mu\nu}_\sigma \frac{\partial}{\partial x^\alpha} \left(\frac{\partial H}{\partial g^{\mu\nu}_\alpha} \right) = \frac{\partial}{\partial x^\alpha} \left(g^{\mu\nu}_\sigma \frac{\partial H}{\partial g^{\mu\nu}_\alpha} \right) - \frac{\partial H}{\partial g^{\mu\nu}_\alpha} \frac{\partial g^{\mu\nu}_\alpha}{\partial x^\sigma}$ Replacing the first term in Eq. (5.47b) (multiplied by $g^{\mu\nu}_\sigma$), the full expression becomes

$$\frac{\partial}{\partial x^\alpha}\left(g_\sigma^{\mu\nu}\frac{\partial H}{\partial g_\alpha^{\mu\nu}}\right) - \frac{\partial H}{\partial g_\alpha^{\mu\nu}}\frac{\partial g_\alpha^{\mu\nu}}{\partial x^\sigma} - g_\sigma^{\mu\nu}\frac{\partial H}{\partial g^{\mu\nu}} = 0$$

The second and third terms can be combined as

$$\frac{\partial H\left(g^{\mu\nu}, g_\alpha^{\mu\nu}\right)}{\partial x^\sigma} = \frac{\partial H}{\partial g_\alpha^{\mu\nu}}\frac{\partial g_\alpha^{\mu\nu}}{\partial x^\sigma} + g_\sigma^{\mu\nu}\frac{\partial H}{\partial g^{\mu\nu}}$$

giving $\frac{\partial}{\partial x^\alpha}\left(g_\sigma^{\mu\nu}\frac{\partial H}{\partial g_\alpha^{\mu\nu}}\right) - \frac{\partial H}{\partial x^\sigma} = 0$

Noting both terms are a derivative, one with respect to x^α and the other with respect to x^σ, by introducing the delta function δ_σ^α, the equation can be written

$$\frac{\partial}{\partial x^\alpha}\left(g_\sigma^{\mu\nu}\frac{\partial H}{\partial g_\alpha^{\mu\nu}} - \delta_\sigma^\alpha H\right) = \frac{\partial}{\partial x^\alpha}\left(t_\sigma^\alpha\right) = 0 \qquad (5.49)$$

For $\sigma = 4$ Einstein integrates Eq. (5.49) over a three-dimensional volume, and identifies the expression as the conservation laws for momentum and energy (l, m, n are the direction cosines of the surface element dS).[146]

$$\frac{d}{dx^4}\int t_\sigma dV = \int\left(lt_\sigma^1 + mt_\sigma^2 + nt_\sigma^3\right)dS \qquad (5.49a)$$

5.4.7.2   Derivation of Eq. (5.50)

By use of Eqs. (5.48), (5.47a), (5.34), and (5.45), this expression for t_σ^α can be manipulated into another form, Eq. (5.50).

$$-2\kappa t_\sigma^\alpha = g_\sigma^{\mu\nu}\frac{\partial H}{\partial g_\alpha^{\mu\nu}} - \delta_\sigma^\alpha H$$

$$= g_\sigma^{\mu\nu}\Gamma_{\mu\nu}^\alpha - \delta_\sigma^\alpha g^{\mu\nu}\Gamma_{\mu\beta}^\lambda\Gamma_{\nu\lambda}^\beta$$

by Eqs (5.48), (5.47a)

$$\kappa t_\sigma^\alpha = +\frac{1}{2}\delta_\sigma^\alpha g^{\mu\nu}\Gamma_{\mu\beta}^\lambda\Gamma_{\nu\lambda}^\beta - \frac{1}{2}g_\sigma^{\mu\nu}\Gamma_{\mu\nu}^\alpha$$

Working with the second term:

$$-\frac{1}{2}g_\sigma^{\mu\nu}\Gamma_{\mu\nu}^\alpha = -\frac{1}{2}(-g^{\mu\tau}\{\tau\sigma,\nu\} - g^{\nu\tau}\{\tau\sigma,\mu\})\Gamma_{\mu\nu}^\alpha$$

by Eq. (5.34)

$$= -\frac{1}{2}(+g^{\mu\tau}\Gamma_{\tau\sigma}^\nu + g^{\nu\tau}\Gamma_{\tau\sigma}^\mu)\Gamma_{\mu\nu}^\alpha$$

by Eq. (5.45)

$$= -\frac{1}{2}(g^{\mu\tau}\Gamma_{\tau\sigma}^\nu\Gamma_{\mu\nu}^\alpha + g^{\nu\tau}\Gamma_{\tau\sigma}^\mu\Gamma_{\mu\nu}^\alpha)$$

Noting all indices with the exception of α and σ are summed over, it is seen the two terms in the parentheses are the same. This can formally be seen by, in the first term, replacing the indices $\nu \to \beta$ and $\tau \to \nu$, and in the second term, by replacing the indices $\mu \to \beta, \nu \to \mu$ and $\tau \to \nu$:

$$-\frac{1}{2}g_\sigma^{\mu\nu}\Gamma^\sigma_{\mu\nu} = -g^{\mu\nu}\Gamma^\beta_{\nu\sigma}\Gamma^\alpha_{\mu\beta}$$

$$\kappa t^\alpha_\sigma = \frac{1}{2}\delta^\alpha_\sigma g^{\mu\nu}\Gamma^\lambda_{\mu\beta}\Gamma^\beta_{\nu\lambda} - g^{\mu\nu}\Gamma^\alpha_{\mu\beta}\Gamma^\beta_{\nu\sigma} \tag{5.50}$$

5.4.7.3 Derivation of Eq. (5.51)

To arrive at the third alternative form of Eq. (5.47), Einstein multiplies Eq. (5.47) by $g^{\nu\sigma}$ and manipulates the resulting equations in the following manner:

$$0 = +g^{\nu\sigma}\left(\frac{\partial\Gamma^\alpha_{\mu\nu}}{\partial x^\alpha} + \Gamma^\alpha_{\mu\beta}\Gamma^\beta_{\nu\alpha}\right) \tag{5.47c}$$

Working initially with the first term on the right-hand side,

$$g^{\nu\sigma}\frac{\partial\Gamma^\alpha_{\mu\nu}}{\partial x^\alpha} = \frac{\partial}{\partial x^\alpha}(g^{\nu\sigma}\Gamma^\alpha_{\mu\nu}) - \left(\frac{\partial g^{\nu\sigma}}{\partial x^\alpha}\right)\Gamma^\alpha_{\mu\nu}$$

$$= \frac{\partial}{\partial x^\alpha}(g^{\nu\sigma}\Gamma^\alpha_{\mu\nu}) - \Gamma^\alpha_{\mu\nu}(-g^{\nu\tau}\{\tau\alpha,\sigma\} - g^{\sigma\tau}\{\tau\alpha,\nu\})$$

by Eq. (5.34)

$$= \frac{\partial}{\partial x^\alpha}(g^{\nu\sigma}\Gamma^\alpha_{\mu\nu}) - \Gamma^\alpha_{\mu\nu}(+g^{\nu\tau}\Gamma^\sigma_{\tau\alpha} + g^{\sigma\tau}\Gamma^\nu_{\tau\alpha})$$

by Eq. (5.45)

$$= \frac{\partial}{\partial x^\alpha}(g^{\nu\sigma}\Gamma^\alpha_{\mu\nu}) - g^{\nu\beta}\Gamma^\alpha_{\mu\nu}\Gamma^\sigma_{\alpha\beta} - g^{\sigma\beta}\Gamma^\alpha_{\mu\nu}\Gamma^\nu_{\beta\alpha}$$

with $\tau \to \beta$ In the third term, exchanging the indices $\nu \leftrightarrow \beta$, it is seen this term is the negative of the second term in Eq. (5.47c). Equation (5.47c) becomes

$$0 = \frac{\partial}{\partial x^\alpha}(g^{\nu\sigma}\Gamma^\alpha_{\mu\nu}) - g^{\nu\beta}\Gamma^\alpha_{\mu\nu}\Gamma^\sigma_{\alpha\beta} \tag{5.47d}$$

We now work with the second term on the RHS. Consider first Eq. (5.50), the expression for t^α_σ. From this one can form the expression for t, defined as $t = t^\alpha_\alpha$.

$$\kappa t^\alpha_\alpha = \frac{1}{2}\delta^\alpha_\alpha g^{\mu\nu}\Gamma^\lambda_{\mu\beta}\Gamma^\beta_{\nu\lambda} - g^{\mu\nu}\Gamma^\alpha_{\mu\beta}\Gamma^\beta_{\nu\alpha}$$

$$= \frac{1}{2}(4)g^{\mu\nu}\Gamma^\lambda_{\mu\beta}\Gamma^\beta_{\nu\lambda} - g^{\mu\nu}\Gamma^\alpha_{\mu\beta}\Gamma^\beta_{\nu\alpha}$$

Excluding the factor of two with the first term, the second term is the same as the first term (with $\alpha \to \lambda$ in the second term):

$$\kappa t = \kappa t^\alpha_\alpha = g^{\mu\nu}\Gamma^\lambda_{\mu\beta}\Gamma^\beta_{\nu\lambda}$$

Using this result and the expression for t^α_σ, we form the quantity,

$$\kappa t^\sigma_\mu - \frac{1}{2}\delta^\sigma_\mu\kappa t = \left(\frac{1}{2}\delta^\sigma_\mu g^{\gamma\nu}\Gamma^\lambda_{\gamma\beta}\Gamma^\beta_{\nu\lambda} - g^{\gamma\nu}\Gamma^\alpha_{\gamma\beta}\Gamma^\beta_{\nu\mu}\right) - \frac{1}{2}\delta^\sigma_\mu g^{\gamma\nu}\Gamma^\lambda_{\gamma\beta}\Gamma^\beta_{\nu\lambda}$$

$$= -g^{\gamma\nu}\Gamma^\alpha_{\gamma\beta}\Gamma^\beta_{\nu\mu}$$

This is the same expression as the second term in Eq. (5.47d) (in the second term in Eq. (5.47d), one can set $\sigma \to \alpha, \beta \to \gamma, \alpha \to \beta$). With this substitution Eq. (5.47d) becomes

$$0 = \frac{\partial}{\partial x^\alpha}(g^{\nu\sigma}\Gamma^\alpha_{\mu\nu}) - g^{\nu\beta}\Gamma^\alpha_{\mu\nu}\Gamma^\sigma_{\alpha\beta} = \frac{\partial}{\partial x^\alpha}(g^{\nu\sigma}\Gamma^\alpha_{\mu\nu}) + \kappa\left(t^\sigma_\mu - \frac{1}{2}\delta^\sigma_\mu t\right)$$

$$\frac{\partial}{\partial x^\alpha}(g^{\nu\sigma}\Gamma^\alpha_{\mu\nu}) = -\kappa\left(t^\sigma_\mu - \frac{1}{2}\delta^\sigma_\mu t\right) \tag{5.51a}$$

$$\sqrt{-g} = 1 \tag{5.51b}$$

5.4.7.4   Derivation of Eq. (5.56)

Contract Eq. (5.52) with respect to the indices μ and σ:

$$\left\{\begin{array}{l} \frac{\partial}{\partial x^\alpha}\left(g^{\beta\sigma}\Gamma^\alpha_{\mu\beta}\right) = -\kappa\left[(t^\sigma_\mu + T^\sigma_\mu) - \frac{1}{2}\delta^\sigma_\mu\,(t+T)\right] \\ \sqrt{-g} = 1 \end{array}\right\} \tag{5.52}$$

$$\frac{\partial}{\partial x^\alpha}(g^{\beta\sigma}\Gamma^\alpha_{\sigma\beta}) = -\kappa[(t^\sigma_\sigma + T^\sigma_\sigma)] - \frac{1}{2}\delta^\sigma_\sigma(t+T)$$

$$= -\kappa\left\{(t+T) - \frac{1}{2}(4)(t+T)\right\} = +\kappa(t+T)$$

In this equation, changing the index $\sigma \to \lambda$, multiplying by $\frac{1}{2}\delta^\sigma_\mu$ and subtracting the result from Eq. (5.52),

$$\frac{\partial}{\partial x^\alpha}\left(g^{\beta\sigma}\Gamma^\alpha_{\mu\beta} - \frac{1}{2}\delta^\sigma_\mu g^{\lambda\beta}\Gamma^\alpha_{\lambda\beta}\right) = -\kappa(t^\sigma_\mu + T^\sigma_\mu) \tag{5.52a}$$

Taking the derivative of Eq. (5.52a) with respect to x^σ,

$$\frac{\partial}{\partial x^\sigma}\frac{\partial}{\partial x^\alpha}\left(g^{\beta\sigma}\Gamma^\alpha_{\mu\beta} - \frac{1}{2}\delta^\sigma_\mu g^{\lambda\beta}\Gamma^\alpha_{\lambda\beta}\right) = -\kappa\frac{\partial}{\partial x^\sigma}(t^\sigma_\mu + T^\sigma_\mu) \tag{5.52b}$$

Working with the first term,

$$\frac{\partial}{\partial x^\sigma}\frac{\partial}{\partial x^\alpha}(g^{\beta\sigma}\Gamma^\alpha_{\mu\beta}) = \frac{\partial^2}{\partial x^\sigma \partial x^\alpha}\left(-g^{\sigma\beta}\{\beta\mu,\alpha\}\right)$$

by Eq. (5.45)

$$= -\frac{\partial^2}{\partial x^\sigma \partial x^\alpha}(g^{\sigma\beta}g^{\alpha\lambda}[\beta\mu,\lambda])$$

by Eq. (5.23)

$$= -\frac{1}{2}\frac{\partial^2}{\partial x^\sigma \partial x^\alpha}\left[g^{\sigma\beta}g^{\alpha\lambda}\left(\frac{\partial g_{\mu\lambda}}{\partial x^\beta} + \frac{\partial g_{\beta\lambda}}{\partial x^\mu} - \frac{\partial g_{\beta\mu}}{\partial x^\lambda}\right)\right]$$

by Eq. (5.21)

The contributions from the first and third terms in the round parentheses, ( ), sum to zero. The two terms are the negative of each other as can be seen by the exchange of the indices $\beta \leftrightarrow \lambda$. In the exchange of

$\beta \leftrightarrow \lambda$ one must also exchange the indices $\sigma \leftrightarrow \alpha$. The remaining term, the second term, is

$$-\frac{1}{2}\frac{\partial^2}{\partial x^\sigma \partial x^\alpha}\left(g^{\sigma\beta}g^{\alpha\lambda}\frac{\partial g_{\beta\lambda}}{\partial x^\mu}\right) = -\frac{1}{2}\frac{\partial^2}{\partial x^\sigma \partial x^\alpha}\left(-\frac{\partial g^{\sigma\alpha}}{\partial x^\mu}\right)$$

by Eq. (5.31) Substituting these results, one obtains

$$\frac{\partial^2}{\partial x^\sigma \partial x^\alpha}(g^{\beta\sigma}\Gamma^\alpha_{\mu\beta}) = \frac{1}{2}\frac{\partial^3 g^{\sigma\alpha}}{\partial x^\sigma \partial x^\alpha \partial x^\mu} = \frac{1}{2}\frac{\partial^3 g^{\alpha\beta}}{\partial x^\alpha \partial x^\beta \partial x^\mu} \qquad (5.54)$$

with $\sigma \to \beta$ in the last expression. Returning to Eq. (5.52b), the second term on the left-hand side becomes

$$\frac{\partial}{\partial x^\sigma}\frac{\partial}{\partial x^\alpha}\left(-\frac{1}{2}\delta^\sigma_\mu g^{\lambda\beta}\Gamma^\alpha_{\lambda\beta}\right) = -\frac{1}{2}\frac{\partial^2}{\partial x^\mu \partial x^\alpha}(g^{\lambda\beta}\Gamma^\alpha_{\lambda\beta})$$

$$= -\frac{1}{2}\frac{\partial^2}{\partial x^\mu \partial x^\alpha}(g^{\lambda\beta}(-\{\lambda\beta,\alpha\}))$$

by Eq. (5.45)

$$= +\frac{1}{2}\frac{\partial^2}{\partial x^\mu \partial x^\alpha}(g^{\lambda\beta}g^{\alpha\delta}[\lambda\beta,\delta])$$

by Eq. (5.23)

$$= \frac{1}{2}\frac{\partial^2}{\partial x^\mu \partial x^\alpha}\left[g^{\lambda\beta}g^{\alpha\delta}\frac{1}{2}\left(\frac{\partial g_{\delta\lambda}}{\partial x^\beta} + \frac{\partial g_{\delta\beta}}{\partial x^\lambda} - \frac{\partial g_{\lambda\beta}}{\partial x^\delta}\right)\right]$$

by Eq. (5.21)

The third term in the round parentheses is zero, since

$$\frac{1}{2}\frac{\partial^2}{\partial x^\mu \partial x^\alpha}\left[-\frac{1}{2}g^{\alpha\delta}\left(g^{\lambda\beta}\frac{\partial g_{\lambda\beta}}{\partial x^\delta}\right)\right] = \frac{1}{2}\frac{\partial^2}{\partial x^\mu \partial x^\alpha}\left[-\frac{1}{2}g^{\alpha\delta}\left(\frac{1}{\sqrt{-g}}\frac{\partial \sqrt{-g}}{\partial x^\delta}\right)\right]$$

by Eq. (5.29)
$= 0$ since $\sqrt{-g} = +1 = $ constant

The first and second terms can be combined as

$$\frac{1}{4}\frac{\partial^2}{\partial x^\mu \partial x^\alpha}\left[g^{\lambda\beta}g^{\alpha\delta}\frac{\partial g_{\delta\lambda}}{\partial x^\beta} + g^{\lambda\beta}g^{\alpha\delta}\frac{\partial g_{\delta\beta}}{\partial x^\lambda}\right]$$

Exchanging the indices $\lambda \leftrightarrow \beta$ in the second term in the square parentheses,[ ], it is seen to be identical to the first term in the parentheses. Using Eq. (5.31),

$$= \frac{1}{2}\frac{\partial^2}{\partial x^\mu \partial x^\alpha}\left[g^{\lambda\beta}g^{\alpha\delta}\frac{\partial g_{\delta\lambda}}{\partial x^\beta}\right] = \frac{1}{2}\frac{\partial^2}{\partial x^\mu \partial x^\alpha}\left[-\frac{\partial}{\partial x^\beta}g^{\alpha\beta}\right]$$

$$= -\frac{1}{2}\frac{\partial^3}{\partial x^\mu \partial x^\alpha \partial x^\beta}(g^{\alpha\beta}) \qquad (5.54a)$$

Assembling these partial results (Eqs. (5.54) and (5.54a)) into Eq. (5.52b),

$$\frac{\partial^3}{\partial x^\alpha \partial x^\beta \partial x^\mu}\left(\frac{1}{2}g^{\alpha\beta} - \frac{1}{2}g^{\alpha\beta}\right) = 0 = -\kappa\frac{\partial}{\partial x^\sigma}\left(t^\sigma_\mu + T^\sigma_\mu\right) \quad (5.56)$$

$$\kappa\frac{\partial}{\partial x^\sigma}\left(t^\sigma_\mu + T^\sigma_\mu\right) = 0$$

Equation (5.56), $\kappa\frac{\partial}{\partial x^\sigma}(t^\sigma_\mu + T^\sigma_\mu) = 0$, is a statement of the conservation of momentum and energy of the total system, whereas Eq. (5.49a) was the same statement but only for the energy components of the gravitational field in the absence of matter.

5.4.7.5   Derivation of Eqs. (5.57), (5.57a)

Derivation of Eq. (5.57):

Multiplying Eq. (5.53) by $\frac{\partial g^{\mu\nu}}{\partial x^\sigma} = g^{\mu\nu}_\sigma$,

$$g^{\mu\nu}_\sigma\left(\frac{\partial \Gamma^\alpha_{\mu\nu}}{\partial x^\alpha} + \Gamma^\alpha_{\mu\beta}\Gamma^\beta_{\nu\alpha}\right) = -\kappa g^{\mu\nu}_\sigma\left(T_{\mu\nu} - \frac{1}{2}g_{\mu\nu}T\right)$$

The left-hand side of Eq. (5.53) is the same as the left-hand side of Eq. (5.47b). This manipulation on Eq. (5.47b) was carried out in Appendix 5.4.7.1, the result being

$$g^{\mu\nu}_\sigma\left(\frac{\partial \Gamma^\alpha_{\mu\nu}}{\partial x^\alpha} + \Gamma^\alpha_{\mu\beta}\Gamma^\beta_{\nu\alpha}\right) = -2\kappa\frac{\partial t^\alpha_\sigma}{\partial x^\alpha}$$

Using this result, and that $g_{\mu\nu}\frac{\partial g^{\mu\nu}}{\partial x^\sigma} = g_{\mu\nu}g^{\mu\nu}_\sigma = 0$ (from Eq. (5.29), $g_{\mu\nu}\frac{\partial g^{\mu\nu}}{\partial x^\sigma} = -2\frac{1}{\sqrt{-g}}\frac{\partial\sqrt{-g}}{\partial x^\sigma} = 0$ since $\sqrt{-g} = +1$),

$$-2\kappa\frac{\partial t^\alpha_\sigma}{\partial x^\alpha} = -\kappa g^{\mu\nu}_\sigma T_{\mu\nu} + \frac{1}{2}\kappa g^{\mu\nu}_\sigma g_{\mu\nu}T = -\kappa g^{\mu\nu}_\sigma T_{\mu\nu}$$

$$\frac{\partial T^\alpha_\sigma}{\partial x^\alpha} + \frac{1}{2}g^{\mu\nu}_\sigma T_{\mu\nu} = \frac{\partial T^\alpha_\sigma}{\partial x^\alpha} + \frac{1}{2}\frac{\partial g^{\mu\nu}}{\partial x^\sigma}T_{\mu\nu} = 0 \quad \text{by Eq. (5.56) (5.57)}$$

Derivation of Eq. (5.57a):

By Eqs. (5.41b) and (5.57), the divergence of the mixed tensor T^σ_μ is (with $\sqrt{-g} = 1$)

$$T_\mu = \frac{\partial T^\sigma_\mu}{\partial x^\sigma} + \frac{1}{2}\frac{\partial g^{\rho\sigma}}{\partial x^\mu}T_{\rho\sigma} = 0$$

by Eq. (5.57)

$$T_\mu = 0 = \frac{\partial T^\sigma_\mu}{\partial x^\sigma} - \{\sigma\mu, \tau\}\,T^\sigma_\tau = \frac{\partial T^\sigma_\mu}{\partial x^\sigma} + \Gamma^\tau_{\sigma\mu}T^\sigma_\tau$$

by Eqs. (5.41), (5.45)

$$\frac{\partial T^\sigma_\mu}{\partial x^\sigma} = -\Gamma^\tau_{\sigma\mu}T^\sigma_\tau$$

By relabeling the indices $\sigma \to \alpha, \mu \to \sigma, \tau \to \beta$,

$$\frac{\partial T^\alpha_\sigma}{\partial x^\alpha} = -\Gamma^\beta_{\alpha\sigma} T^\alpha_\beta \tag{5.57a}$$

5.4.8 Calculation of the Bending of Starlight

The total deflection of starlight passing near to the sun will be deflected by an amount $B = \int\limits_{-\infty}^{+\infty} \frac{\partial\gamma}{\partial x^1} dx^2$. From Section 5.2.22, we have that

$$\gamma = 1 - \frac{1}{2}\frac{\alpha}{r}\left(1 + \frac{(x^2)^2}{r^2}\right). \tag{5.74}$$

$$\frac{\partial\gamma}{\partial x^1} = -\frac{\alpha}{2}\left[-\frac{1}{r^2}\left(1 + \frac{(x^2)^2}{r^2}\right) + \frac{1}{r}\left(-\frac{2(x^2)^2}{r^3}\right)\right]\left(\frac{\partial r}{\partial x^1}\right)$$

$$= -\frac{\alpha}{2}\left[-\frac{1}{r^2}\left(1 + \frac{3(x^2)^2}{r^2}\right)\right]\left(\frac{x^1}{r}\right) = +\frac{\alpha}{2}\left[\frac{x^1}{r^3} + \frac{3x^1(x^2)^2}{r^5}\right]$$

The integral for B becomes

$$B = \int\limits_{-\infty}^{+\infty} \frac{\partial\gamma}{\partial x^1} dx^2 = \frac{\alpha}{2}\int\limits_{-\infty}^{+\infty}\left(\left[\frac{x^1}{r^3} + \frac{3x^1(x^2)^2}{r^5}\right]\right) dx^2$$

Making the following substitutions:

$$x^1 = \Delta; r = \frac{\Delta}{\cos\phi}; x^2 = r\sin\phi = \Delta\frac{\sin\phi}{\cos\phi} \Rightarrow dx^2 = \frac{\Delta}{\cos^2\phi} d\phi$$

$$B = \frac{\alpha}{2\Delta}\int\limits_{\phi=-\frac{\pi}{2}}^{\phi=+\frac{\pi}{2}} (\cos\phi + 3\sin^2\phi\cos\phi)\, d\phi = \frac{\alpha}{2\Delta}\left[\sin\phi + \sin^3\phi\right]_{\phi=-\frac{\pi}{2}}^{\phi=+\frac{\pi}{2}} = \frac{2\alpha}{\Delta}$$

5.4.9 Calculation of the Precession of the Perihelion of Mercury

To obtain the expression for the precession of the perihelion of the planet Mercury, Einstein began with the expression for the motion of a point mass in a gravitational field (the point mass eventually to become the planet Mercury in the gravitational field of the sun), Eq. (5.46):[147]

$$\frac{d^2 x^\tau}{ds^2} = \Gamma^\tau_{\mu\nu}\frac{dx^\mu}{ds}\frac{dx^\nu}{ds} \tag{5.75}$$

A solution to Eq. (5.75) is obtained through a series of successive approximations in the parameter α. At the outset, Einstein said he was looking for a solution "without discussing the question whether the solution might be unique."[148] In the derivation, he uses the following:

1. Energy conservation per unit mass

$$\frac{1}{2}u^2 + \Phi = constant = A \tag{5.76}$$

2. Kepler's Second Law

$$r^2\frac{d\phi}{ds} = constant = B \tag{5.77}$$

3. Gravitational potential (per mass)

$$\Phi = -\frac{GM_{sun}}{r} = -\frac{\alpha}{2r} \tag{5.78}$$

4. $u =$ speed in polar coordinates

$$u^2 = \frac{dr^2 + r^2 d\phi^2}{ds^2} = \left(\frac{dr}{ds}\right)^2 + r^2\left(\frac{d\phi}{ds}\right)^2 \tag{5.79}$$

Using the assumed solutions for $g_{\mu\nu}$, Eqs. (5.70), and the definition of $\Gamma^\tau_{\mu\nu}$, Eqs. (5.45), (5.23), and (5.21), to a first approximation,[149]

$$\begin{aligned}\Gamma^\tau_{\rho\sigma} &= -\frac{\alpha}{2}\left(2\delta_{\rho\sigma}\frac{x^\tau}{r^3} - 3\frac{x^\mu x^\nu x^\tau}{r^5}\right) \\ \Gamma^\nu_{44} &= \Gamma^4_{4\nu} = -\frac{\alpha}{2}\frac{x^\nu}{r^3}\end{aligned} \tag{5.80}$$

Those components for which the index 4 appears one or three times vanish.

Substituting Eqs. (5.80) into Eq. (5.75) (and using item h in Section 5.2.21),

$$\begin{aligned}\frac{d^2x^\nu}{ds^2} &= \Gamma^\nu_{44}\frac{dx^4}{ds}\frac{dx^4}{ds} = \Gamma^\nu_{44} = -\frac{\alpha}{2}\frac{x^\nu}{r^3} \quad \text{for } \nu = 1,2,3 \\ \frac{d^2x^4}{ds^2} &= 0 \qquad\qquad\qquad\qquad\qquad \text{for } \nu = 4\end{aligned} \tag{5.81}$$

Equation (5.81), the first approximation, is Newton's second law for motion in a gravitational field.

The equations are then evaluated to the second order in the parameter α. Using Eqs. (5.47) and (5.80), and remembering $\Gamma^\tau_{\rho\sigma} \neq 0$ only when two of the indices $= 4$,

$$\begin{aligned}\frac{\partial\Gamma^\nu_{44}}{\partial x^\nu} &= -\Gamma^\nu_{4\mu}\Gamma^\mu_{4\nu} = -2\left(\Gamma^4_{41}\Gamma^1_{44} + \Gamma^4_{42}\Gamma^2_{44} + \Gamma^4_{43}\Gamma^3_{44}\right) \\ &-2\left(-\frac{\alpha}{2r^3}\right)^2\left[\left(x^1\right)^2 + \left(x^2\right)^2 + \left(x^3\right)^2\right] = -\frac{\alpha^2}{2r^4}\end{aligned} \tag{5.82}$$

Einstein says, from Eq. (5.82), the following expression can be deduced for Γ^ν_{44}:

$$\Gamma^\nu_{44} = -\frac{\alpha}{2}\frac{x^\nu}{r^3}\left(1 - \frac{\alpha}{r}\right) \tag{5.83}$$

It will be sufficient for our purposes to show that this is "a solution" to Eq. (5.82).

$$\frac{\partial \Gamma^{\nu}_{44}}{\partial x^{\nu}} = \frac{\partial}{\partial x^{\nu}}\left[-\frac{\alpha}{2}\frac{x^{\nu}}{r^3}\left(1-\frac{\alpha}{r}\right)\right]$$

$$= -\frac{\alpha}{2}\left[\frac{1}{r^3}\left(1-\frac{\alpha}{r}\right)+x^{\nu}\left(\frac{-3}{r^4}+\frac{4\alpha}{r^5}\right)\frac{x^{\nu}}{r}\right]$$

Summing over $\nu = 1, 2, 3$,

$$= -\frac{\alpha}{2}\left[\frac{3}{r^3}\left(1-\frac{\alpha}{r}\right)+r\left(\frac{-3}{r^4}+\frac{4\alpha}{r^5}\right)\right]$$

$$= -\frac{\alpha}{2r^3}\left[\frac{-3\alpha}{r}+\frac{4\alpha}{r}\right] = -\frac{\alpha}{2r^4}$$

This shows that Eq. (5.83) is a solution to Eq. (5.82). Using this revised expression for Γ^{ν}_{44}, the expression for the equation of motion is recalculated:

$$\frac{d^2 x^{\nu}}{ds^2} = \Gamma^{\nu}_{\sigma\tau}\frac{dx^{\sigma}}{ds}\frac{dx^{\tau}}{ds} \tag{5.84}$$

Remembering $\Gamma^{\tau}_{\rho\sigma} \neq 0$ only when two of the indices $= 4$ or when none of the indices $= 4$, we consider first the case when the two indices $\sigma, \tau = 4$. By Eqs. (5.83) and (5.72), the right-hand side of Eq. (5.84) becomes

$$\Gamma^{\nu}_{44}\frac{dx^4}{ds}\frac{dx^4}{ds}$$

$$= -\frac{\alpha}{2}\frac{x^{\nu}}{r^3}\left(1-\frac{\alpha}{r}\right)\left(1+\frac{\alpha}{r}\right)\left(1+\frac{\alpha}{r}\right)$$

$$= -\frac{\alpha}{2}\frac{x^{\nu}}{r^3}\left(1+\frac{\alpha}{r}\right) \text{ to order } \alpha^2 \tag{5.85}$$

When neither σ nor τ is $= 4$, i.e., $\sigma \neq 4 \neq \tau$, using Eq. (5.80), the right-hand side of Eq. (5.84) becomes

$$\Gamma^{\nu}_{\sigma\tau}\frac{dx^{\sigma}}{ds}\frac{dx^{\tau}}{ds}$$

$$= -\alpha\left(\delta_{\sigma\tau}\frac{x^{\nu}}{r^3}-\frac{3}{2}\frac{x^{\sigma}x^{\tau}x^{\nu}}{r^5}\right)\frac{dx^{\sigma}}{ds}\frac{dx^{\tau}}{ds}$$

$$= -\frac{\alpha x^{\nu}}{r^3}\left(\delta_{\sigma\tau}-\frac{3}{2}\frac{x^{\sigma}x^{\tau}}{r^2}\right)\frac{dx^{\sigma}}{ds}\frac{dx^{\tau}}{ds} \tag{5.86}$$

Using $r\frac{dr}{ds} = r\frac{\partial r}{\partial x^{\nu}}\frac{dx^{\nu}}{ds} = r\left(\frac{x^{\nu}}{r}\frac{dx^{\nu}}{ds}\right) = x^{\nu}\frac{dx^{\nu}}{ds}$ and noting $x^{\sigma}\left(\frac{dx^{\sigma}}{ds}\right)x^{\tau}$ $\left(\frac{dx^{\tau}}{ds}\right) = \left(r\frac{dr}{ds}\right)\left(r\frac{dr}{ds}\right) = \left(r\frac{dr}{ds}\right)^2$, the sum over σ, τ in Eq. (5.86) gives

$$= -\frac{\alpha x^{\nu}}{r^3}\left[u^2-\frac{3}{2}\left(\frac{dr}{ds}\right)^2\right] \tag{5.87}$$

Using Eqs. (5.85) and (5.87),

$$\frac{d^2x^\nu}{ds^2} = -\frac{\alpha}{2}\frac{x^\nu}{r^3}\left[1 + \frac{\alpha}{r} + 2u^2 - 3\left(\frac{dr}{ds}\right)^2\right] \tag{5.88}$$

Using Eqs. (5.79) and (5.77), $\left(\frac{dr}{ds}\right)^2 = u^2 - r^2\left(\frac{d\phi}{ds}\right)^2 = u^2 - \frac{B^2}{r^2}$,

$$\frac{d^2x^\nu}{ds^2} = -\frac{\alpha}{2}\frac{x^\nu}{r^3}\left[1 + \frac{\alpha}{r} - u^2 + 3\left(\frac{B^2}{r^2}\right)\right]$$

Using Eqs. (5.76) and (5.78), $u^2 = 2(A - \Phi) = 2\left(A + \frac{\alpha}{2r}\right)$,

$$\frac{d^2x^\nu}{ds^2} = -\frac{\alpha}{2}\frac{x^\nu}{r^3}\left[1 - 2A + \frac{3B^2}{r^2}\right]$$

Replacing the variable s with $s' = s\sqrt{1 - 2A}$,

$$\frac{d^2x^\nu}{ds'^2}(1 - 2A) = -\frac{\alpha}{2}\frac{x^\nu}{r^3}\left[1 - 2A + \frac{3B^2}{r^2}\right]$$

Relabeling s' as s, setting $B'^2 = \frac{1}{1-2A}B^2$ and then relabeling B' as B, Eq. (5.88) becomes

$$\frac{d^2x^\nu}{ds^2} = -\frac{\partial\Phi_E}{\partial x^\nu} \text{ with } \Phi_E = \Phi_{Einstein} = -\frac{\alpha}{2r}\left(1 + \frac{B^2}{r^2}\right)$$

With this, the expression for energy conservation becomes

$$\frac{1}{2}u^2 + \Phi_E = A$$

$$u^2 = 2A - 2\Phi_E$$

$$\left(\frac{dr}{ds}\right)^2 + r^2\left(\frac{d\phi}{ds}\right)^2 = 2A - 2\Phi = 2A + \frac{\alpha}{r} + \frac{\alpha B^2}{r^3}$$

Using Kepler's Second Law, Eq. (5.77), to eliminate $d\phi$, and substituting $r = \frac{1}{x}$,

$$B^2\left(\frac{dx}{d\phi}\right)^2 + B^2x^2 = 2A + \alpha x + \alpha B^2 x^3$$

$$\left(\frac{dx}{d\phi}\right)^2 = \frac{2A}{B^2} + \frac{\alpha}{B^2}x - x^2 + \alpha x^3 \tag{5.89}$$

Einstein points out, "This equation differs from the corresponding one in Newtonian theory only in the last term on the right side."[150] Excluding the last term, the solutions to Eq. (5.89) are the elliptical orbits of Newton. Including the last term will give a solution that can be described as the elliptical orbits of Newton, but orbits that are precessing. Inverting Eq. (5.89),

$$d\phi = \frac{dx}{\sqrt{\frac{2A}{B^2} + \frac{\alpha}{B^2}x - x^2 + \alpha x^3}}$$

The angle ϕ between the perihelion and the aphelion is

$$\phi = \int_{\alpha_1}^{\alpha_2} \frac{dx}{\sqrt{\frac{2A}{B^2} + \frac{\alpha}{B^2}x - x^2 + \alpha x^3}}$$

where $\alpha_1 = \frac{1}{r_1} = perihelion$ and $\alpha_2 = \frac{1}{r_2}aphelion$. Upon integration, the angle ϕ between perihelion and aphelion is (a is the semi-major axis of the ellipse and e is the eccentricity of the ellipse)

$$\phi = \pi \left(1 + \frac{3}{2}\frac{\alpha}{a(1+e^2)} \right) \tag{5.90}$$

In traveling in orbit from perihelion to aphelion, Eq. (5.90) indicates the angle ϕ increases by more than π, by an amount equal to $\frac{3\pi}{2}\frac{\alpha}{a(1+e^2)}$. Completing the orbit back to the perihelion will increase the angle by this amount once again, for a total advance per orbit of $3\pi\frac{\alpha}{a(1+e^2)}$. This calculation yields for Mercury a perihelion advance of $43''$ per century.

5.4.10 The Bending of Starlight Experiment

When a star is far from the vicinity of the sun, its light travels in a straight line to the earth. As we look out from the earth along that straight line, we see where the star is. See Figure 5.15.

If the starlight passes near to the sun, light coming from the star would be bent around the sun and travel to the earth as indicated in Figure 5.16. Looking back along this line from the earth, the star would appear to be in a slightly different position, relative to all of the other stars in the sky, than when the starlight was not passing near to the sun.

Measuring the position of the particular star, relative to the other stars in the background of stars, allows one to measure how far that star appears to have been displaced (relative to the other background stars) when the light ray from the star passes near to the sun.

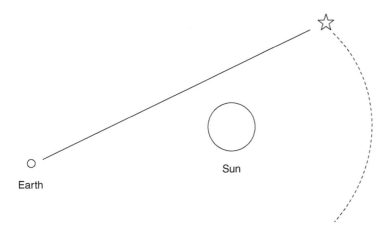

Fig. 5.15 Starlight traveling to the earth passing far from the sun.

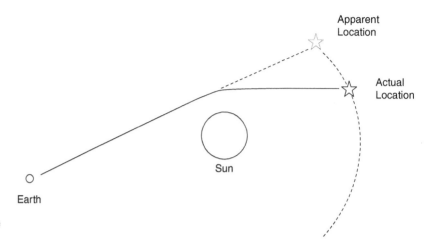

Fig. 5.16 Starlight traveling to the earth passing near to the sun.

5.4.11   Newton's Bucket

To distinguish absolute motion (motion relative to absolute space) from relative motion, Newton proposed an experiment that has become known as "Newton's Bucket." First, Newton describes a situation familiar to most people, that water in a rotating bucket will rise up the sides of the bucket and be depressed in the center of the bucket. In Newton's words:

> ...If a bucket is hanging from a very long cord and is continually turned around until the cord becomes twisted tight, and if the bucket is thereupon filled with water and is at rest along with the water and then, by some sudden force, is made to turn around in the opposite direction and, as the cord unwinds, perseveres for a while in this motion; then the surface of the water will at first be level, just as it was before the vessel began to move. But after the vessel, by the force gradually impressed upon the water, has caused the water also to begin revolving perceptibly, the water will gradually recede from the middle and rise up on the sides of the vessel, assuming a concave shape (as experience has shown me), and, with an ever faster motion, will rise further and further until, when it completes its revolutions in the same times as the vessel, it is relatively at rest in the vessel.[151]

In his analysis of this phenomenon, Newton arrives at the conclusion of the "true and absolute circular motion of the water," i.e., the motion of the water relative to absolute space. Continuing in his analysis, Newton writes:

> The rise of the water reveals its endeavor to recede from the axis of motion, and from such an endeavor one can find out and measure the true and absolute circular motion of the water, which here is the direct opposite of its relative motion. In the beginning, when the relative motion of the water in the vessel was greatest, that motion was not giving rise to any endeavor to recede from the axis; the water did not seek the circumference by rising up the sides of the vessel but remained level, and therefore its true circular motion had not yet begun. But afterward, when the relative motion of the water decreased, its rise up the sides of the vessel revealed its endeavor to recede from the axis, and

this endeavor showed the true circular motion of the water to be continually increasing and finally becoming greatest when the water was relatively at rest in the vessel. Therefore, that endeavor does not depend on the change of position of the water with respect to surrounding bodies, and thus true circular motion cannot be determined by means of such changes of position. The truly circular motion of each revolving body is unique, corresponding to a unique endeavor as its proper and sufficient effect ... [152]

In this experiment, Newton distinguished three situations:

1. At the start of the experiment, before the bucket has begun spinning, the water is at rest relative to the bucket and the surface of the water is flat.
2. Midway through the experiment, when the bucket is spinning rapidly but the water has just begun to move, the water has its greatest motion relative to the bucket and the surface of the water is still nearly flat.
3. Near the end of the experiment, when the bucket is spinning rapidly and the water is rotating at the same rate as the bucket, the water is at rest relative to the bucket and the surface of the water is concave.

By this example, Newton showed there are effects that depend on more than the relative motion of two objects, i.e., the relative motion of the bucket and water are the same (no relative motion) in situations 1 and 3, yet the physical result (the shape of the surface of the water) is not the same. This indicated to Newton the rotation is relative to absolute space.

5.5 Notes

1. Einstein, Albert, The Foundation of the General Theory of Relativity, 20 March, 1916, *Annalen der Physik* 354 (7), (1916), pp. 769–822; Kox, A. J., Klein, Martin, J., and Schulmann, Robert, editors, *The Collected Papers of Albert Einstein*, Volume 6, [CPAE6], Princeton University Press, Princeton, NJ, 1996, pp. 283–339; English translation by Alfred Engel, [CPAE6 ET], Princeton University Press, Princeton, NJ, 1997, pp. 146–200.
2. Newton, Isaac, *The Principia*, A new translation by I. Bernard Cohen and Anne Whitman, University of California Press, Berkeley, CA, 1999, paperback version, p. 218.
3. Eötvös, B., *Mathematische und Naturwissenschaftliche Berichte aus Ungarn* 8 (1890); Wiedemann's *Beiblätter* 15 (1891), 688; citation from Einstein, Albert, and Grossmann, Marcel, *Outline of a Generalized Theory of Relativity and of a Theory of Gravitation*, Teubner, Leipzig, 1913; Klein, Martin, Kox, A. J., Renn, Jürgen, and Schulmann, Robert, editors, *The Collected Papers of Albert Einstein*, Volume 4, [CPAE4], Princeton University Press, Princeton, NJ, 1995, p. 304, English translation by Anna Beck, [CPAE4 ET], p. 151.
4. Pais, Abraham, *Subtle is the Lord*, Oxford University Press, New York, 1982, p. 216; Miller, Arthur, Albert Einstein's 1907 Jahrbuch Paper:

The First Step from SRT to GRT, in *Studies in the History of General Relativity*, Volume 3, Birkhäuser, Boston, MA, 1988, p. 326.

5. I thank Don Howard for bringing this to my attention.

6. Ishiwara, J., *Einstein Koén-Roku*, (Kyoto lecture), Tokyo-Tosho, Tokyo, 1977; citation from Pais, Abraham, *Subtle is the Lord*, pp. 179, 183.

7. Einstein, Albert, On the Relativity Principle and the Conclusions Drawn from It, *Jahrbuch der Radioaktivität und Elektronik* 4 (1907), pp. 411–462; Stachel, John, editor, *The Collected Papers of Albert Einstein*, Volume 2, [CPAE2], Princeton University Press, Princeton, NJ, 1989, pp. 432–489; English translation by Anna Beck, [CPAE2 ET], 1989, pp. 252–311.

8. Einstein, Albert, The Relativity Principle; [CPAE2, p. 476; CPAE2 ET, pp. 301–302].

9. Hoffmann, Banesh, *Relativity and Its Roots*, Scientific American Books, W. H. Freeman and Company, New York, 1983, p. 131.

10. Hoffmann, Banesh, *Relativity and Its Roots*, pp. 132–133.

11. Einstein, Albert, The Relativity Principle; [CPAE2, pp. 483–484; CPAE2 ET, p. 310].

12. Lorentz, Hendrik, The Radiation of Light, *Nature* 113 (26 April, 1924), 608–611.

13. Einstein, Albert, The Relativity Principle; [CPAE2, pp. 483–484; CPAE2 ET, p. 310].

14. Pais, Abraham, *Subtle is the Lord*, p. 152.

15. Einstein, A., On the Influence of Gravitation on the Propagation of Light, *Annalen der Physik* 340 (10), (1911), Klein, Martin J., Kox, A. J., Renn, Jürgen, and Schulmann, Robert, editors, *The Collected Papers of Albert Einstein*, Volume 3, [CPAE3], Princeton University Press, Princeton, NJ, 1993, pp. 486, 496; English translation by Anna Beck, [CPAE3 ET], pp. 379, 387.

16. Stachel, John, *Einstein from 'B' to 'Z'*, Birkhäuser, Boston, MA, 2002, pp. 228, 229; [CPAE3, p. 494, CPAE3 ET, p. 385].

17. I thank the reviewers of the manuscript for this description.

18. Einstein, Albert, Propagation of Light; [CPAE3, p. 492; CPAE3 ET, p. 384].

19. Pais, Abraham, *Subtle is the Lord*, p. 303.

20. Pais, Abraham, *Subtle is the Lord*, p. 217.

21. Stachel, John, *Einstein from 'B' to 'Z'*, pp. 229, 230.

22. Stachel, John, *Einstein from 'B' to 'Z'*, p. 230.

23. Einstein, Albert, and Grossmann, Marcel, Outline of a Generalized Theory of Relativity and of a Theory of Gravitation, *Zeitschrift für Mathematik und Physik* 62 (1914), p. 22, Eq. (5.259); [CPAE4. pp. 302–343; CPAE4 ET, pp. 151–188].

24. Stachel, John, *Einstein from 'B' to 'Z'*, p. 231.

25. Pais, Abraham, *Subtle is the Lord*, 303, 304.

26. Hoffmann, Banesh, *Relativity and Its Roots*, p. 157; Pais, Abraham, *Subtle is the Lord*, p. 250.

27. Lorentz, Hendrik, The Radiation of Light, *Nature* 113 (26 April, 1924), pp. 608–611.

28. Einstein, Albert, On the General Theory of Relativity, 4 November, 1915, Report to the Prussian Academy of Sciences, Königlich Preussische Akademie der Wissenschaften (Berlin), *Sitzungsberichte* (1915); Kox *et al.* [CPAE6, pp. 215–216, 221; CPAE6 ET, pp. 98, 104].

29. Einstein, Albert, On the General Theory of Relativity (Addendum), 11 November, 1915, Report to the Prussian Academy of Sciences, Königlich Preussische Akademie der Wissenschaften (Berlin), *Sitzungsberichte* (1915); Kox *et al.* [CPAE6, p. 227, CPAE6 ET, pp. 108, 109].

30. Einstein, Albert, Explanation of the Perihelion Motion of Mercury from the General Theory of Relativity, 18 November, 1915, Report to the Prussian Academy of Sciences, [CPAE6, pp. 234, 237, 242; CPAE6 ET, pp. 113, 114, 116].

31. Hoffmann, Banesh, *Relativity and Its Roots*, p. 157.

32. Einstein, Albert, The Field Equations of Gravitation, 25 November, 1915, Report to the Prussian Academy of Sciences, Königlich Preussische Akademie der Wissenschaften (Berlin), *Sitzungsberichte* (1915); Kox *et al.* [CPAE6, pp. 245, 248; CPAE6 ET, pp. 117, 120].

33. Einstein, Albert, letter to Paul Ehrenfest 26 December, 1915; Schulmann, Robert, Kox, A. J., and Jansenn, Michael, *The Collected Papers of Albert Einstein*, Volume 8, [CPAE8], Princeton University Press, Princeton, NJ, 1998, p. 228; English translation by Ann M. Hentschel, [CPAE8 ET], p. 167.

34. Einstein, Albert, letter to Hendrik Lorentz, 17 January, 1916; Schulmann *et al.* [CPAE8, p. 245; CPAE8 ET, p. 179].

35. Einstein, Albert, Foundation of General Relativity; [CPAE6, pp. 283–339; CPAE6 ET, p. 146–200]. With the exception of the first page, the English version of the paper in [CPAE6 ET] is the translation by H. A. Lorentz. The translation of the first page was furnished by the editors of the [CPAE6] since it was missing in the translation by Lorentz.

36. Einstein, Albert, The Formal Foundation of the General Theory of Relativity; [CPAE6, pp. 72–130; CPAE6 ET, pp. 30–84].

37. Einstein, Albert, and Grossmann, Marcel, Outline of a Generalized Theory of Relativity; Klein *et al.* [CPAE4. pp. 302–343; CPAE4 ET, pp. 151–188].

38. Einstein, Albert, Foundation of General Relativity; [CPAE6, p. 297; CPAE6 ET, pp. 158, 159].

39. Einstein, Albert, Foundation of General Relativity; [CPAE6, pp. 283–294; CPAE6 ET, pp. 147–156].

40. Einstein, Albert, Foundation of General Relativity; [CPAE6, p. 291; CPAE6 ET, p. 153].

41. Einstein, Albert, Foundation of General Relativity; [CPAE6, pp. 294–316; CPAE6 ET, pp. 156–178].

42. Einstein, Albert, Foundation of General Relativity; [CPAE6, pp. 316–325; CPAE6 ET, pp. 178–187].

43. Einstein, Albert, Foundation of General Relativity; [CPAE6, pp. 325–337; CPAE6 ET, pp. 187–200].

44. Einstein, Albert, Foundation of General Relativity; [CPAE6, p. 285; CPAE6 ET, p. 148].

45. Einstein, Albert, Foundation of General Relativity; [CPAE6, p. 286; CPAE6 ET, pp. 148, 149].

46. Einstein, Albert, Foundation of General Relativity; [CPAE6, p. 287; CPAE6 ET, p. 149].

47. Einstein, Albert, Foundation of General Relativity; [CPAE6, p. 287; CPAE6 ET, p. 150].

48. Einstein, Albert, Foundation of General Relativity; [CPAE6, p. 288; CPAE6 ET, p. 150].

49. Einstein, Albert, Foundation of General Relativity; [CPAE6, p. 288; CPAE6 ET, pp. 150, 151].
50. Einstein, Albert, Foundation of General Relativity; [CPAE6, p. 290; CPAE6 ET, p. 152].
51. Einstein, Albert, Foundation of General Relativity; [CPAE6, p. 290; CPAE6 ET, p. 153].
52. Einstein, Albert, Foundation of General Relativity; [CPAE6, p. 291; CPAE6 ET, p. 153].
53. Einstein, Albert, Foundation of General Relativity; [CPAE6, p. 291; CPAE6 ET, p. 153].
54. Einstein, Albert, Foundation of General Relativity; [CPAE6, p. 294; CPAE6 ET, p. 156].
55. Einstein, Albert, and Grossmann, Marcel, Outline of a Generalized Theory of Relativity; [CPAE4. pp. 302–343; CPAE4 ET, pp. 151–188].
56. Einstein, Albert, Foundation of General Relativity; [CPAE6, p. 300; CPAE6 ET, p. 162].
57. Einstein, Albert, Foundation of General Relativity; [CPAE6, p. 304; CPAE6 ET, p. 166].
58. Einstein, Albert, Foundation of General Relativity; [CPAE6, p. 304; CPAE6 ET, p. 166].
59. Einstein, Albert, Foundation of General Relativity; [CPAE6, p. 305; CPAE6 ET, p. 167].The method is similar to the derivation of Lagrange's equations using Hamilton's principle. See, for example, Goldstein, Herbert, Poole, Charles, and Safko, John, *Classical Mechanics*, 3rd edition, Addison-Wesley, San Francisco, CA, 2002, pp. 44–45.
60. Einstein, Albert, Foundation of General Relativity; [CPAE6, p. 307; CPAE6 ET, p. 168].
61. Einstein, Albert, Foundation of General Relativity; [CPAE6, p. 314; CPAE6 ET, p. 176].
62. Einstein, Albert, Foundation of General Relativity; [CPAE6, p. 295; CPAE6 ET, p. 157].
63. Einstein, Albert, Foundation of General Relativity; [CPAE6, p. 318; CPAE6 ET, p. 180].
64. Einstein, Albert, Foundation of General Relativity; [CPAE6, p. 319; CPAE6 ET, pp. 180, 181].
65. Einstein, Albert, Foundation of General Relativity; [CPAE6, p. 319; CPAE6 ET, p. 181].
66. Einstein, Albert, On the General Theory of Relativity, 4 November, 1915, Report to the Prussian Academy of Sciences; [CPAE6, pp. 221, 222; CPAE6 ET, pp. 104, 105].
67. Einstein, Albert, Foundation of General Relativity; [CPAE6, p. 322; CPAE6 ET, p. 184].
68. Einstein, Albert, Manuscript on Special Relativity, [CPAE4, pp. 91–98; CPAE4 ET, pp. 78–85].
69. Einstein, Albert, Generalized Theory of Relativity, [CPAE 4, p. 322; CPAE4 ET, p. 170].
70. Einstein, Albert, Foundation of General Relativity; [CPAE6, p. 323; CPAE6 ET, p. 185].
71. I thank the reviewers of the manuscript for pointing this out.
72. Einstein, Albert, Foundation of General Relativity; [CPAE6, pp. 324, 325; CPAE6 ET, p. 187].

73. Einstein, Albert, The Formal Foundation of the General Theory of Relativity; [CPAE6, pp. 104, 105; CPAE6 ET, pp. 60, 61].
74. Einstein, Albert, The Formal Foundation of the General Theory of Relativity; [CPAE6, p. 104; CPAE6 ET, p. 61].
75. I thank the reviewers of this manuscript for this description.
76. Einstein, Albert, Foundation of General Relativity; [CPAE6, p. 329; CPAE6 ET, p. 191].
77. These equations are contained in the November 18, 1915, report to the Prussian Academy of Science, Explanation of the Perihelion Motion of Mercury from the General Theory of Relativity, [CPAE6, p. 236; CPAE6 ET, p. 113]. On p. 113 of the English translation, Einstein reports that these equations are an "assumed solution."
78. Einstein, Albert, Propagation of Light; [CPAE3, pp. 494–496; CPAE3 ET, pp. 385–387].
79. I thank Gintaras Duda for this description.
80. The calculation is contained in the November 18, 1915, report to the Prussian Academy of Science, Explanation of the Perihelion Motion of Mercury from the General Theory of Relativity, [CPAE6, pp. 233–243; CPAE6 ET, pp. 112–116].
81. Einstein, Albert, Report to the Prussian Academy of Science, Explanation of the Perihelion Motion of Mercury from the General Theory of Relativity, [CPAE6, p. 242; CPAE6 ET, p. 116].
82. French, A. P., editor, *Einstein: A Centenary Volume*, Harvard University Press, Cambridge, MA, 1979, pp. 94–96.
83. Pais, Abraham, *Subtle is the Lord*, p. 253.
84. Einstein, Albert, letter to Paul Ehrenfest, 17 January, 1916; [CPAE.8, p. 244; CPAE8 ET, p. 179].
85. Wheeler, John Archibald, *A Journey Into Gravity and Spacetime*, Scientific American Books, 1990, p. 182.
86. Pais, Abraham, *Subtle is the Lord*, pp. 303, 304.
87. Hoffmann, Banesh, *Relativity and Its Roots*, p. 156; Pais, Abraham, *Subtle is the Lord*, p. 304.
88. Pais, Abraham, *Subtle is the Lord*, p. 304; Clark, Ronald W., *Einstein: The Life and Times*, The World Publishing Company, New York, 1971, pp. 228, 229.
89. Lorentz, Hendrik, telegram to Einstein, 22 September, 1919; Buchwald, Diana Kormos, Schulmann, Robert, Illy, József, Kennefick, Daniel J., and Sauer, Tilman, editors, *The Collected Papers of Albert Einstein*, Volume 9, [CPAE9], p. 167; English translation [CPAE9 ET], p. 95.
90. Thomson, J. J., *Proc. Roy. Soc.*, 96A, 311 (1919); citation from Pais, Abraham, *Subtle is the Lord*, pp. 305, 324.
91. Holton, Gerald, and Elkana, Yehuda, *Albert Einstein: Historical and Cultural Perspectives*, Princeton University Press, Princeton, NJ, 1982, p. 207; also in Holton, Gerald, *Thematic Origins of Scientific Thought*, revised edition, Harvard University Press, Cambridge, MA, 1988 (1973), pp. 254–255. Reference is to a manuscript by Ilse Rosenthal-Schneider, "Reminiscences of Conversation with Einstein," dated 23 July, 1957.
92. Pais, Abraham, *Subtle is the Lord*, p. 177.
93. Earman, J., and Janssen, M, Einstein's Explanation of the Motion of Mercury's Perihelion; Earman, J., Janssen, M., and Norton, J. D., editors, *Einstein Studies*, Volume 5, Birkhäuser, Boston, MA, 2002, p. 129.

94. Hartle, James B., *Gravity: An Introduction to Einstein's General Relativity*, Addison-Wesley, San Francisco, CA, 2003, p. 118.

95. Urani, John, and Gale, George, E.A. Milne and the Origins of Modern Cosmology: An Essential Presence; Earman *et al.*, *Einstein Studies*, Volume 5, p. 391; Clark, Ronald W., *Einstein*, p. 212.

96. Einstein, Albert, Cosmological Considerations in the General Theory of Relativity, Königlich Preussische Akademie der Wissenschaften (Berlin), *Sitzungsberichte* (1917); citation from Kox *et al.* [CPAE6, pp. 541–552, CPAE6 ET, pp. 421–432].

97. Pais, Abraham, *Subtle is the Lord*, pp. 342, 343.

98. Pais, Abraham, *Subtle is the Lord*, p. 9.

99. Definition from the dictionary in Microsoft Word.

100. Clark, Ronald W., *Einstein*, p. 212.

101. Clark, Ronald W., *Einstein*, p. 213.

102. Newton, Isaac, *The Principia*, (see the Scholium after Definition 8), pp. 412, 413.

103. Mach, Ernst, *The Science of Mechanics*, translation by Thomas J. McCormack, 2nd edition, Open Court, Chicago, IL, 1907, p. 299; citation taken from Casper, Barry M., and Noer, Richard J., *Revolutions in Physics*, W. W. Norton & Company, New York, 1972, pp. 405, 471.

104. Ellis, George F. R., *The Expanding Universe: A History from 1917 to 1960*; Howard, Don, and Stachel, John, editors, *Einstein Studies*, Volume 1, Birkhäuser, Boston, MA, 1989, p. 370.

105. Einstein, Albert, Cosmological Considerations; [CPAE6, pp. 540–552; CPAE6 ET, pp. 421–432].

106. Einstein, Albert, Cosmological Considerations; [CPAE6, pp. 541–542; CPAE6 ET, pp. 421–422].

107. Einstein, Albert, Cosmological Considerations; [CPAE6, pp. 542, 543; CPAE6 ET, p. 423].

108. Pais, Abraham, *Subtle is the Lord*, pp. 286, 287; Einstein, Albert, Cosmological Considerations; Kox *et al.* [CPAE6, p. 550; CPAE6 ET, p. 430].

109. Einstein, Albert, Cosmological Considerations; [CPAE6, p. 551; CPAE6 ET, p. 431].

110. Pais, Abraham, *Subtle is the Lord*, p. 287.

111. Einstein, Albert, Cosmological Considerations; [CPAE6, p. 551; CPAE6 ET, pp. 431–432].

112. Ellis, George, Cosmology from 1917 to 1960; Howard, Don, and Stachel, John, *Einstein Studies*, Vol. 1, p. 371.

113. Pais, Abraham, *Subtle is the Lord*, p. 286.

114. Pais, Abraham, *Subtle is the Lord*, p. 287.

115. Ellis, George, Cosmology from 1917 to 1960; Howard, Don, and Stachel, John, *Einstein Studies*, Vol. 1, p. 370.

116. Pais, Abraham, *Subtle is the Lord*, p. 287; Ellis, George, *Einstein Studies*, Vol. 1, pp. 371–372.

117. Ellis, George, Cosmology from 1917 to 1960; Howard, Don, and Stachel, John, *Einstein Studies*, Vol. 1, pp. 371–372, 375.

118. Gamow, George, *My World Line*, Viking New York, 1970; citation from Clark, Ronald W., *Einstein*, p. 215.

119. Ellis, George, Cosmology from 1917 to 1960; Howard, Don, and Stachel, John, *Einstein Studies*, Vol. 1, p. 379.

120. Ellis, George, Cosmology from 1917 to 1960; Howard, Don, and Stachel, John, *Einstein Studies*, Vol. 1, pp. 381–382.

121. Urani, John, and Gale, George, Origins of Modern Cosmology; Earman *et al.*, *Einstein Studies*, Vol. 5, pp. 391–392.

122. I thank Don Howard for this translation.

123. Einstein, Albert, *The Meaning of Relativity*, 5th edition, Princeton University Press, Princeton, NJ, 1956 (1922), p. 127, footnote to item 1 in "Summary and Other Remarks."

124. Gamow, George, *My World Line*, New York, 1970; citation from Clark, Ronald W., *Einstein*, p. 215.

125. Smith, R. W., private communication from Don Howard; citation from Ellis, George, *Einstein Studies*, Vol. 1, p. 383.

126. Ellis, George, Cosmology from 1917 to 1960; Howard, Don, and Stachel, John, *Einstein Studies*, Vol. 1, pp. 383, 385, 393.

127. Einstein, Albert, *Relativity: The Special and General Theory*, 17th edition. Crown Publishers, Inc., New York, 1952 (1916), p. 155.

128. Jammer, M., In *Concepts of Space*, Harvard University Press, Cambridge, MA, 1994, pp. xiii–xvi; citation from Stachel, John, *Einstein from 'B' to 'Z'*, pp. 423, 425. (Einstein 1954b.)

129. Pais, Abraham, *Subtle is the Lord*, pp. 22, 23.

130. Vizgen, Vladimir, *Einstein, Hilbert, and Weyl: The Genesis of the Geometrical Unified Field Theory Program;* Howard, Don, and Stachel, John, *Einstein Studies*, Vol. 1, p. 304.

131. Pais, Abraham, *Subtle is the Lord*, p. 287.

132. Pais, Abraham, *Subtle is the Lord*, p. 20.

133. Bergia, Silvio, Attempts at Unified Field Theories (1919–1955). Alleged Failure and Intrinsic Validation/Refutation Criteria; Earman *et al.*, *Einstein Studies*, Vol. 5, p. 286.

134. Kaluza, Theodor (1921). "Zum Unitätsproblem der Physik." Königlich Preussische Akademie der Wissenschaften (Berlin), *Sitzungsberichte*, p. 967. Citation from Bergia, Silvio in *Einstein Studies*, Earman *et al.*, Vol. 5, pp. 282, 304.

135. Bergia, Silvio, Attempts at Unified Field Theories (1919–1955); Earman *et al.*, *Einstein Studies*, Vol. 5, p. 286.

136. Biezunski, Michel, Inside the Coconut: The Einstein-Cartan Discussion on Distant Parallelism; Howard, Don, and Stachel, John, *Einstein Studies*, Vol. 1, p. 316.

137. Bergia, Silvio, Attempts at Unified Field Theories (1919–1955); Earman *et al.*, *Einstein Studies*, Vol. 5, p. 292.

138. Einstein, Albert, Foundation of General Relativity; [CPAE6, p. 300; CPAE6 ET, p. 162].

139. Einstein, Albert, Foundation of General Relativity; [CPAE6, p. 305; CPAE6 ET, p. 167]. The method is similar to the derivation of Lagrange's equations using Hamilton's principle. See, for example, Goldstein, Herbert, Poole, Charles, and Safko, John, *Classical Mechanics*, 3rd edition, Addison-Wesley, San Francisco, CA, 2002, pp. 44–45.

140. Einstein, Albert, Foundation of General Relativity; [CPAE6, p. 306; CPAE6 ET, p. 168].

141. Einstein, Albert, Foundation of General Relativity; [CPAE6, p. 306].

142. Einstein, Albert, Foundation of General Relativity; [CPAE6, p. 306; CPAE6 ET, p. 168].

143. Mathews, Jon, and Walker, J. L., *Mathematical Methods of Physics*, 2nd edition, W. A. Benjamin, New York, 1970, p. 419.

144. Einstein, Albert, Foundation of General Relativity; [CPAE6, p. 319; CPAE6 ET, p. 181].

145. Goldstein, Herbert, *Classical Mechanics*, p. 44.

146. Einstein, Albert, Foundation of General Relativity; [CPAE 6, p. 321; CPAE6 ET, p. 183].

147. The derivation in this section follows closely the derivation in Einstein, Albert, Perihelion Motion of Mercury, [CPAE6, pp. 234, 237, 242; CPAE6 ET, pp. 113–116]; and in Earman, John, and Janssen, Michel, Einstein's Explanation of the Motion of Mercury's Perihelion, *The Attraction of Gravitation*, Birkhäuser, Boston, MA, 1993, pp. 129–172.

148. Einstein, Albert, Perihelion Motion of Mercury; [CPAE6, p. 235; CPAE6 ET, p. 113].

149. Einstein, Albert, Perihelion Motion of Mercury; [CPAE6, p 237; CPAE6 ET, p. 114]; Earman, John, and Janssen, Michel, Einstein's Explanation, p. 21.

150. Einstein, Albert, Perihelion Motion of Mercury; [CPAE6, p. 240; CPAE ET, p. 115].

151. Newton, Isaac, *The Principia*, (see the Scholium after Definition 8), pp. 412, 413.

152. Newton, Isaac, *The Principia*, (see the Scholium after Definition 8), p. 413.

5.6   Bibliography

Bergia, Silvio, Attempts at Unified Field Theories (1919–1955). Alleged Failure and Intrinsic Validation/Refutation Criteria, in *Einstein Studies*, Volume 5, Birkhäuser, Boston, MA, 2002.

Biezunski, Michel, Inside the Coconut: The Einstein–Cartan Discussion on Distant Parallelism, in *Einstein Studies*, Volume 1, Birkhäuser, Boston, MA, 1989.

Buchwald, Diana Kormos, Schulman, Robert, Illy, József, Kennefick, Daniel J., and Sauer, Tilman, editors, *The Collected Papers of Albert Einstein*, Volume 9 [CPAE9], Princeton University Press, Princeton, 2004; English translation by Ann Hentschel [CPAE9 ET].

Casper, Barry M., and Noer, Richard J., *Revolutions in Physics*, W. W. Norton & Company, New York, 1972.

Clark, Ronald W., *Einstein: The Life and Times*, The World Publishing Company, New York, 1971.

Earman, J., Janssen, M., and Norton, J. D., editors, *Einstein Studies*, Volume 5, Birkhäuser, Boston, MA, 2002.

Einstein, Albert, Manuscript on the Special Theory of Relativity [CPAE4, pp. 9–108; CPAE ET, pp. 33–88].

Einstein, Albert, The Formal Foundation of the General Theory of Relativity, Königlich Preussische Akademie der Wissenschaften (Berlin). *Sitzungsberichte* (1914) [CPAE6, pp. 72–130; CPAE ET, pp. 30–84].

Einstein, Albert, On the Relativity Principle and the Conclusions Drawn from It, *Jahrbuch der Radioaktivität und Elektronik* 4 (1907), pp. 411–462 [CPAE2, pp. 432–489; CPAE2 ET, pp. 252–311].

Einstein, A., On the Influence of Gravitation on the Propagation of Light, *Annalen der Physik*, 35 (1911) [CPAE3; CPAE3 ET].

Einstein, Albert, and Grossmann, Marcel, Outline of a Generalized Theory of Relativity and of a Theory of Gravitation, *Zeitschrift für Mathematik und Physik* 62, 1914 [CPAE4. pp. 302–343; CPAE4 ET, pp. 151–188].

Einstein, Albert, On the General Theory of Relativity, 4 November, 1915, Report to the Prussian Academy of Sciences, Königlich Preussische Akademie der Wissenschaften (Berlin). *Sitzungsberichte* (1915) [CPAE6; CPAE6 ET].

Einstein, Albert, On the General Theory of Relativity (Addendum), 11 November, 1915, Report to the Prussian Academy of Sciences, Königlich Preussische Akademie der Wissenschaften (Berlin). *Sitzungsberichte* (1915) [CPAE6; CPAE6 ET].

Einstein, Albert, Explanation of the Perihelion Motion of Mercury from the General Theory of Relativity, 18 November, 1915, Report to the Prussian Academy of Sciences, Königlich Preussische Akademie der Wissenschaften (Berlin). *Sitzungsberichte* (1915) [CPAE6; CPAE6 ET].

Einstein, Albert, The Field Equations of Gravitation, 25 November, 1915, Report to the Prussian Academy of Sciences, Königlich Preussische Akademie der Wissenschaften (Berlin). *Sitzungsberichte* (1915) [CPAE6; CPAE6 ET].

Einstein, Albert, The Foundation of the General Theory of Relativity, *Annalen der Physik* 354 (7), (1916), pp. 769–822; [CPAE6, pp. 283–339; CPAE6 ET, pp. 146–200].

Einstein, Albert, *Relativity: The Special and General Theory*, 17th edition, Crown Publishers, Inc., New York, 1952 (1916).

Einstein, Albert, Cosmological Considerations in the General Theory of Relativity, Königlich Preussische Akademie der Wissenschaften (Berlin). *Sitzungsberichte* (1917) [CPAE6; CPAE6 ET].

Einstein, Albert, *The Meaning of Relativity*, 5th edition, Princeton University Press, Princeton, NJ, 1956 (1922).

Ellis, George F. R., The Expanding Universe: A History from 1917 to 1960, in *Einstein Studies*, Volume 1, Birkhäuser, Boston, MA, 1989.

Eötvös, B., *Mathematische und Naturwissenschaftliche Berichte aus Ungarn* 8 (1890); Wiedemann's *Beiblätter* 15 (1891). (Citation in Einstein, Albert, Outline of a Generalized Theory of Relativity [CPAE4; CPAE4 ET].)

French, A. P., editor, *Einstein: A Centenary Volume*, Harvard University Press, Cambridge, MA, 1979.

Gamow, George, *My World Line*, New York, 1970. (Citation in Clark, Ronald, *Einstein: The Life and Times*.)

Goldstein, Herbert, Poole, Charles, and Safko, John, *Classical Mechanics*, 3rd edition, Addison-Wesley Publishing Company, Reading, MA, 2002.

Hartle, James B., *Gravity: An Introduction to Einstein's General Relativity*, Addison-Wesley, San Francisco, CA, 2003.

Hoffmann, Banesh, *Relativity and Its Roots*, Scientific American Books, W. H. Freeman and Company, New York, 1983.

Holton, Gerald, *Thematic Origins of Scientific Thought*, revised edition, Harvard University Press, Cambridge, 1988 (1973).

Holton, Gerald, and Elkana, Yehuda, *Albert Einstein: Historical and Cultural Perspectives*, Princeton University Press, Princeton, NJ, 1982, p. 207.

Howard, D., Stachel, J., editors, *Einstein Studies*, Volume 1, Birkhäuser, Boston, 1989.

Ishiwara, J., *Einstein Koén-Roku*, Kyoto lecture, Tokyo-Tosho, Tokyo, 1977. (Citation in Pais, Abraham, *Subtle is the Lord*.)

Jammer, M., *Concepts of Space*, Harvard University Press, Cambridge, MA, 1954, 1963, 1993 (Reference from John Stachel.)

Kaluza, Theodor, *Zum Unitätsproblem der Physik*, Königlich Preussische Akademie der Wissenschaften (Berlin). *Sitzungsberichte* (1921).

Klein, Martin, Kox, A. J., Renn, Jürgen, and Schulmann, Robert, editors, *The Collected Papers of Albert Einstein*, Volume 4 [CPAE4], Princeton University Press, Princeton, NJ, 1995; English translation by Anna Beck [CPAE4 ET, p. 151].

Klein, Martin J., Kox, A. J., Renn, Jürgen, and Schulmann, Robert, editors, *The Collected Papers of Albert Einstein*, Volume 3 [CPAE3], Princeton University Press, Princeton, NJ, 1993; English translation by Anna Beck [CPAE3 ET, pp. 486, 496].

Kox, A. J., Klein, Martin, J., and Schulmann, Robert, editors, *The Collected Papers of Albert Einstein*, Volume 6 [CPAE6], Princeton University Press, Princeton, NJ, 1996; English translation by Alfred Engel [CPAE6 ET].

Lorentz, Hendrik, The Radiation of Light, *Nature* 113 (26 April, 1924).

Mach, Ernst, *The Science of Mechanics*, 2nd edition, English translation by Thomas J. McCormack, Open Court, Chicago, IL, 1907. (Citation in Casper, Barry and Noer, Richard, *Revolutions in Physics*.)

Mathews, Jon, and Walker, R. L., *Mathematical Methods of Physics*, W. A. Benjamin, New York, 1970.

Miller, Arthur I., Albert Einstein's 1907 Jahrbuch Paper: The First Step from SRT to GRT, in *Einstein Studies*, Volume 3, Birkhäuser, Boston, MA, 1988.

Newton, Isaac, *The Principia*, A new translation by I. Bernard Cohen and Anne Whitman, University of California Press, Berkeley, CA, 1999.

Pais, Abraham, *Subtle is the Lord*, Oxford University Press, New York, 1982.

Schulmann, Robert, Kox, A. J., and Jansenn, Michael, *The Collected Papers of Albert Einstein*, Volume 8 [CPAE8], Princeton University Press, Princeton, NJ, 1998; English translation by Ann M. Hentschel [CPAE8 ET, p. 228].

Stachel, John, *Einstein from 'B' to 'Z'*, Birkhäuser, Boston, MA, 2002.

Stachel, John, editor, *The Collected Papers of Albert Einstein*, Volume 2 [CPAE2], Princeton University Press, Princeton, NJ, 1989; English translation by Anna Beck [CPAE2 ET].

Thomson, J. J., *Proc. Roy. Soc.* 96A, 311 (1919). (Citation in Pais, Abraham, *Subtle is the Lord*.)

Urani, John, and Gale, George, E. A. Milne and the Origins of Modern Cosmology: An Essential Presence, in *Einstein Studies*, Volume 5, Birkhäuser, Boston, MA, 2002.

Vizgen, Vladimir, The Genesis of the Geometrical Unified Field Theory Program, in *Einstein Studies*, Volume 1, Birkhäuser, Boston, MA, 1989.

Wheeler, John Archibald, *A Journey Into Gravity and Spacetime*, Scientific American Books, New York, 1990.

Einstein and Quantum Mechanics

6

The topics presented in this chapter are intended to give an overview of Einstein's contribution to the evolution of the ideas of quantum mechanics, not to present detailed calculations as was done for his 1905 papers and for the general theory of relativity. Any one of the topics would be worthy of an entire book, or at least a chapter, to present his contribution. The intent of this chapter is to show how strongly Einstein was involved in the development of quantum mechanics by highlighting several of his contributions and to show how, in fact, he remained alone for nearly twenty years in his defense of the electromagnetic field being composed of quanta.

6.1 Historical Background   265

6.2 The Evolution of Quantum Mechanics   267

6.3 Discussion and Comments   281

6.4 Appendices   282

6.5 Notes   283

6.6 Bibliography   287

6.1 Historical Background

In the 1800s, spectroscopists were studying the light emitted by different elements. Unlike the spectra of sunlight in which there was a continuous distribution of the colors from red to violet, different elements emitted discrete spectra, that is, only certain lines of color (or, equivalently, certain wavelengths or frequencies), with gaps between the lines. See Figure 6.1.

By looking at the discrete spectra, an experienced spectroscopist could identify from which element the spectra had been obtained. Scientists began to look for some pattern in the emitted discrete spectra.

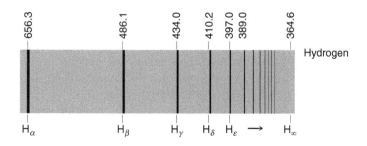

Fig. 6.1 The discrete spectra of Hydrogen.

Source: (From Tipler, Paul A., and Llewellyn, Ralph A., *Modern Physics*, Fifth Edition.[1] Reprinted with permission.)

Some of the spectra were quite complicated. However, the spectrum of hydrogen was relatively simple. For this reason, the emission spectra of hydrogen became one of the most studied. Also, the "reverse" of this effect was noted, i.e., when white light was shone through hydrogen gas and the transmitted light was analyzed, it was noted that dark bands appeared (bands where light of certain wavelengths were missing – this was called the absorption spectra). The dark bands of the hydrogen absorption spectra exactly matched the discrete emission spectra. Because of the extreme precision in the measurement of these spectra, any empirical formula for the discrete spectra would need to be as precise. In 1885, Johann Balmer, a Swiss mathematician and secondary school teacher in Basel, found the lines in the visible portion of the hydrogen spectra could be represented by the empirical formula,[2]

$$\lambda_m = 364.6 \frac{m^2}{m^2 - 2^2} \text{ nm, with } m \text{ an integer equal to } 3, 4, 5, \dots$$

This became known as the Balmer series. Subsequently, it was found that replacing 2^2 by $3^2, 4^2, \dots$ led to other spectral lines outside the visible spectrum. Although this basically was numerology, it was hoped that having such a precise formula for the wavelengths would aid in obtaining a physical explanation for the emission and absorption spectra. But none could be found.

For nearly two decades after 1905, Einstein remained alone as the defender of the electromagnetic field being composed of the elementary quanta.[3] The concept of the quantum generally was accepted by the scientific community, but as due to the interaction of the radiation field with matter, not as an element of the radiation itself. In 1913, eight years after Einstein's 1905 paper on the quanta, in his nomination of Einstein for membership in the Prussian Academy of Sciences, Planck reflected the outlook of the time when he said of Einstein, "That he may sometimes have missed the target in his speculations, as, for instance, in his hypothesis of light-quanta, cannot really be held too much against him, ..."[4] In 1918, thirteen years after the 1905 paper, in a letter to Besso, Einstein wrote, "Here I have been pondering for countless hours about the quantum problem again, naturally without making any headway. But I no longer have doubts about the *reality* of quanta in radiation, even though I'm still quite alone in this conviction."[5] Additionally, the independence of Einstein's quanta was valid only in the Wien (high ν) region of the blackbody radiation; the quanta were not independent in the low ν region.[6] But before these concerns were answered, in the early 1920s a number of other advances were forthcoming.

Until the mid 1920s, Einstein was the leading scientist in matters regarding the quantum. With the contributions of Compton, de Broglie, and Schrödinger, the development of quantum mechanics went in a direction that parted ways with Einstein.

6.2 The Evolution of Quantum Mechanics

6.2.1 The Theory of Specific Heat (1906)

The specific heat capacity, denoted c, is the amount of heat needed to raise the temperature of one mole of a substance by one degree (the phrase "specific heat capacity" is usually shortened to the term "specific heat"). In the early 1800s, Dulong and Petit discovered that for several metals the specific heat had nearly the same value of 6 cal/mole · deg (see Appendix 6.4.1 for a derivation of this result).[7] Although generally true, it was not universally so; carbon was found to be a significant exception. In 1840, measurements of diamond dust gave a specific heat of 1.4,[8] while 1841 measurements gave a value of 1.8.[9] In the 1870s, Heinrich Weber noted the different readings for the specific heat of diamond were for different temperature ranges. Proceeding to retake measurements for diamond over temperatures ranging from 0° C to 200° C, Weber found the specific heat increasing with temperature by a factor of three. Later, extending the range of his measurements, he found the specific heat to increase by a factor of 15 in the temperature range from –100° C to +1000° C. As the temperature rose, the value of the specific heat became closer to the value of Dulong and Petit. Extending the materials measured, Weber's data grew to include boron, silicon, graphite, and diamond, each in the range from –100° C to +1000° C.[10]

In his 1907 paper, "Planck's Theory of Radiation and the Theory of Specific Heat,"[11] Einstein re-derives the Planck blackbody radiation distribution function, starting from the quantization of the radiation field. He then says:

I believe that we must not content ourselves with this result. For the question arises: If the elementary structures that are to be assumed in the theory of energy exchange between radiation and matter cannot be perceived in terms of the current molecular-kinetic theory, are we then not obliged also to modify the theory for the other periodically oscillating structures considered in the molecular theory of heat? In my opinion the answer is not in doubt. If Planck's radiation theory goes to the root of the matter, then contradictions between the current molecular-kinetic theory and experience must be expected in other areas of the theory of heat as well, which can be resolved along the lines indicated. In my opinion this is actually the case, as I shall now attempt to show.[12]

Assuming each atom in a solid undergoes oscillations about its equilibrium position, classically the specific heat was calculated to be 5.94 cal/mole · deg (times the number of atoms in a molecule). This is in agreement with the law of Dulong and Petit. But the exceptions of carbon, boron and silicon remained puzzling.[13]

For blackbody radiation, Planck had obtained the relation (see Section 2.1.3):

$$U(\nu, T) = \frac{h\nu}{e^{\frac{h\nu}{kT}} - 1}$$

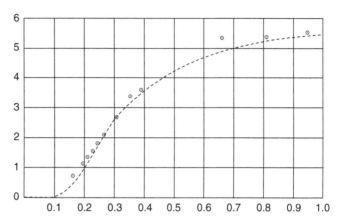

Fig. 6.2 Weber's specific heat data for diamond.

Source: (From Einstein, Albert, *The Collected Papers of Albert Einstein.*[17] Reproduced with permission.)

The first published graph dealing with the quantum theory of the solid state: Einstein's expression for the specific heat of solids [given in Eq. 20.4] plotted versus $h\nu/kT$. The little circles are Weber's experimental data for diamond. Einstein's best fit to Weber's measurements corresponds to $h\nu/k \cong 1300$ K.

For the oscillating atoms, Einstein used the same equation for the energy. For one mole of a substance with each atom having three degrees of freedom, the total internal energy of the system is $3N_A U(\nu, T) = 3N_A \frac{h\nu}{e^{\frac{h\nu}{kT}} - 1}$. Using the thermodynamic relation $c_V = (\partial U/\partial T)_V$, Einstein obtained for the specific heat:[14]

$$c_V = 3R \frac{\left(\frac{h\nu}{kT}\right)^2 e^{\frac{h\nu}{kT}}}{\left(e^{\frac{h\nu}{kT}} - 1\right)^2} = 5.94 \frac{\left(\frac{h\nu}{kT}\right)^2 e^{\frac{h\nu}{kT}}}{\left(e^{\frac{h\nu}{kT}} - 1\right)^2} = 5.94 \frac{\left(\frac{h\nu}{kT}\right)^2 e^{\frac{-h\nu}{kT}}}{\left(1 - e^{\frac{-h\nu}{kT}}\right)^2}$$

For low temperatures, i.e., $T \to 0$, the ratio $(h\nu/kT) \to \infty$ and $c_V \to 0$. Similarly, for high temperatures, i.e., $T \to \infty$, the ratio $(h\nu/kT) \to 0$ and $c_V \to 5.94$.[15]

There was one parameter, the frequency ν or, equivalently, by setting the ratio $(h\nu/kT) = 1$, the Einstein temperature, T_E. Comparing this expression for c_V to Weber's data for diamond, Einstein found an almost exact fit for $T_E = 1300$ K. See Figure 6.2. This is well above the typical room temperature of 300 K, and thus might be expected to show quantum effects at room temperature. On the other hand, the element lead has an Einstein temperature of 70 K, well below the typical room temperature.[16]

Planck had introduced the quantum to explain the interaction between blackbody radiation and the walls of the container; Einstein used it as a fundamental constituent of the radiation field and, with its application to specific heat, extended it to the solid state of materials.

6.2.2   The Dual Nature of Radiation (1909)

In the 1909 paper, "On the Present Status of the Radiation Problem,"[18] Einstein derived an expression for the energy fluctuations of blackbody radiation in the frequency range ν to $\nu + d\nu$:[19]

$$\langle \varepsilon^2(\nu, T) \rangle = kT^2 V \left(\frac{\partial \rho}{\partial T} \right) d\nu$$

where V is the volume of the container.

In the low frequency limit, where the wave description of radiation suffices, ρ is given by the Rayleigh–Jeans law. From the Rayleigh–Jeans expression, Einstein obtained (actually, it was Pais who obtained this expression, and the succeeding one. See the previous endnote):

$$\langle \varepsilon^2(\nu, T) \rangle = \frac{c^3}{8\pi\nu^2} \rho^2 V d\nu$$

In the high frequency limit, where the wave description was not sufficient, ρ is given by Wien's law. In his 1905 paper, Einstein had shown in the Wien region the radiation behaved as if it were composed of n independent particles.[20] From the Wien expression, Einstein (again, Pais) obtained:

$$\langle \varepsilon^2(\nu, T) \rangle = h\nu \rho V d\nu$$

And using the Planck expression for the radiation density ρ Einstein obtained:[21]

$$\langle \varepsilon^2(\nu, T) \rangle = \left(h\nu\rho + \frac{c^3}{8\pi\nu^2} \rho^2 \right) V d\nu$$

From Planck's expression, the two terms in the above expression for $\langle \varepsilon^2(\nu, T) \rangle$ emerge naturally, the first indicating a particle aspect of the radiation, the second indicating a wave aspect of the radiation. Einstein described the meaning this held for him:

I have already tried to show that our current foundations of the radiation theory must be abandoned....[22]

Once it had been recognized that light exhibits the phenomena of interference and diffraction, it seemed hardly doubtful any longer that light is to be conceived as wave motion. Since light can also propagate through vacuum... it was necessary to assume... [a] luminiferous ether...

However, today we must regard the ether hypothesis as an obsolete standpoint.... It is therefore my opinion that the next stage in the development of theoretical physics will bring us a theory of light that can be understood as a kind of fusion of the wave and emission theories of light.[23]

...

All I wanted is briefly to indicate... that the two structural properties (the undulatory structure and the quantum structure) simultaneously displayed by radiation according to the Planck formula should not be considered as mutually incompatible.[24]

...we are only dealing with a *modification* of our current theory, not with its complete *abolition*.[25]

6.2.3 The Bohr Atom (1913)

Einstein's 1905 papers on the size of the atom and Brownian motion provided the last piece of evidence to tip the scales solidly in favor of

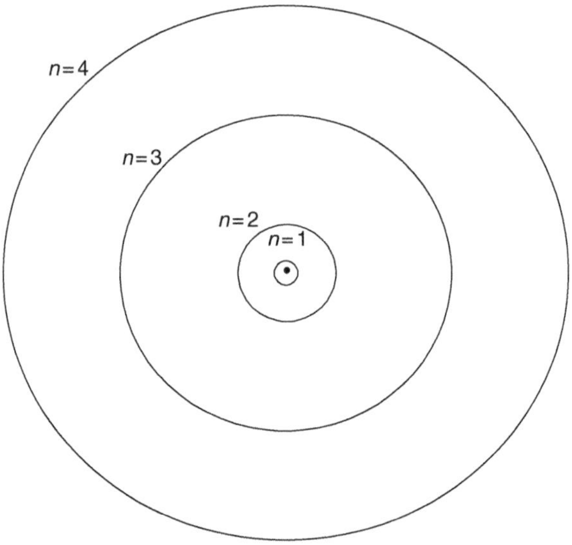

Fig. 6.3 The Bohr atom.

the reality of atoms. Then followed more focused investigations of the atom and its composition. In 1905, Einstein (along with a variety of other investigators) had determined the radius of the atom to be about 10^{-10} m. In 1911, Ernest Rutherford determined the preponderance of mass in the atom was associated with a positive charge and was concentrated in a "nucleus" with a radius of 10^{-14} m, 10,000 times smaller than the atom. Around this positive nucleus, the negative electrons were placed into an orbit of 10^{-10} m, held in orbit by the Coulomb force. But the electrons in circular orbits were accelerating (out of straight-line motion) and would radiate energy, quickly spiraling into the nucleus.

In 1913, Niels Bohr suggested that on the atomic level there might be new laws in play that we had not observed previously. The suggestion was that there were certain preferred orbits (termed stationary states by Bohr) where the electron, even though accelerating, would not radiate energy. Each of these preferred orbits would correspond to a particular value of the energy, and to a particular value of angular momentum (the angular momentum of these orbits was a multiple of $\hbar$).[26] As an electron moved from one orbit to another, the difference in energy, ΔE, would be given off as radiation, with the frequency of the radiation determined by the Planck–Einstein relation $\Delta E = h\nu$. See Figure 6.3.

This model not only predicted discrete spectra from atoms, but predicted the correct values of the frequencies of the experimental spectra, in particular producing the Balmer formula that had defied a physical explanation for nearly 30 years.[27] By the end of 1914, these ideas had been confirmed by the experiments of James Franck and Gustav Ludwig Hertz (referred to today as the Franck–Hertz experiment).[28]

The unanswered question now was, why should there be such preferred orbits? Niels Bohr responded that, perhaps, entirely new and different laws are needed to describe nature on the atomic scale. In the succeeding years a variety of explanations were offered, but none of them was totally satisfactory.

6.2.4 Spontaneous and Induced Transitions (1916)

Building on the Bohr assumption that atoms have a set of discrete energy states Einstein, looked at the interaction between a gas of these atoms and electromagnetic radiation.[29] He focused on two of the discrete energy levels of the atoms, E_m and E_n, with $E_m > E_n$. For the gas in equilibrium with the radiation the number of atoms in state m is $N_m = p_m e^{-E_m/kT}$, with a similar expression for N_n (p is a weighting factor). In time dt, Einstein hypothesized that the number of transitions dW from state $m \to n$ and from state $n \to m$ is given by[30]

$$dW_{mn} = N_m(\rho B_{mn} + A_{mn})dt$$

$$dW_{nm} = N_n \rho B_{nm} dt$$

The A coefficient corresponds to the probability of a spontaneous transitions, while the B terms refer to the probability of an induced emission and absorption. ρ is the spectral density of the radiation present. In order that these equations lead to Planck's law, "[i]t is necessary that the transitions $m \underset{\rightarrow}{\leftarrow} n$ are accompanied by a single monochromatic radiation quantum. By this remarkable reasoning, Einstein therefore established a bridge between blackbody radiation and Bohr's theory of spectra."[31] However, even more, Einstein was "suggesting that probabilities themselves might have to be regarded as fundamental, basic, physical properties of atomic systems."[32] This work is often referred to as the basis for the laser.[33]

6.2.5 The Compton Scattering Experiment (1923)

In 1923, Arthur Compton[34] and Peter Debye[35] each derived the relativistic kinematics for an x-ray photon scattering from an electron. They treated the collision between an incoming photon and an electron, initially free and at rest, as a hard sphere (billiard ball) collision, assuming energy and momentum conservation, but using the relativistic expressions for the energy and momentum. See Figure 6.4.

The electrons, even if bound, could be considered free and initially stationary since the energy of the incoming x-rays was so much greater than the binding energy. (From the photoelectric effect, an estimate of the binding energy would be on the order of the incoming ultraviolet photon. The energy of an x-ray is several orders of magnitude greater than the energy of an ultraviolet photon.)[37]

Since the photon has zero mass, the relativistic energy equation, $E^2 = p^2c^2 + (m_0c^2)^2$, becomes $E = pc$. For a photon of momentum $\vec{p}_0$ before

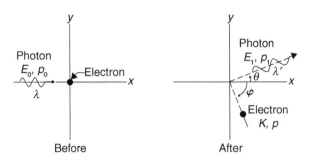

Fig. 6.4 The Compton scattering experiment.

Source: (From Eisberg, Robert, and Resnick, Robert, *Modern Physics*.[36] Reproduced with permission.)

the collision, and $\vec{p}_1$ after the collision, the equations for conservation of momentum and energy are,

$$\vec{P}_o = \vec{p}_e + \vec{p}_1$$

$$|\vec{p}_0|c + m_0c^2 = |\vec{p}_1|c + (p_e^2 c^2 + m_0 c^2)^{1/2}$$

Solving these equations in a manner similar to the classical billiard ball collision, the momentum of the outgoing photon, p_1, is determined as a function of the photon scattering angle θ:[38]

$$\frac{1}{p_1} - \frac{1}{p_0} = \frac{1}{m_0c}(1 - \cos\theta)$$

Writing $E = h\nu = pc$ and using the wave speed relation, $\lambda\nu = c$, the momentum of the photon can be written $p = h/\lambda$. Multiplying by h, the previous equation relating the change in momentum of the photon to the scattering angle θ the change in wavelength of the photon is

$$\lambda_1 - \lambda_0 = \Delta\lambda = \frac{h}{m_0c}(1 - \cos\theta)$$

Experiments by Compton showed this relation to be satisfied within experimental error.[39]

Thus, finally, an experiment showed the photon as an entity itself; that it was a basic constituent of the electromagnetic field and existed independently of interactions between the field and the walls of the container. From this time, the idea of the photon became widely accepted.[40] No longer was Einstein "still quite alone in [his] conviction."[41] The reality of the photon, Einstein's quanta of the radiation field, now received general acceptance within the scientific community.

6.2.6   Bose–Einstein Statistics (1924)

In Boltzmann's derivation of the entropy expression, $S = k \log W$, W is the probability of the state occurring, calculated by counting the number of "complexions," i.e., by counting the number of ways the given state could occur.[42] Planck used Boltzmann's definition of probability in terms of complexions in his derivation of the radiation density equation for blackbody radiation. Einstein had misgivings about counting complexions to determine the probability of the state. To Einstein, the time-ensemble definition of the probability of the state of a physical system was primary. The probability a system is in a given region Γ of the state variables $p_1, p_2, p_3, \ldots p_n$ is given by the quantity τ/T in the limit of $T \to \infty$, where τ is the time spent in Γ out of the total time T.[43] Calculations of probability based on the counting of complexions assume the complexions are equally probable, an assumption that bothered Einstein. Planck, Boltzmann, and others, assumed all complexions to be equally probable, an assumption justified a posteriori by its obvious success in obtaining Planck's formula.[44]

Satyendra N. Bose, a very bright young Indian physicist, advised his students, "Never accept an idea as long as you are yourself not satisfied with its consistency and the logical structure upon which the concepts are based."[45] Following his own advice when studying derivations of Planck's blackbody radiation formula he found, "In every case, the derivations do not appear to me to be sufficiently logically justified."[46] Working through the derivations to satisfy himself, Bose developed a derivation of Planck's radiation law with no mention of the electromagnetic theory. Pais describes this achievement:

The paper by Bose is the fourth and last of the revolutionary papers of the old quantum theory (the other three being by, respectively, Planck, Einstein, and Bohr). Bose's arguments divest Planck's law of all supererogatory elements of electromagnetic theory and base its derivation on the bare essentials. It is the thermal equilibrium law for particles with the following properties: they are massless, they have two states of polarization, the number of particles is not conserved, and the particles obey a new statistics.[47]

In 1924, Bose sent these ideas in a paper to Einstein, with an accompanying note saying, "I have ventured to send you the accompanying article for your perusal and opinion. I am anxious to know what you think of it.... If you think the paper worth publication, I shall be grateful if you arrange for its publication in *Zeitschrift für Physik*."[48] Einstein submitted the article, with a note appended, saying, "Bose's derivation of Planck's law appears to me to be an important step forward."[49] In 1925, Bose traveled to Berlin to work with Einstein for a year. During this collaboration, one of the questions they worked on was whether the new statistics of Bose implied a new type of interaction.

Einstein applied Bose's new ideas to his thoughts on the quantum gas. He notes that his theory "is based on the hypothesis of a far-reaching formal relationship between radiation and gas. According to this theory, the degenerate gas deviates from the gas of [classical] statistical mechanics in a way that is analogous to that in which radiation obeying Planck's law deviates from radiation obeying Wien's law.... [I]f it is justified to regard radiation as a quantum gas, then the analogy between quantum gas and gas of molecules must be complete."[50] In a subsequent paper he adds, "This theory seems justified if one starts with

Distribution	Particles in Cell 1	Particles in Cell 2
1	A and B	None
2	A	B
3	B	A
4	None	A and B

Fig. 6.5 Two distinguishable particles in two cells.

the conviction that (apart from its property of being polarized) a light quantum differs essentially from a monatomic molecule only in that its rest mass is vanishingly small..."[51]

Stachel summarizes the situation regarding complexions and probability:

... The counting of complexions is based on the assumption that, while the material oscillators or the normal modes of the radiation field are *distinguishable*, the quanta of energy of a given frequency ... are *indistinguishable*. If there are seven quanta of energy associated with one oscillator and five with another, it makes no sense to ask *which* seven and *which* five quanta of energy are involved.

... While elements of the total energy may be indistinguishable, in the sense that it makes no sense to ask *which* quanta of that energy are in a given state, how can one avoid this question for energy quanta, i.e., particles? Classically, one *cannot*, and it must have seemed by the nature of the particle concept that such a question makes sense. No wonder that even Einstein paused before entering such a statistical morass.

But Bose did so, with a courage perhaps born in part from an incomplete awareness of the perils ... [H]e turned the problem of partitioning *quanta of energy* among normal modes of the field into one of distributing *energy quanta* – that is photons, considered as particles – among cells in phase space; but continued to count with the *old* method to solve the *new* problem. Now the cells, like the normal modes, are certainly independent of each other; but, if the old counting method is used for the new problem, then the photons *cannot* be statistically independent. For if they were, as Einstein had known since 1905, Boltzmann's standard counting method for independent particles, accepted by everyone, would lead to Wien's radiation formula rather than Planck's.[52]

For a concrete example, consider the simple case of two particles and two cells.[53] Consider first the case of two distinguishable particles, labeled A and B. These can be distributed into the two cells in four distinct ways, each way equally probable to the others. See Figure 6.5.

From the table it is seen in two of the four cases, particles A and B are in the same cell. If the particles are not distinguishable, one can only say how many are in a given cell (for photons, this leads to the statement that one can only say how much energy is in a given cell, not

Distribution	Total in Cell 1	Total in Cell 2
1	Two Particles	None
2	One Particle	One Particle
3	None	Two Particles

Fig. 6.6 Two indistinguishable particles in two cells.

which photons are in a given cell). For two indistinguishable particles the distribution is shown in Figure 6.6.

For the second case (indistinguishable particles), it is seen in two of the three cases there are two particles in the same cell. The first case (distinguishable particles) corresponds to the usual classical way of distributing distinguishable particles among cells. The second case corresponds to the counting introduced by Bose and Einstein, which is justified a posteriori since it leads to the correct Planck formula.[54]

From the Bose method of counting, the quanta are more likely to be together (two out of three cases = 67%) than if they were statistically independent of one another (two out of four cases = 50%). The Bose statistics introduces some sort of dependence between the light quanta, a feature peculiar to quantum theory. By making the (apparently) simple shift from "quanta of energy to energy quanta, [Bose] introduced a type of statistical dependence between the latter peculiar to quantum mechanics."[55] The term now used to talk of this peculiar dependence of quantum particles is *quantum entanglement*. Thus, two indistinguishable particles are more likely to be found near one another (67% of the time) than are two distinguishable particles (50% of the time).[56]

Einstein saw Bose's new statistics as

... mysterious and disturbing ... it was clear to Einstein that something more fundamental must lie behind Bose statistics ....

Einstein was looking for some way of assimilating *entanglement* of quantum systems ... to some more traditional type of *interaction* between particles ....

... The work of Einstein, Podolsky and Rosen and subsequent tests of Bell-type inequalities show that, once two quantum systems of any type interact (in the ordinary sense of the word), they are quantum-entangled forever ...

... even more than the probabilistic element it was entanglement, and the attendant non-locality that it introduced into the heart of physics, that most bothered Einstein about quantum mechanics.[57]

6.2.7 Einstein, de Broglie (1924), and Schrödinger (1926)

Louis de Broglie, while contemplating Einstein's wave–particle duality for electromagnetism (fields (waves) and photons (particles)) "suddenly had the idea, during the year 1923, that the discovery made by Einstein in 1905 should be generalized by extending it to all material particles and notably to electrons."[58]

In a report to the French Academy of Sciences, Louis de Broglie put forth the idea that the Planck–Einstein relation relating the frequency of electromagnetic waves to the energy of the constituent photons, $E = h\nu$, should hold also for material particles such as the electron. This idea was a natural extension of the idea that if waves exhibit particle characteristics, the converse should also be true. Inverting the equation for the energy and momentum of the photon of the electromagnetic field ($E = h\nu$ and $p = h/\lambda$), de Broglie determined the frequency and wavelength of the associated electron field ($\nu = E/h$ and $\lambda = h/p$). These ideas were developed in de Broglie's thesis, which he defended in November, 1924.[59]

Einstein was intrigued and impressed with de Broglie's thesis, terming it "a very notable publication."[60] In his 1909 work on the radiation of the electromagnetic field, Einstein noted the expression for $\langle \varepsilon^2(\nu, T) \rangle$ was composed of two terms, one indicating a wave aspect, the other a particle aspect (see Section 6.2.2). In 1924, he had obtained a similar expression for the quantum gas, but was uncertain how to interpret the term corresponding to the wave characteristics. The de Broglie waves fit neatly into this picture.[61] Both de Broglie and Einstein talked of experimental confirmation by suitable diffraction experiments with electrons. In 1927, Davisson and Germer, working at the Bell Laboratories, performed the experiment showing the wave nature of the electron.[62]

De Broglie's theory showed the preferred orbits of Bohr are those into which an integral number of wavelengths will fit, i.e., the preferred orbits of Bohr corresponded to the orbits of standing de Broglie waves. De Broglie waves are the "reason" for the preferred orbits of Bohr. See Figure 6.7.

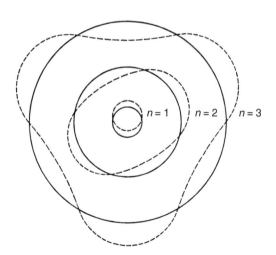

Fig. 6.7 De Broglie standing waves.

Source: (Eisberg, Robert, and Resnick, Robert, *Modern Physics*.[63] Reproduced with permission.)

Now the task was to determine what these waves of de Broglie correspond to in the physical world, if they do, indeed, correspond to anything in the physical world.

Einstein's comment that de Broglie's ideas "involve more than merely an analogy,"[64] drew the attention of Erwin Schrödinger. Schrödinger was not enthused with the new Bose–Einstein statistics, and set about starting with the wave picture of the gas and superimposing on that the quantization conditions.[65] He recalled that, "My theory was stimulated by de Broglie's thesis and by short but infinitely far-seeing remarks by Einstein."[66] He comments, also, that his theory "means nothing else but taking seriously the de Broglie–Einstein wave theory of moving particles, according to which the particles are nothing more than a kind of 'wave crest' on a background of waves."[67] Out of the de Broglie–Einstein wave theory was born the Schrödinger equation in 1926:

$$i\hbar \frac{\partial \Psi}{\partial t} = -\frac{\hbar^2}{2m}\nabla^2 \Psi + U\Psi$$

Initially, Schrödinger interpreted the square of the wave function, $|\Psi|^2$, to be the matter density or, possibly, the charge density of the particle. As Cushing points out, the problem with this interpretation is that at a barrier, "Since part of the wave function is reflected ... and part transmitted, an electron would have to split up at the barrier, part being reflected and part transmitted. However, experimentally, one always detects a whole electron or none at all, but never a piece of an electron."[68] Later, in 1926, Max Born gave us the presently accepted interpretation, that $|\Psi|^2$ represents the probability density of finding the electron at location $\vec{r}$ at time t.[69]

6.2.8 Einstein and Bohr (1927, 1930)

In 1925, Heisenberg proposed a theoretical description for quantum mechanics that replaced the kinematical and dynamical variables of classical mechanics "by symbols subjected to a non-commutative algebra."[70] This is the matrix formulation of quantum mechanics. From this comes the uncertainty principle for any two conjugate variables, such as (p, q) or (E, t),

$$\Delta q \cdot \Delta p \geq \hbar/2$$

$$\Delta t \cdot \Delta E \geq \hbar/2$$

At the Como conference in September of 1927, Niels Bohr stated his view

... it is decisive to recognize that, *however far the phenomena transcend the scope of classical physical explanation, the account of all evidence must be expressed in classical terms* .... [This] implies the *impossibility of any sharp separation between the behavior of atomic objects and the interaction with the measuring instruments which serve to define the conditions under which the phenomena appear* .... Consequently, evidence obtained under different

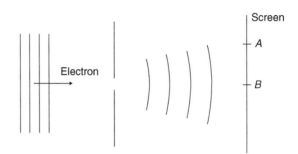

Fig. 6.8 Einstein's thought experiment for wave function diffraction.
Source: (From Cushing, James T., *Philosophical Concepts in Physics.*[72] Reproduced with permission.)

experimental conditions cannot be comprehended within a single picture, but must be regarded as *complementary* in the sense that only the totality of the phenomena exhausts the possible information about the objects.

Under these circumstances an essential element of ambiguity is involved in ascribing conventional physical attributes to atomic objects, as is at once evident in the dilemma regarding the corpuscular and wave properties of electrons and photons, where we have to do with contrasting pictures, each referring to an essential aspect of empirical evidence.

... It must here be remembered that ... we are dealing with an implication of the formalism which defies unambiguous expression in words suited to describe classical physical pictures.[71]

Einstein had not been at the Como conference, but he attended the Fifth Solvay conference the following month (in October of 1927). At the conference he tried to point out inconsistencies within quantum mechanics. He considered the example of a single electron passing through a very small hole and hitting a screen, where it is detected. As Cushing describes it:

Near the beginning of the general discussion session of the 1927 meeting, he [Einstein] considered an electron passing through a small hole and being detected on a screen, as shown in [Figure 6.8].

Before any observation is made, there is, according to quantum mechanics, at nearly every point on the screen a nonzero probability of detecting an electron. However, once the electron has been detected at, say, point A, there will then be absolutely zero probability of finding the electron at any other point B. Einstein argued that, if one were to say that the electron had been virtually present everywhere over an appreciable portion of the screen before the observation but that the probability at B had been instantaneously affected by an observation at A, then this would require an action at a distance that relativity is usually taken to rule out (since an effect ought not be propagated instantaneously between two spatially separated points). On the other hand, he contended, if there did exist some actual trajectory along which the electron proceeded through the slit to point A on the screen and if quantum mechanics were incapable of yielding that information, then quantum mechanics would be an incomplete theory.[73]

Einstein considered this "ghostlike remote effect" (it also has been referred to as a "spooky action-at-a-distance force") an unacceptable part of any physical theory.[74] Cushing continues,

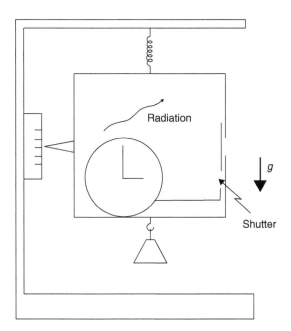

Fig. 6.9 Einstein's thought experiment to circumvent the uncertainty principle.

Source: (Cushing, James T., *Philosophical Concepts in Physics*.[76] Reproduced with permission.)

In the view of Bohr, Born, Dirac, Heisenberg and nearly all of the other quantum theorists, Einstein's argument missed the point since the wave function did not, in their opinion, represent anything like an ordinary wave propagating in a space-time background.[75]

At the sixth Solvay conference, held in 1930, Einstein came prepared with another example, this one apparently violating the uncertainty principle in the form $\Delta t \cdot \Delta E \geq \hbar/2$. Again using Cushing's description,

Basically what he proposed was to have a box filled with low-density electromagnetic radiation and equipped with a shutter driven by a clock inside the box, as illustrated in [Figure 6.9].

The clock would be set to open and close the shutter very quickly in a time Δt so that only a single photon would escape from the box. This time interval Δt could be set as accurately as desired. Since $E = mc^2$, simple accurate weighings of the radiation filled box before and after the photon had been emitted would determine the energy difference ΔE that would be the energy of the ejected photon. Because Δt and ΔE could each be independently determined to any degree of accuracy, the simultaneous values of Δt and ΔE could be made so small that $\Delta t \cdot \Delta E < \hbar/2$, in contradiction to Heisenberg's uncertainty relation.[77]

Initially, Bohr was unable to find a flaw in Einstein's proposed experiment. After a sleepless night, he realized the flaw was from Einstein's own general theory of relativity. The loss of the photon produced a change in the gravitational field in the box because of the loss of mass, $\Delta m = \Delta E/c^2$. The change in the gravitational field affects the rate at which the clock keeps time, which, Bohr showed, exactly restored the Heisenberg uncertainty principle. After this, Einstein ceased his search for inconsistencies in quantum mechanics and focused his efforts on the incompleteness question.[78]

In 1935, Albert Einstein, Boris Podolsky, and Nathan Rosen collaborated on a paper (usually referred to as the EPR paper) titled "Can Quantum-Mechanical Description of Reality Be Considered Complete?"[79] The paper defines a complete theory as follows:

In a complete theory there is an element corresponding to each element of reality. A sufficient condition for the reality of a physical quantity is the possibility of predicting it with certainty, without disturbing the system. In quantum mechanics in the case of two physical quantities described by non-commuting operators, the knowledge of one precludes the knowledge of the other.[80]

Consider a system of two particles with position and momentum (q_1, p_1) and (q_2, p_2), respectively. The Heisenberg uncertainty principle holds for each pair of non-commuting variables (q_1, p_1) and (q_2, p_2). But the relative position of the two particles, $Q = q_1 - q_2$, and the total momentum of the system, $P = p_1 + p_2$, form a pair of commuting variables (see Appendix 6.4.2). P and Q, being commuting variables, can be known simultaneously to any desired degree of accuracy.

The two particles are allowed to interact. Much later, when p_1 is measured, the value of p_2 will also be known without disturbing particle 2. Thus p_2 is an element of reality. Then one measures q_1, giving the value of q_2 without disturbing the system. Thus q_2 also is an element of reality. But, according to quantum mechanics, p_2 and q_2 cannot simultaneously be elements of reality because (p_2, q_2) is a non-commuting pair of variables. Quantum mechanics must, therefore, be incomplete.

Bohr responded quickly. He admitted to the correctness of the analysis in the EPR paper. But he stressed again that, "because of the finite value of the quantum of action [Planck's constant] it was impossible – in contrast to classical physics – to speak of 'physical reality' without including the measuring process. If that is borne in mind, the contradiction highlighted by EPR is only apparent, and quantum mechanics is a complete description of what physicists can discover about nature."[81]

One interpretation of the wave function (actually $|\Psi|^2$) giving the probability of an electron being at location $\vec{r}$ at time t is that this is fundamentally the best one can do in light of the uncertainty principle. A second interpretation is that, due to our incomplete knowledge of the system, there are additional variables of which we are not yet aware. These theories are termed "hidden variable" theories, similar to the idea of unseen atoms underlying the structure of thermodynamics. Although many place Einstein in the hidden variable theory camp, John Stachel does not, based partially on a statement by Einstein in 1953, "I think that it is not possible to get rid of the statistical character of the present quantum theory by merely adding something to the latter without changing the fundamental concepts about the whole structure."[82]

Einstein continued to look for the fusion of the wave and particle theories. In 1939, he wrote:

I do not believe that the light-quanta have reality in the same immediate sense as the corpuscles of electricity. Likewise I do not believe that the particle-waves have reality in same sense as the particles themselves. The wave-character of particles and the particle-character of light will – in my opinion – be understood in a more indirect way, not as immediate physical reality.[83]

In 1952, he wrote:

In present-day physics there is manifested a kind of battle between the particle-concept and the field-concept for leadership, which will probably not be decided for a long time. It is even doubtful if one of the two rivals finally will be able to maintain itself as a fundamental concept.[84]

6.3 Discussion and Comments

Of the two types of physical theories, theories of principle and theories of construction, we spoke of Einstein's special theory of relativity being a theory of principle, while Lorentz's theory of relativity was a theory of construction. Quantum theory is a theory of construction, being constructed piece by piece. In this construction, Einstein was a major player, at times making contributions of his own, at other times supporting the work of others or raising points of concern regarding their work.

In 1905, Einstein brought forth a radically different view of light, the idea of energy packets (later to be called photons) rather than light being a wave. He was uncomfortable with the dualism of light being described some times as particles and at others as a wave. Rather than seeing the wave and particle aspects of radiation as being at odds with one another, Einstein began looking for "a theory of light that can be understood as a kind of fusion of the wave and emission theories of light,"[85] He extended the quantum ideas beyond gases and liquids to the solid state with his work on the specific heat of solids, explaining the anomalous behavior of some materials not following the law of Dulong and Petit. With the success of the Bohr model of the hydrogen atom, Einstein used the principle of discrete energy states in his work on spontaneous and induced transitions. Nevertheless, there remained concerns regarding the determination of probabilities by simply counting the number of complexions, assuming the complexions were equally probable. Einstein's work with Bose addressed this concern. To this point in the early 1920s, Einstein had remained alone as the defender of the electromagnetic field being composed of elementary quanta, rather than the quanta arising from the interaction of the electromagnetic field with the walls of the container.[86] In 1923, the Compton scattering experiment showed the photon as an entity itself, that it was a basic constituent of the electromagnetic field.

After this time, Einstein remained convinced of his interpretation of quantum theory and what conditions it must satisfy. At the fifth Solvay conference in 1927, and at the sixth Solvay conference in 1930, he pointed out what he considered inconsistencies in the interpretation of quantum theory as presented by Niels Bohr. From this point (1930), Einstein ceased his search for inconsistencies in the quantum theory and focused his efforts on the incompleteness question.[87]

6.4   Appendices

6.4.1   The Specific Heat of Dulong and Petit

The first law of thermodynamics is $\Delta U = \Delta Q - \Delta W$, with ΔU the change in the internal energy of the system, ΔQ the amount of heat added to the system, and ΔW the amount of work done by the system. The amount of work done by the system is $p\Delta V$. Rewriting the first law as $\Delta U = \Delta Q - p\Delta V$, it can be seen that in any process in which $\Delta V = 0$, the change in internal energy of the system, ΔU, is equal to the heat added to the system, ΔQ.

The specific heat, c, is the amount of heat needed to raise the temperature of one mole of a substance by one degree Kelvin, $c = \Delta Q/\Delta T$. If the process takes place at constant volume, the specific heat at constant volume is $c_V = (\Delta Q/\Delta T)_V = (\Delta U/\Delta T)_V = (\partial U/\partial T)_V$. From kinetic theory, the kinetic energy of an atom is $\frac{3}{2}kT$. Boltzmann had shown that "the average kinetic energy equals the average potential energy for a system of particles each one of which oscillates under the influence of external harmonic forces."[88] The average total energy of one atom of a system is $\langle U \rangle = \langle KE \rangle + \langle PE \rangle = 3kT$. The average total internal energy of one mole of the system is $U = N_A 3kT = 3RT$,

$$c_V = (\partial U/\partial T) = 3R = 3 \times 1.98 \frac{\text{calories}}{\text{K}} = 5.94 \frac{\text{calories}}{\text{K}}$$

6.4.2   The Commutator of P and Q

The commutator of two variables, x and y, is defined as $[x, y] = xy - yx$. For typical situations such as x and y being numbers the commutator $[x, y]$ is equal to zero. However, some entities do not commute, i.e., their commutator is not equal to zero (one common example of this is matrices). In quantum mechanics, the position and momentum variables of a particle do not commute, i.e., the commutator of those variables is not zero, $[q, p] = qp - pq = i\hbar$, with q the position of the particle, p the momentum of the particle, $i = \sqrt{-1}$, and $\hbar = h/2\pi$.

For a system of two particles, the position and momentum of particle 1, q_1 and p_1, are non-commuting variables, as are the position and momentum of particle 2, q_2 and p_2. But the position and momentum variables of particle 1 commute with the position and momentum variables of particle 2:

$$[q_1, p_1] = q_1 p_1 - p_1 q_1 = i\hbar$$

$$[q_2, p_2] = i\hbar$$

$$[q_1, q_2] = 0 = [q_1, p_2] = [p_1, q_2] = [p_1, p_2]$$

But the relative position of the two particles, $Q = q_1 - q_2$, and the total momentum of the system, $P = p_1 + p_2$, are commuting variables:[89]

$$[Q, P] = QP - PQ = (q_1 - q_2)(p_1 + p_2) - (p_1 + p_2)(q_1 - q_2)$$

$$= (q_1 p_1 + q_1 p_2 - q_2 p_1 - q_2 p_2) - (p_1 q_1 - p_1 q_2 + p_2 q_1 - p_2 q_2)$$

$$= (q_1 p_1 - p_1 q_1) + (q_1 p_2 - p_2 q_1) + (-q_2 p_1 + p_1 q_2) + (-q_2 p_2 + p_2 q_2)$$

$$= i\hbar + 0 + 0 - i\hbar$$

$$= 0$$

6.5 Notes

1. Tipler, Paul A., and Llewellyn, Ralph A., *Modern Physics*, Fifth Edition, W. H. Freeman and Company, New York, 2008, p. 149.
2. Tipler, Paul A., and Llewellyn, Ralph A., *Modern Physics*, pp. 149–150; Howard, Don, *Albert Einstein: Physicist, Philosopher, Humanitarian*, Part 2, The Teaching Company, Chantilly, VA, 2008, p.26; Eisberg, Robert, and Resnick, Robert, *Quantum Physics*, Second Edition, John Wiley & Sons, New York, 1985, pp. 96–98.
3. Stachel, John, editor, Einstein's Early Work on the Quantum Hypothesis, *The Collected Papers of Albert Einstein*, Volume 2, [CPAE2], Princeton University Press, Princeton, NJ, 1989, p. 142.
4. Kirsten, G., and Körber, H., *Physiker über Physiker*, p. 201, Akademie Verlag, Berlin, 1975; citation from Pais, Abraham, *Subtle is the Lord*, Oxford University Press, Oxford, 1982, pp. 382, 387.
5. Letter from Einstein to Besso, July 29, 1918; Schulmann, Robert, Kox, A. J., and Jansenn, Michael, editors, *The Collected Papers of Albert Einstein*, Volume 8, [CPAE8], Princeton University Press, Princeton, NJ, 1998, p. 836, English translation by Ann M. Hentschel, [CPAE8 ET], p. 613.
6. Howard, Don, and Stachel, John, editors, *Einstein: The Formative Years, 1879–1909*, Birkhäuser, Boston, MA, 2000, p. 244.
7. Petit, S. T., and Dulong, P. L., *Ann. Chim. Phys.* 10, 395 (1819); citation from Pais, *Subtle is the Lord*, pp. 390, 401.
8. de la Rive, A., and Marcet, F., *Ann. Chim. Phys.* 75, 113 (1840); citation from Pais, Abraham, *Subtle is the Lord*, pp. 391, 401.
9. Regnault, H. V., *Ann. Chim. Phys.* 1, 129 (1841), especially pp. 202–5; citation from Pais, Abraham, *Subtle is the Lord*, pp. 391, 401.
10. Pais, Abraham, *Subtle is the Lord*, pp. 391–392.
11. Einstein, Albert, Planck's Theory of Radiation and the Theory of Specific Heat, *Annalen der Physik* 327 (1), (1907), 180–190; [CPAE2. pp. 378–391]; English translation by Anna Beck, [CPAE2 ET], 1989, pp. 214–224.
12. Einstein, Albert, Planck's Theory of Specific Heat; [CPAE2. p. 383; CPAE2 ET, pp. 218–219].

13. Einstein, Albert, Planck's Theory of Specific Heat; [CPAE2. pp. 383–384; CPAE2 ET, p. 219].

14. Note the subscript V has been added to the specific heat, c. c_V is the specific heat at constant volume.

15. Pais, Abraham, *Subtle is the Lord*, pp. 395–396.

16. Pais, Abraham, *Subtle is the Lord*, p. 396.

17. Einstein, Albert, *The Collected Papers of Albert Einstein*, Volume 2, Hebrew University and Princeton University Press, Princeton, NJ, 1989, p. 220.

18. Einstein, Albert, On the Present Status of the Radiation Problem, *Physikalische Zeitschrift* 10 (1909), 185–193; [CPAE2, pp. 541–553; CPAE2 ET, pp. 357–376].

19. Pais, Abraham, *Subtle is the Lord*, p. 403; the following three equations for energy fluctuations are implicitly in Einstein's papers. For the convenience of the reader, Pais writes them out explicitly.

20. Einstein, Albert, On a Heuristic Point of View Concerning the Production and Transmission of Light, *Annalen der Physik* 17 (1905), 132–148; [CPAE2, p. 161; CPAE2 ET, p. 97].

21. Einstein, Albert, The Radiation Problem; [CPAE2, p. 547; CPAE2 ET, p. 369]. For consistency in notation the form of the equation is that obtained in Pais, Abraham, *Subtle is the Lord*, p. 403.

22. Einstein, Albert, On the Development of Our Views Concerning the Nature and Constitution of Radiation, *Deutsche Physikalische Gesellschaft Verhandlungen* 7 (1909), 482–500; [CPAE2, p. 577; CPAE2 ET, p. 390].

23. Einstein, Albert, The Nature and Constitution of Radiation; [CPAE2, p. 564; CPAE2 ET, p. 379].

24. Einstein, Albert, The Nature and Constitution of Radiation; [CPAE2, p. 582; CPAE2 ET, p. 394].

25. Einstein, Albert, The Radiation Problem; [CPAE2, p. 549; CPAE2 ET, p. 372].

26. Tipler, Paul A., and Llewelyn, Ralph A., *Modern Physics*, p. 161.

27. Further details can be found in a modern physics text, for example, Paul Tipler and Ralph Llewelyn's *Modern Physics*, pp. 159–164.

28. A treatment of this can be found in a modern physics text, for example, Paul Tipler and Ralph Llewelyn's *Modern Physics*, pp. 174–175.

29. Klein, Martin J., Einstein and the Development of Quantum Physics, p. 147, in French, A. P., editor, *Einstein, A Centenary Volume*, Harvard University Press, Cambridge, MA, 1979, and Pais, Abraham, *Subtle is the Lord*, p. 405.

30. Einstein, Albert, Emission and Absorption of Radiation in Quantum Theory, in Kox, A. J., Klein, Martin, J., and Schulmann, Robert, editors, *The Collected Papers of Albert Einstein*, Volume 6, [CPAE6], Princeton University Press, Princeton, NJ, 1996, p. 367; English translation by Alfred, Engel, [CPAE6 ET], Princeton University Press, Princeton, NJ, 1997, p. 214. Also in Pais, Abraham, *Subtle is the Lord*, p. 406.

31. Pais, Abraham, *Subtle is the Lord*, p. 407.

32. Howard, Don, *Albert Einstein: Physicist, Philosopher, Humanitarian*, p. 27.

33. Klein, Martin J., *Einstein and the Development of Quantum Physics*, p. 147.

34. Compton, A. H., *Phys, Rev.* 21, 483 (1923); citation from Pais, Abraham, *Subtle is the Lord*, pp. 413, 414.

35. Debye, P., *Phys. Zeitschr.* 24, 161 (1923); citation from Pais, Abraham, *Subtle is the Lord*, pp. 413, 414.

36. Eisberg, Robert, and Resnick, Robert, *Quantum Physics*, p. 36.

37. Eisberg, Robert, and Resnick, Robert, *Quantum Physics*, p. 36.

38. See, for example, Eisberg, Robert, and Resnick, Robert. *Quantum Physics*, pp. 34–38.

39. Eisberg, Robert, and Resnick, Robert, *Quantum Physics*, p. 37; Pais, Abraham, *Subtle is the Lord*, pp. 413, 414.

40. Pais, Abraham, *Subtle is the Lord*, pp. 413–414.

41. Letter from Einstein to Besso, July 29, 1918; [CPAE8, p. 836; CPAE8 ET, p. 613].

42. Kuhn, Thomas, *Black-Body Theory and the Quantum Discontinuity, 1894–1912*, University of Chicago Press, Chicago, Il, 1987, pp. 48–49.

43. Einstein, Albert, A Theory of the Foundations of Thermodynamics, *Annalen der Physik* 316 (5), (1903), 170–187; [CPAE2, pp. 78–79, CPAE2 ET, pp. 49–50].

44. Howard, Don, and Stachel, Stachel, *The Formative Years*, p. 243.

45. Mehra, Jagdish, and Rechenberg, Helmut, *The Historical Development of Quantum Theory*, Volume 1, Part 2: *The Quantum Theory of Planck, Einstein, Bohr and Sommerfeld: Its Foundation and the Rise of its Difficulties 1900–1925*, Springer-Verlag, New York/Heidelberg/Berlin, 1982, p. 564; citation from Stachel, John, Einstein and Bose, in *Einstein from 'B' to 'Z'*, Birkhäuser, Boston, MA, 2002, pp. 520, 538.

46. Bose, Satyendra Nath, Planck's Gesetz und Lichquantenhypothese, *Zeitschrift für Physik* 26, (1924), 179; citation from Stachel, John, Einstein and Bose, in *Einstein from 'B' to 'Z'*, pp. 521, 537.

47. Pais, Abraham, *Subtle is the Lord*, p. 425.

48. Letter from Bose to Einstein, Einstein Archive, Doc. 6-127; citation from Stachel, John, Einstein and Bose, in *Einstein from 'B' to 'Z'*, p. 524.

49. Bose, Satyendra Nath, Planck's Gesetz und Lichquantenhypothese, *Zeitschrift für Physik* 26 (1924), 179; citation from Stachel, John, Einstein and Bose, in *Einstein from 'B' to 'Z'*, pp. 524, 537.

50. Einstein, Albert, On the Quantum Theory of the Monatomic Ideal Gas. Second Paper, 1924, p. 3; translation from Stachel, John, in *Einstein from 'B' to 'Z'*, p. 528.

51. Einstein, Albert, On the Theory of the Ideal Gas, translation from Stachel, John, in *Einstein from 'B' to 'Z'*, p. 528.

52. Stachel, John, *Einstein from 'B' to 'Z'*, p. 531.

53. This example follows closely that given in Stachel, John, *Einstein from 'B' to 'Z'*, p. 532, which, in turn, cites the work of Howard, Don, The Prehistory of EPR, 1909–1935: Einstein's Early Worries About Quantum Mechanics of Composite Systems, in *Sixty-Two Years of Uncertainty*, Arthur Miller, editor, Plenum, New York, p. 67.

54. Einstein, On the Theory of the Ideal Gas, p. 18; citation from Stachel, John, *Einstein from 'B' to 'Z'*, p. 528.

55. Stachel, John, *Einstein from 'B' to 'Z'*, p. 533.

56. Howard, Don, *Albert Einstein: Physicist, Philosopher, Humanitarian*, p. 30.

57. Stachel, John, *Einstein from 'B' to 'Z'*, pp. 533–535.

58. de Broglie, Louis, preface to his reedited 1924 Ph.D. thesis, *Recherches sur la Théorie des Quanta*, Masson, Paris, 1963, p. 4; citation from Pais, *Subtle is the Lord*, pp. 436, 439.

59. Pais, Abraham, *Subtle is the Lord*, p. 436.

60. Pais, Abraham, *Subtle is the Lord*, p. 437.

61. Pais, Abraham, *Subtle is the Lord*, pp. 436, 437.

62. Pais, Abraham, *Subtle is the Lord*, pp. 436–437; for a discussion of the Davisson–Germer experiment, see a modern physics text such as Paul Tipler and Ralph Llewelyn's *Modern Physics*, pp. 188–195.

63. Eisberg, Robert, and Resnick, Robert, *Modern Physics*, John Wiley & Sons, Inc., New York, 1985, p. 113.

64. Jammer, M., *The Conceptual Development of Quantum Mechanics*, 2nd edition, Tomash Publishing Company, New York, 1989, p. 258; citation from Cushing, James T., *Philosophical Concepts in Physics*, Cambridge University Press, Cambridge, 1998, pp. 286, 391, 406.

65. Pais, Abraham, *Subtle is the Lord*, pp. 438–439.

66. Schrödinger, E., *AdP.* 79, 734 (1926); footnote on p. 735; citation from Pais, Abraham, *Subtle is the Lord*, p. 439; see also Klein, M. J., Einstein and the Wave-Particle Duality, in Gershenon, D.E. and Greenberg, D.A., editors, 1964, *The Natural Philosopher*, Vol. 3, Blaisdell Publishing Co., New York, p. 4; citation from Cushing, James T., *Philosophical Concepts*, pp. 286, 391, 407.

67. Schrödinger, E., *Phys. Zeitschr.* 27, 95 (1926); citation from Pais, Abraham, *Subtle is the Lord*, p. 439; see also Klein, M. J., Einstein and the Wave-Particle Duality, in Gershenon, D.E. and Greenberg, D.A., editors, 1964, *The Natural Philosopher*, Vol. 3, Blaisdell Publishing Co., New York, p. 43; citation from Cushing, James T., *Philosophical Concepts*, pp. 286, 391, 407.

68. Cushing, James T., *Philosophical Concepts*, p. 293.

69. Born, M., Zur Quantenmechanik der Stossvorgänge, *Zeitschrift für Physik* 37 (1926), p. 863; citation from Cushing, James T., *Philosophical Concepts*, p. 293, 391, 401.

70. Bohr, Niels, Discussion with Einstein on Epistemological Problems in Atomic Physics, in *Albert Einstein: Philosopher–Scientist*, Schilpp, Paul A., editor, Cambridge University Press, London, 1949 (1970), p. 207.

71. Bohr, Niels, Discussion with Einstein, in Schilpp, Paul A., *Albert Einstein: Philosopher–Scientist*, pp. 209–211.

72. Cushing, James T., *Philosophical Concepts*, p. 308.

73. Cushing, James, T., *Philosophical Concepts*, p. 307.

74. Fölsing, Albrecht, *Albert Einstein*, Viking Penguin, New York, 1997, p. 699.

75. Cushing, James T., *Philosophical Concepts*, pp. 307–308.

76. Cushing, James T., *Philosophical Concepts*, p. 308.

77. Cushing, James T., *Philosophical Concepts*, pp. 308–309.

78. Cushing, James T., *Philosophical Concepts*, p. 309; Pais, Abraham, *Subtle is the Lord*, p. 448.

79. Einstein, A., Podolsky, B. and Rosen, N., Can Quantum-Mechanical Description of Physical Reality Be Considered Complete? *Phys. Rev.*, 47, 777 (1935); citation taken from Pais, Abraham, *Subtle is the Lord*, pp. 455, 458.

80. Einstein, A., Podolsky, B. and Rosen, N., Can Quantum-Mechanical Description of Physical Reality Be Considered Complete? *Phys. Rev.*

47, 777 (1935); citation and translation taken from Cushing, James T., *Philosophical Concepts*, pp. 319. 393, 404.

81. Fölsing, Albrecht, *Albert Einstein*, p. 698.

82. Einstein to Aron Kupperman, 14 November 1953, Item 8-036 in the Einstein Archive; citation from Stachel, John, Einstein and Quantum Mechanics, in *Einstein from 'B' to 'Z'*, p. 412. The competing interpretations of underlying deterministic hidden variables and of the variables of the system being undetermined until a measurement was made on the system was put to the test in a series of tests developed by John Bell. See, for example, Cushing, James T., *Philosophical Concepts*, for a discussion of the Bell theorem and Bell experiments, pp. 319–330.

83. Einstein to Paul Bonogield, September 18, 1939, Item 6-118.1 in the Einstein Archive; citation from Stachel, John, Einstein and the Quantum: Fifty Years of Struggle, in *Einstein from 'B' to 'Z'*, p. 389.

84. Einstein to Herbert Kondo, 11 August 1952, Item 15-408 in the Einstein Archive; citation from Stachel, John, Einstein and the Quantum, in *Einstein from 'B' to 'Z'*, p. 395.

85. Einstein, Albert, The Nature and Constitution of Radiation; [CPAE2, p. 565; CPAE2 ET, p. 379].

86. Stachel, John, editor, Einstein's Early Work on the Quantum Hypothesis; [CPAE2, p. 142].

87. Cushing, James T., *Philosophical Concepts*, p. 309; Pais, Abraham, *Subtle is the Lord*, p. 448.

88. Boltzmann, L., *Wiener Ber.* 63, 679 (1871); reprinted in *Wissenschaftliche Abhandlugen von L. Boltzmann* (F. Hasenohrl, ed.), Volume 1, p. 259; Reprinted by Chelsea, New York, 1968; citation from Pais, Abraham, *Subtle is the Lord*, pp. 392, 400.

89. Bohr, Niels, Discussion with Einstein, in *Philosopher–Scientist*, p. 233; see also Pais, Abraham, *Subtle is the Lord*, pp. 455–456.

6.6 Bibliography

Bohr, Niels, Discussion with Einstein on Epistemological Problems in Atomic Physics, in *Albert Einstein: Philosopher–Scientist*.

Boltzmann, L., *Wiener Ber.* 63, 679 (1871); reprinted in *Wissenschaftliche Abhandlugen von L. Boltzmann* (F. Hasenohrl, ed.) Vol. 1, p. 259; Reprinted by Chelsea, New York, 1968. (Citation in Pais, Abraham, *Subtle is the Lord*.)

Born, M. Zur Quantenmechanik der Stossvorgänge, *Zeitschrift für Physik* 37 (1926) (Citation in Cushing, James, *Philosophical Concepts in Physics*.)

Bose, Satyendra Nath, Planck's Gesetz und Lichquantenhypothese, *Zeitschrift für Physik* 26 179 (1924). (Citation in Stachel, John, *Einstein from 'B' to 'Z'*.)

Compton, A. H., *Phys., Rev.* 21, 483 (1923). (Citation in Pais, Abraham, *Subtle is the Lord*.)

Cushing, James T., *Philosophical Concepts in Physics*, Cambridge University Press, Cambridge, 1998.

de Broglie, Louis, Preface to his reedited 1924 Ph. D. thesis, *Recherches sur la Théorie des Quanta*, Masson, Paris, 1963, p. 4. (Citation in Pais, Abraham,*Subtle is the Lord*.)

Debye, P., *Phys. Zeitschr.* 24, 161 (1923). (Citation in Pais, Abraham, *Subtle is the Lord*.)

Einstein, Albert, A Theory of the Foundations of Thermodynamics, *Annalen der Physik* 316 (5), (1903), pp. 170–187; [CPAE2; CPAE2 ET].

Einstein, Albert, On a Heuristic Point of View Concerning the Production and Transmission of Light, *Annalen der Physik* 322 (6), (1905), pp. 132–148; [CPAE2; CPAE2 ET].

Einstein, Albert, Planck's Theory of Radiation and the Theory of Specific Heat, *Annalen der Physik* 327 (1), (1907), pp. 180–190; [CPAE2; CPAE2 ET].

Einstein, Albert, On the Present Status of the Radiation Problem, *Phys. Zeitschr.* 10 (1909) [CPAE2; CPAE2 ET].

Einstein, Albert, On the Development of Our Views Concerning the Nature and Constitution of Radiation, *Deutsche Physikalische Gesellschaft Verhandlungen* 7 (1909) [CPAE2; CPAE2 ET].

Einstein, Albert, The Nature and Constitution of Radiation [CPAE2; CPAE ET].

Einstein, Albert, Emission and Absorption of Radiation in Quantum Theory [CPAE6].

Einstein, Albert, On the Quantum Theory of the Monatomic Ideal Gas. Second Paper, 1924. (Citation from Stachel, John, *Einstein from 'B' to 'Z'*.)

Einstein, A., Podolsky, B., and Rosen, N., Can Quantum-Mechanical Description of Physical Reality Be Considered Complete? *Phys. Rev.* 47, 777 (1935). (Citation in Pais, Abraham, *Subtle is the Lord*.)

Eisberg, Robert, Resnick, Robert, *Quantum Physics*, Second Edition, John Wiley & Sons, New York, 1985.

French, A. P., (editor), *Einstein, A Centenary Volume*, Harvard University Press, Cambridge, MA, 1979.

Fölsing, Albrecht, *Albert Einstein*, Viking Penguin, New York, 1997.

Howard, Don, *Albert Einstein: Physicist, Philosopher, Humanitarian*, Part 2, The Teaching Company, Chantilly, VA, 2008.

Howard, Don, The Prehistory of EPR, 1909–1935: Einstein's Early Worries About Quantum Mechanics of Composite Systems, in *Sixty-Two Years of Uncertainty*, Miller, Arthur, editor, Plenum, New York, 1990.

Howard, Don, and Stachel, John, editors, *Einstein: The Formative Years, 1879–1909*, Birkhäuser, Boston, 2000.

Jammer, M., *The Conceptual Development of Quantum Mechanics*, 2nd edition, Tomash Publishing Company, New York, 1989, p. 258. (Citation in Cushing, James, *Philosophical Concepts in Physics*.)

Kirsten, G., and Körber, H., *Physiker über Physiker*, p. 201, Akademie Verlag, Berlin, 1975. (Citation in Pais, Abraham, *Subtle is the Lord*.)

Klein, Martin J., *Einstein and the Development of Quantum Physics*, in French, A.P. (ed.), *Einstein, A Centenary Volume*.

Klein, M. J., Einstein and the Wave-Particle Duality, in Gershenon, D.E. and Greenberg, D.A., editors, 1964, *The Natural Philosopher*, Vol. 3, Blaisdell Publishing Co., New York. (Citation in Cushing, James, *Philosophical Concepts in Physics*.)

Kox, A. J., Klein, Martin, J., and Schulmann, Robert, editors, *The Collected Papers of Albert Einstein*, Volume 6 [CPAE6], Princeton University Press, Princeton, NJ, 1996; English translation by Alfred Engel [CPAE6 ET, 1997].

Kuhn, Thomas, *Black-Body Theory and the Quantum Discontinuity, 1894–1912*, University of Chicago Press, Chicago, IL, 1987.

Mehra, Jagdish, and Rechenberg, Helmut, *The Historical Development of Quantum Theory*, Volume 1, Part 2: *The Quantum Theory of Planck, Einstein, Bohr and Sommerfeld: Its Foundation and the Rise of its Difficulties*

1900–1925, Springer-Verlag, New York/Heidelberg/Berlin. (Citation in Stachel, John, *Einstein from 'B' to 'Z'*.)

Miller, Arthur, editor, *Sixty-Two Years of Uncertainty*, Plenum, New York, 1990. (Citation in Stachel, John, *Einstein from 'B' to 'Z'*.)

Pais, Abraham, *Subtle is the Lord*, Oxford University Press, Oxford, 1982.

Petit, S. T., Dulong, P. L., *Ann. Chim. Phys.* 10, 395 (1819). (Citation in Pais, Abraham, *Subtle is the Lord*.)

Regnault, H.V., *Ann. Chim. Phys.* 1, 129 (1841). (Citation in Pais, Abraham, *Subtle is the Lord*.)

de la Rive, A., Marcet, F., *Ann. Chim. Phys.* 75, 113 (1840). (Citation in Pais, Abraham, *Subtle is the Lord*.)

Schilpp, Paul A., editor, *Albert Einstein: Philosopher–Scientist*, Cambridge University Press, London, 1949 (1970).

Schrödinger, E., *AdP.* 79, 734 (1926). (Citation in Pais, Abraham, *Subtle is the Lord*.)

Schrödinger, E., *Phys. Zeitschr.* 27, 95 (1926). (Citation in Pais, Abraham, *Subtle is the Lord*.)

Schulman, Robert, Kox, A. J., Janssen, Michel, editors, *The Collected Papers of Albert Einstein*, Volume 8 [CPAE8], Princeton University Press, Princeton, NJ, 1998; English translation by Ann M. Hentschel. [CPAE8 ET].

Stachel, John, *Einstein from 'B' to 'Z'*, Birkhäuser, Boston, MA, 2002.

Stachel, John, Einstein and Quantum Mechanics, in *Einstein from 'B' to 'Z'*, Birkhäuser, Boston, MA, 2002.

Stachel, John, Einstein and the Quantum: Fifty Years of Struggle, in *Einstein from 'B' to 'Z'*, Birkhäuser, Boston, MA, 2002.

Stachel, John, editor, *Einstein's Early Work on the Quantum Hypothesis* [CPAE2].

Stachel, John, Einstein and Bose, in *Einstein from 'B' to 'Z'*, Birkhäuser, Boston, MA, 2002.

Stachel, John, editor, *The Collected Papers of Albert Einstein*, Volume 2 [CPAE2], Princeton University Press, Princeton, NJ, 1989; English translation by Anna Beck [CPAE2 ET].

Tipler, Paul A., and Llewellyn Ralph A., *Modern Physics*, Fifth Edition, W. H. Freeman and Company, New York, 1999, 2008.

Epilogue

7.1 The Inflexible Boundary Condition

7.1 The Inflexible
    Boundary Condition   290
7.2 Notes   293
7.3 Bibliography   295

In describing Albert Einstein's approach to science, he frequently is described as more of an artist than a scientist, that his guiding principles were order and beauty. In 1933, Einstein said, "Our experience up to date justifies us in feeling sure that in nature is actualized the ideal of mathematical simplicity."[1] And, in 1947, he said, "My views are near those of Spinoza: admiration for the beauty of and belief in the logical simplicity of the order and harmony which we can grasp humbly and only imperfectly."[2] In 1970, Banesh Hoffmann said of Einstein, "One might almost say that he was not so much a scientist as an artist of science."[3]

However, even though it may be proper to describe him as an artist because of his æsthetic sense and search for beauty and order, Einstein remained acutely attuned to the "inflexible boundary condition of agreeing with physical reality."[4] This sense was honed in his father's electrical engineering company, in Weber's courses on experimental physics at the ETH, and in examining patents in the patent office, leading to a discrimination of what was essential from what was incidental. After detailing many of the mathematical details of his papers in the previous chapters, we conclude with some reflection on his focus on experimental results to provide him insights, to set the direction, to correct his path, and to verify his results.

In Munich, as the Einstein home was on the grounds of the factory, Albert grew up in daily contact with electromechanical equipment,[5] with first-hand experience of electromagnetism, a cutting edge technology at the time, and its application in the real world. In college, Professor Weber's laboratory courses stressed the importance of measurement. Einstein's attitude toward experiment can be seen in his comment on his enjoyment of the laboratory experience, that he was "...fascinated by the direct contact with experience..."[6] Even prior to 1905, in commenting on the work of Boltzmann, Einstein was concerned that Boltzmann had not more closely tied his work to experimental results. In a 1901 letter to Mileva Marić, he wrote, "At present I am again studying Boltzmann's theory of gases. Everything is very nice, but there is too little stress on the comparison with reality."[7]

In the paper, "On a Heuristic Point of View Concerning the Production and Transformation of Light,"[8] Einstein began by noting that in the blackbody radiation experiment, in the high frequency regime (Wien's distribution law regime), "the theoretical principles...fail completely."[9] In sections 7, 8, and 9 of the paper, Einstein shows how the concept of light quanta can be used to give simple explanations of three experimental phenomena: Stokes' rule for photoluminescence; the photoelectric effect; and the generation of cathode rays by photo-ionization.

In the paper, "A New Determination of Molecular Dimensions,"[10] Einstein arrived at the size of the atom by comparing his theoretical results to experimental data on diffusion and viscosity. In the paper "On the Movement of Small Particles Suspended in Stationary Liquids Required by the Molecular-Kinetic Theory of Heat,"[11] he concludes it with a request for experimental confirmation (or falsification), "Let us hope that a researcher will soon succeed in solving the problem posed here, which is of such importance in the theory of heat."[12] And, after the confirmation of the Brownian motion equation by Perrin, Einstein recalled that "...the agreement of these considerations [theory of Brownian motion] with experience...convinced the skeptics, who were quite numerous at that time (Ostwald, Mach) of the reality of atoms."[13]

The paper, "On the Electrodynamics of Moving Bodies,"[14] opens with a comment on the electrodynamic induction experiment, on "asymmetries [in electrodynamic induction] that do not seem to attach to the phenomena."[15], and "...the failure of attempts to detect a motion of the earth relative to the 'light medium' [the experimental fact of the constant value of the speed of light]..."[16] Starting with the principle of relativity, Einstein obtained the Lorentz transformations and obtained the same transformations of the electric and magnetic fields as were obtained by Lorentz. As the Lorentz transformations were obtained specifically to be consistent with experiment, Einstein's results also would be consistent with experiment.

In this paper, Einstein considered an electron at rest at the origin of K' (the moving reference frame). In K', the electromagnetic force on the electron will be purely electric: $\vec{F} = q\vec{E}$. In K, the electron is moving at speed v and, transforming to the reference frame K,[17] the force becomes $\vec{F} = q(\vec{E} + \frac{\vec{v}}{c} \times \vec{B})$. This is the Lorentz force on a charged particle moving in an electromagnetic field, obtained in a natural way as a consequence of the principle of relativity, not as a separate postulate as was done by Lorentz.

Of interest, as it was in apparent contradiction to Einstein's strong reliance on experiment to guide his theories and to verify his results, was his response to the experiments of Walter Kaufmann on the mass of high speed electrons. Stachel describes his response as "... cautious about accepting Kaufmann's results as definitive, perhaps because of his familiarity with Planck's critical analysis of the experiments."[18] In 1905, after a series of experiments measuring the mass of high speed electrons, Kaufmann had written that, "The prevalent results decidedly speak against the correctness of Lorentz's assumptions as well as Einstein's."[19]

Einstein waited two years before responding. In his 1907 paper, "On the Relativity Principle and the Conclusions Drawn from It," Einstein wrote, "It should also be mentioned that ... [other] ... theories of the motion of the electron yield curves that are significantly closer to the observed curve than the curve obtained from the theory of relativity. However, the probability that their theories are correct is rather small, in my opinion, because their basic assumptions concerning the dimensions of the moving electron are not suggested by theoretical systems that encompass larger complexes of phenomena."[20] Eventually, after two years of investigation, Planck found an inconsistency in Kaufmann's data that shifted the conclusions slightly in favor of Einstein's theory.[21]

As John Stachel notes, "The earliest widely accepted empirical evidence for the quantum hypothesis came not from radiation phenomena, but from data on specific heats of solids."[22] In his 1907 paper, "Planck's Theory of Radiation and the Theory of Specific Heat,"[23] Einstein obtained an expression for the specific heat of a solid.[24] Comparing this expression for c_V to Weber's data for diamond, Einstein found an almost exact fit for $T_E = 1300$ K.

In his 1907 review of the special theory of relativity, entitled "On the Relativity Principle and the Conclusions Drawn from It,"[25] in the last section of the paper, Einstein already commented on the possibility of a general theory of relativity and its verification by measurement of a gravitational redshift and the bending of light in a gravitational field.

The paper, "The Foundation of the General Theory of Relativity,"[26] begins with the statement, "This view is made possible for us by the teaching of experience as to the existence of a field of force, namely the gravitational field, which possesses the remarkable property of imparting the same acceleration to all bodies."[27] He then references the Eötvös experiment,[28] "Eötvös has proved experimentally that the gravitational field has this property to great accuracy."[29] He concludes the paper with reference to three possible verifications (or falsifications) of the theory: the precession of the perihelion of the orbit of Mercury (a known, but not explained, fact); a gravitational redshift (a prediction of a new phenomenon); and the bending of starlight passing near to the sun (a prediction of a new phenomenon).

In the cosmology paper, "Cosmological Considerations in the General Theory of Relativity," Einstein introduced the cosmological constant to be in agreement with a static universe – the accepted knowledge of the day.[30] Years later, after the Hubble expansion of the universe was known, looking back on his introduction of the cosmological constant, he said, "If Hubble's expansion had been discovered at the time of the creation of the general theory of relativity, the cosmological member never would have been introduced."[31]

The unified field theory had no physical insight or conflict to guide him; it was purely a quest for harmony. But even here, in his exchanges with Cartan, it could be seen Einstein was mostly concerned with possible physical interpretation, whereas Cartan would be guided by considerations of logical necessity.[32]

In a 1921 lecture in London, Einstein said, "The abandonment of a certain concept . . . must not be regarded as arbitrary . . . the justification for a physical concept lies exclusively in its clear and unambiguous relation to facts that can be experienced."[33] In 1924, describing his breakthrough to the special theory of relativity, he said, ". . . our concepts and laws of space and time can only claim validity insofar as they stand in a clear relation to our experiences; and that experience could very well lead to the alteration of these concepts and laws."[34] In 1933, apparently moving closer to pure mathematical constructions, in a lecture delivered at Oxford, he said, "I am convinced that we can discover by means of purely mathematical constructions the concepts and the laws connecting them with each other, which furnish the key to the understanding of natural phenomena . . . the creative principle resides in mathematics."[35]

However, as Michel Biezunski described it in a 1989 article, Albert Einstein was a person who ". . . wanted a theory that is complete, general, aesthetically satisfying, and unified. The means were less important in his eyes. He used a mathematical theory as a tool, and never more. He was concerned with physical problems, and refused to be caught in the trap of formal, mathematical structures."[36]

7.2   Notes

1. Einstein, Albert, *On the Method of Theoretical Physics*, Oxford University Press, New York, 1933. Reprinted in *Phil. Sci.* 1, 162 (1934). Citation from Pais, Abraham, *Subtle is the Lord*, Oxford University Press, Oxford, 1982, pp. 466–468.
2. Hoffmann, Banesh, *Albert Einstein: Creator and Rebel*, New American Library, New York, 1973, p. 95.
3. Hoffmann, Banesh, author March 24, 1970. Citation from Clark, Ronald, *Albert Einstein: The Life and Times*, The World Publishing Company, New York, pp. 535, 688.
4. Cushing, James T., *Philosophical Concepts in Physics*, Cambridge University Press, Cambridge, 1998, p. 360.
5. Howard, Don, *Albert Einstein: Physicist, Philosopher, Humanitarian*, The Teaching Company, Virginia, VA, 2008, pp. 10–12.
6. Howard, Don, and Stachel, John, editors, *Einstein: The Formative Years, 1879–1909*, Birkhäuser, Boston, MA, 2000, pp. 63, 64, 66.
7. Stachel, John, editor, *The Collected Papers of Albert Einstein*, [CPAE1], Princeton University Press, Princeton, NJ, 1987, pp. 293–295; English translation by Anna Beck, [CPAE1 ET], p. 168.
8. Einstein, Albert, On a Heuristic Point of View Concerning the Production and Transformation of Light, *Annalen der Physik* 322 (6), (1905), pp. 132–148; Stachel, John, editor, *The Collected Papers of Albert Einstein*, [CPAE2], Princeton University Press, Princeton, NJ, 1989, pp. 150–166; English translation by Anna Beck, [CPAE2 ET], pp. 86–103.
9. Stachel, John, ed., [CPAE2, pp. 139, 155; CPAE2 ET, p. 91].
10. Einstein, Albert, A New Determination of Molecular Dimensions. Dissertation, University of Zürich, 1905; [CPAE2, pp. 183–202; CPAE2 ET, pp. 104–122].

11. Einstein, Albert, On the Movement of Small Particles Suspended in Stationary Liquids Required by the Molecular-Kinetic Theory of Heat, *Annalen der Physik* 322 (8), (1905): pp. 549–560; [CPAE2, pp. 223–235; CPAE2 ET, pp. 123–134].

12. Einstein, Albert, Movement of Small Particles; [CPAE2, p. 235; CPAE2 ET, p. 134].

13. Schilpp, Paul Arthur, editor, *Albert Einstein: Philosopher–Scientist*, Cambridge University Press, London, 1970, p. 49.

14. Einstein, Albert, On the Electrodynamics of Moving Bodies, *Annalen der Physik* 17 (1905); [CPAE2, pp. 275–306; CPAE2 ET, pp. 140–171].

15. Einstein, Albert, Electrodynamics of Moving Bodies; [CPAE2, p. 276; CPAE2 ET, p. 140].

16. Einstein, Albert, Electrodynamics of Moving Bodies; [CPAE2, p. 276; CPAE2 ET, p. 140].

17. Einstein, Albert, Electrodynamics of Moving Bodies; [CPAE2, pp. 303–304; CPAE2 ET, p. 168].

18. Stachel, John, editor, *Einstein's Miraculous Year*, Princeton University Press, Princeton, NJ, 1998, p. 120.

19. Cushing, James T., *Philosophical Concepts*, p. 215.

20. Einstein, Albert, On the Relativity Principle and the Conclusions Drawn from It, *Jahrbuch der Radioaktivität und Elektronik* 4 (1907); [CPAE2, p. 461; CPAE2 ET, p. 284].

21. An excellent re-creation of Planck's analysis can be found in James T. Cushing's article, Electromagnetic Mass, Relativity, and the Kaufmann Experiments, *Am. J. Phys.* 49 (1981), 1133–1149.

22. Stachel, John, editor, *Einstein's Miraculous Year*, p. 174.

23. Einstein, Albert, Planck's Theory of Radiation and the Theory of Specific Heat, *Annalen der Physik* 327 (1), (1907), 180–190; [CPAE2. pp. 378–391]; English translation, pp. 214–224.

24. Note the subscript V has been added to the specific heat, c. c_V is the specific heat at constant volume.

25. Einstein, Albert, On the Relativity Principle and the Conclusions, pp. 411–462; Stachel, John, ed., [CPAE2, pp. 432–489; CPAE2 ET, pp. 252–311].

26. Einstein, Albert, The Foundation of the General Theory of Relativity, 20 March, 1916, *Annalen der Physik* 354 (7), (1916), 769–822; Kox, A. J., Klein, Martin, J., and Schulmann, Robert, editors, [CPAE6], Princeton University Press, Princeton, NJ, 1996, pp. 283–339; English translation by Alfred Engel, [CPAE6 ET], Princeton University Press, Princeton, NJ, 1997, pp. 146–200.

27. Einstein, Albert, The Foundation of the General Theory of Relativity, [CPAE6, p. 288; CPAE ET, p. 150].

28. Eötvös, B., *Mathematische und Naturwissenschaftliche Berichte aus Ungarn* 8 (1890); Wiedemann's *Beiblätter* 15 (1891), 688; citation from Einstein, Albert, *Outline of a Generalized Theory of Relativity and of a Theory of Gravitation*, Teubner, Leipzig, 1913; Klein, Martin, Kox, A. J., Renn, Jürgen, and Schulmann, Robert, editors, [CPAE4], Princeton University Press, Princeton, NJ, 1995, p. 304, English translation by Anna Beck, [CPAE4 ET], p. 151.

29. Einstein, Albert, The Foundation of the General Theory of Relativity, [CPAE6, p. 288; CPAE6 ET, p. 150].

30. Einstein, Albert, Cosmological Considerations in the General Theory of Relativity, Königlich Preussische Akademie der Wissenschaften (Berlin). *Sitzungsberichte* (1917); citation from Kox *et al.*, [CPAE6, pp. 542–543; CPAE ET, p. 423].
31. Einstein, Albert, *The Meaning of Relativity*, 5th edition, Princeton University Press, Princeton, NJ, 1956 (1922), p. 127, footnote to item 1 in "Summary and Other Remarks."
32. Biezunski, Michel, Inside the Coconut: The Einstein–Cartan Discussion on Distant Parallelism, in *Einstein Studies*, Volume 1, Birkhäuser, Boston, 1989, p. 315.
33. Einstein, Albert, *Essays in Science*, Philosophical Library, 1934, p. 48. Citation from Schwinger, Julian *Einstein's Legacy*, Scientific American Library, New York, 1986, pp. 238, 239.
34. Einstein, Albert, Recording transcribed in Friedrich Herneck, "Zwei Tondokumente Einsteins sur Relativitätstheorie," *Forschung und Fortschritte* 40 (1966), 134. Citation 5 in *Einstein's Miraculous Year*, Stachel, John, editor, Princeton University Press, Princeton, NJ, 1998, pp. 112, 121.
35. Einstein, Albert, *Essays in Science*, p. 48. Citation from Schwinger, Julian, *Einstein's Legacy*, pp. 238, 239.
36. Biezunski, Michel, Inside the Coconut: The Einstein-Cartan Discussion, in *Einstein Studies*, Volume 1, p. 321.

7.3 Bibliography

Biezunski, Michel, Inside the Coconut: The Einstein–Cartan Discussion on Distant Parallelism, in *Einstein Studies*, Volume 1, Birkhäuser, Boston, MA, 1989.
Clark, Ronald W., *Einstein, The Life and Times*, The World Publishing Company, New York, 1971.
Cushing, James T., Electromagnetic Mass, Relativity, and the Kaufmann Experiments, *Am. J. Phys.* 49 (1981).
Cushing, James T., *Philosophical Concepts in Physics*, Cambridge University Press, Cambridge, 1998.
Einstein, Albert, *On the Method of Theoretical Physics*, Oxford University Press, New York, 1933.
Einstein, Albert, On a Heuristic Point of View Concerning the Production and Transformation of Light, *Annalen der Physik* 322 (6), (1905), pp. 132–148.
Einstein, Albert, A New Determination of Molecular Dimensions, Dissertation, University of Zürich, 1905.
Einstein, Albert, On the Movement of Small Particles Suspended in Stationary Liquids Required by the Molecular-Kinetic Theory of Heat, *Annalen der Physik* 322 (8), (1905), pp. 549–560.
Einstein, Albert, On the Electrodynamics of Moving Bodies, *Annalen der Physik* 322 (10), (1905), pp. 891–921.
Einstein, Albert, On the Relativity Principle and the Conclusions Drawn from It, *Jahrbuch der Radioaktivität und Elektronik* 4 (1907).

Einstein, Albert, Planck's Theory of Radiation and the Theory of Specific Heat, *Annalen der Physik* 327 (1), (1907), pp. 180–190.

Einstein, Albert, The Foundation of the General Theory of Relativity, 20 March, 1916, *Annalen der Physik* 354 (7), (1916), pp. 769–822.

Einstein, Albert, Cosmological Considerations in the General Theory of Relativity, Königlich Preussische Akademie der Wissenschaften (Berlin). *Sitzungsberichte* (1917).

Einstein, Albert, *The Meaning of Relativity*, 5th edition, Princeton University Press, Princeton, NJ, 1956 (1922).

Einstein, Albert, *Outline of a Generalized Theory of Relativity and of a Theory of Gravitation*, Teubner, Leipzig, 1913.

Einstein, Albert, *Essays in Science*, Philosophical Library, 1934.

Einstein, Albert, Recording transcribed in Herneck, Friedrich, Zwei Tondokumente Einsteins sur Relativitätstheorie, *Forschung und Fortschritte* 40 (1966).

Eötvös, B., *Mathematische und Naturwissenschaftliche Berichte aus Ungarn* 8 (1890); Wiedemann's *Beiblätter* 15 (1891).

Hoffmann, Banesh, *Albert Einstein: Creator and Rebel*, New American Library, New York, 1973.

Howard, Don, *Albert Einstein: Physicist, Philosopher, Humanitarian*, The Teaching Company, Chantilly, VA, 2008.

Howard, Don, and Stachel, John, editors, *Einstein: The Formative Years, 1879–1909*, Birkhäuser, Boston, 2000.

Klein, Martin, Kox, A. J., Renn, Jürgen, and Schulmann, Robert, editors, [CPAE4], Princeton University Press, Princeton, NJ, 1995; English translation by Anna Beck, [CPAE4 ET].

Kox, A. J., Klein, Martin, J., and Schulmann, Robert, editors, *The Collected Papers of Albert Einstein*, Volume 6 [CPAE6], Princeton University Press, Princeton, NJ, 1996; English translation by Alfred Engel, [CPAE6 ET, 1997]. Pais, Abraham, *Subtle is the Lord*, Oxford University Press, Oxford, 1982.

Pais, Abraham, *Subtle is the Lord*, Oxford University Press, Oxford, 1982.

Schilpp, Paul Arthur, editor, *Albert Einstein: Philosopher–Scientist*, Cambridge University Press, London, 1970.

Schwinger, Julian *Einstein's Legacy*, Scientific American Library, New York, 1986.

Stachel, John, editor, *The Collected Papers of Albert Einstein*, Volume 1, [CPAE1], Princeton University Press, Princeton, NJ, 1987; English translation by Anna Beck [CPAE2 ET].

Stachel, John, editor, *The Collected Papers of Albert Einstein*, Volume 2, [CPAE2], Princeton University Press, Princeton, NJ, 1989; English translation by Anna Beck, [CPAE2 ET].

Stachel, John, editor, *Einstein's Miraculous Year*, Princeton University Press, Princeton, NJ, 1998.

Index

Aarau 17
aberration 124
Abraham, Max 20, 62
absolute rest 114
absorption 35, 45, 47, 48, 216, 271
  coefficient 34
  spectra 266
acceleration, linear 164, 173
adiabatic 94, 199
aesthetic 161, 163, 165, 290, 293
aether (ether):
  absolute space 14, 21, 108, 109, 123, 131, 138, 139
  Aristotle 4, 13
  Einstein 24, 33, 40
  ether 13
  electromagnetic 13, 14, 33, 62, 108, 109, 112, 113, 122, 123
  gravity 13
  Lorentz 14, 21, 111, 112
  luminiferous 13, 14, 24, 108, 112, 113, 114, 131, 138, 139, 140, 269
  Maxwell 21, 108, 122, 123
Almagest 5
Annalen der Physik 33, 56, 105, 161, 171
annus mirabilis 11
Aristotle 3, 4, 5, 9, 10, 13, 57, 58
astronomer 4, 5, 6, 170, 215
astronomy (astronomical):
  Copernicus 6
  Eddington 214, 215, 220
  Einstein 215, 219
  Greek 2, 3, 4
  Ptolemy 5, 6
asymmetries 105, 291
atom (atomic, atomist, atomism, atomistic) 4, 21, 40, 56, 57, 58, 59
  Avogadro 59
  Bernoulli 58
  Bohr 269, 270, 271, 277, 278, 281
  Boltzmann 33, 59, 77, 282
  Boyle 58, 61
  Dalton 58
  Democritus 57
  Einstein 20, 40, 44, 45, 47, 56, 70, 72, 76–78, 216, 267–271, 274, 280, 281, 291

Epicurus 57
Gassendi 57, 58
Gay-Lussac 58
Mach 59, 77, 78, 91
Newton 58, 61
Ostwald 60, 62, 77, 78, 291
Perrin 77
Poincaré 77
Rutherford 270
Rynasiewicz 111
Avogadro, Amedeo 58, 59
Avogadro's number 63, 68, 71, 76, 77

Balmer formula 266, 270
Balmer, Johann 266
Balmer series 266
bending of light 165, 167, 171, 199, 208, 215, 292
bending of starlight 169, 192, 210, 214, 215, 222, 249, 253, 292
Bernoulli, Daniel 58
Besso, Michele 266
Bianchi identity 198
Biezunski, Michel 293
big bang 221
blackbody 34, 35, 42
blackbody radiation 34–39
  Bose 273
  Einstein 18, 19, 33, 39, 40, 44, 47, 51, 266, 267, 268, 271, 290, 291
  Planck 33, 40, 42, 50, 267, 268, 272, 273
  Rayleigh–Jeans 41
  Weber 18, 19
  Wien 18, 39, 41, 50
Bohr atom 269, 270, 281
Bohr, Niels 270, 273, 276, 277, 279, 280, 282
Bohr theory 271
Boltzmann's constant 34, 41, 44, 72
Boltzmann, Ludwig 18, 20, 33, 34, 41, 44, 47, 56, 59, 273, 274
  Einstein 44, 45, 290
  entropy 34, 44, 45, 50, 59, 272
  kinetic theory 20, 33, 34, 41, 59, 282, 290
  Planck 272

statistical mechanics 20, 34, 40, 59, 77
Boorse, Henry, and Motz, Lloyd 58, 61
Born, Max 277, 279
Bose, Satyendra Nath 49, 273, 274, 275, 281
Bose–Einstein statistics 49, 272, 275, 277
boundary condition
  inflexible 1, 23, 290
  sphere in a liquid 64, 65, 85
Boyle, Robert 58, 61
Boyle's law 58
Brahe, Tycho 6, 7, 9, 10, 13
Brown, Robert 57, 60
Brownian motion 56, 57, 60, 70, 71, 76, 77, 78, 269, 291

c (speed of light) 114
calculus:
  Einstein 18
  Newton 11
  tensor 168, 169
carbon 267
  carbon, boron, silicon 18, 267
  carbon filament 40
Cartan, Elie 292
cathode rays 45, 46, 291
cavity 35, 36, 41, 42, 48
cavity radiation 35, 38, 41, 42
celeritas 114
Christoffel 168, 169, 186, 187, 238
  notation (symbol) 186, 189, 222, 228, 234
  tensor 179, 191, 239, 240
circular motion 2, 4, 5, 6, 7, 10, 12, 107, 254, 255
circular orbits 9, 270
coefficient of absorption, *see* absorption coefficient
coefficient of diffusion 63, 68, 69, 70, 74, 75, 98
coefficient of viscosity 63, 66, 67, 68, 69, 74, 90, 93
Como conference 276, 278
complexions 34, 40, 49, 50, 272, 273, 274, 281

Compton scattering (Compton effect) 49, 271, 272, 281
Compton, Arthur 266, 272
configuration 34, 94, 96, 139
conservation laws:
  angular momentum 25, 26
  energy 33, 45, 130, 169, 195, 198, 199, 202, 205, 248, 250, 252, 271, 272
  momentum 33, 65, 169, 195, 198, 199, 202, 248, 271, 272
convection currents 128
Copernican system 6, 8, 10
Copernicus, Nicolaus 6, 10, 217
cosmological constant 218, 219, 220, 221, 292
cosmology (cosmological) 216, 218, 219, 220, 221, 292
Coulomb's law 162, 164, 270
covariant (co-variant)
  equations 168–171
  tensors 179–184
CPAE xvi
curved space 167, 217
curved space-time 168, 200, 214
Cushing, James T. 1, 22, 24, 38, 61, 62, 111, 113, 277, 278, 279

Dalton, John 58, 59
Davisson, Clinton J. 276
Davisson–Germer experiment 276
de Broglie, Louis 266, 275, 276, 277
de Broglie waves 276, 277
de Sitter, Willem 214, 219, 220, 221
Debye, Peter 271
deferent 6, 10
degrees of freedom 34, 36, 268
Democritus 57, 58
diamond 267, 268, 292
diffusion 62, 68, 73, 74, 75, 98, 291, *see also* coefficient of diffusion
dilatational motion (dilatation) 63, 64, 91
Dirac, P. A. M. 279
discrete energy elements 50
discrete energy levels (states) 271, 281
dissolved (dissolved molecules) 62, 67, 68, 69, 71, 72, 93, 95
distinguishable 50, 274, 275
Doppler 124, 126, 150, 216
dual nature of radiation 268, 269, 275, 276, 281
Duhem, Pierre 2, 4, 20, 25
Dulong, S. T., and Petit, P. L. 18, 267, 281, 282
dynamic equilibrium 42, 74, 75
dynamics 3, 106, 128

$E = mc^2$ 129, 131, 279
eccentric (eccentricity) 5, 6, 253

Eddington, Sir Arthur 23, 214, 215, 220
Ehrenfest, Paul 170, 214
Einstein convention 179
Einstein temperature 268, 292
Einstein, Albert:
  pre-college years 15, 16, 17
  college years 17, 18, 19
  1900–1905 years 19, 20
  scientific outlook 1, 21, 25, 132, 221, 290, 293
  experiment 19, 21, 33, 40, 41, 76, 78, 105, 113, 116, 123, 132, 133, 165, 167, 168, 170, 173, 214, 267, 268, 276, 290, 291, 292, 293
  1905 paper "On a Heuristic Point of View Concerning the Production and Transformation of Light" 39–47
  1905 paper "A New Determination of Molecular Dimensions" 62–70
  1905 paper "On the Movement of Small Particles Suspended in Stationary Liquids Required by the Molecular-Kinetic Theory of Heat" 70–76
  1905 paper "On the Electrodynamics of Moving Bodies" 113–129
  1905 paper "Does the Inertia of a Body Depend on Its Energy Content?" 129–131
  1916 paper "The Foundation of the General Theory of Relativity" 171–213
  see also topics such as relativity, general theory; relativity, special theory; Lorentz transformations; etc.
Einstein, Hans Albert 19
Einstein, Hermann 15, 17
Einstein, Jakob 15, 17
Einstein, Marie (Maja) 15, 17
Einstein, Pauline (Koch) 15
Einstein's Odyssey 168
electric:
  charge 14, 128
  field 108, 110, 111, 122–126, 129, 132, 136, 149–153, 291
  force 128, 153, 162, 291
  force vector 122
  mass 203
electrical 14, 15, 17, 20, 162, 290
electricity 13, 14, 19, 222, 281
electromagnetic 21, 40, 61, 113
  aether 14, 33
  field 14, 16, 17, 61, 111, 128, 132, 178, 193, 200, 202, 203, 205, 216, 222, 265, 266, 272, 276, 281, 291
  field transformations 46, 149, 150, 153
  force 122, 8, 153, 178, 291

foundation 15, 61, 62
  induction 14, 105, 114, 132
  phenomena 21, 61, 62, 132
  potential 200, 222
  radiation 34, 40, 271, 279
  tensor 222, 223
  theory 15, 18, 21, 40, 61, 62, 131, 132, 199, 273
  waves 14, 21, 40, 42, 126, 152, 276
  worldview 15, 20, 21, 40, 61, 62, 131, 132, 199, 273
electromagnetism 13, 14, 21, 33, 40, 61, 62, 108, 128, 200, 222, 223, 275, 276
electrostatics 290
electron:
  accelerated 128
  Bohr orbits 270
  bound 41, 42
  Compton effect 271, 272
  Davisson–Germer experiment 276
  de Broglie 276
  diffraction 278
  energy 41, 48, 129
  free, bound electrons 41, 42, 271
  Kaufmann experiments 132, 133, 291
  Lorentz force 128, 153, 291
  Lorentz theory of electrons 14, 21, 61, 111
  Maxwell's theory 61
  photoemission 46, 47, 48
  relativistic 128, 129, 132, 291, 292
  resonators 48
  Schrödinger wave function 277, 280
Elements:
  Aristotle 4, 57, 58
  atomic number 111
  Democritus 57
  energy 50
  lead 268
  spectra 265
ellipse 7, 12, 213, 252
ellipsoid 121, 126, 127, 173
elliptical orbits 7, 8, 9, 12, 13, 25, 26, 213, 252
empirical (empirically) 2, 3, 4, 18, 23, 113, 161, 266, 278, 292
empiricism 2
energetics world view, *see* worldview, energetics
energy:
  binding 271
  conservation of 33, 45, 130, 169, 195, 198, 199, 202, 205, 248, 250, 252, 271, 272
  density 35, 37, 38, 39, 41, 42, 50
  discrete (states) 39, 50, 270, 271, 281
  distribution 18, 38
  electromagnetic 40, 202, 203, 205, 276
  elements 50

fluctuations 268
equipartition of 20, 34, 35, 37, 39
free 72, 73
internal 268, 282
kinetic 41, 46 48, 129, 130, 131, 282
quanta 274, 275
quanta of 18, 38, 39, 45, 46, 47, 48, 50, 274, 275, 281
radiation 34, 37, 38, 39, 41, 42, 45, 46, 51, 126, 127, 270
relativistic 129, 271
tensor 169, 170, 195–199, 203, 205, 207
work function 46, 47
enforced motion 3, 6
entanglement 275
entropy:
Boltzmann 34, 44, 45, 59, 272
Einstein 20
ideal gas 43, 44, 45
radiation 40, 42, 43, 44, 45, 47, 51
thermodynamic 33, 49, 50, 60, 72, 93, 95
Wien 51
statistical 60, 72, 93, 95
Entwurf 162, 168, 178, 179, 197
Eötvös, Baron Roland von 162, 292
Epicurus 57, 58
epicycle 5, 6, 10
epistemological 57, 172, 173
EPR paper 280
equal reaction 11, 106
equant 5, 6
equation of continuity 65, 78, 79
equation of state 71
equipartition of energy 20, 34, 35, 37, 39
equivalence 20, 105, 164, 165
equivalence principle 167, 216
ether *see* aether
ETH 17–20, 40, 168, 290
Eudoxus 3, 4, 5
Euler's equations 199, 200

factitious 173
falsification (falsifiable, falsifiability) 1, 22, 24, 291, 292
Faraday, Michael 14, 25
field equations:
electromagnetism 200
gravity 168–171, 175, 192, 193, 196, 198, 207, 218, 219, 221
fifth Solvay conference *see* Solvay conference, fifth
FitzGerald, George 112, 140
Fizeau, Hyppolyte 109, 123
fluid 56, 62, 63, 65, 66, 78, 199
fluorescence 45
four dimensional:
region 175, 195, 242

space (continuum, world) 18, 167, 176, 178, 179, 184, 185, 186, 222, 223, 225
space-time 168
straight line 193
volume 177, 184, 224
Franck, James 270
Franck–Hertz experiment 270
free electrons *see* electrons, free
Fresnel, Augustin-Jean 13
Freundlich, Erwin 170, 215
Friedmann, Alexander 220
fusion 269, 281

Galilean principle of relativity 10, 105, 106
Galilean transformations 107, 108, 111–113, 122, 133, 140
Galileo, Galilei 1, 6, 9, 10–13, 105, 106, 165, 173
Gamow, George 220, 221
Gassendi, Pierre 57, 58
Gay-Lussac, Joseph 58
generally covariant
(equations) 168–171, 178, 179, 185, 187
geocentric 3, 5
geodetic 187, 188, 193, 225, 229
geodetic equation 187, 193
geodetic line 179, 186, 187, 193, 225–228
Germer, Lester 276
gravitational:
acceleration 12, 161–165
constant 12
energy 197, 198
force 12, 13, 162, 164, 172, 178, 213
field 18, 96, 162–165, 167–170, 172, 173, 175–179, 184, 186, 192–196, 198–200, 206–211, 213, 216, 218, 219, 221, 222, 241, 248, 249, 250, 279, 292
Mass, *see* mass, gravitational
potential 168, 170, 207, 208, 218, 222, 250
redshift 167, 168, 199, 210, 215, 219, 292
gravity 161, 164, 165, 167
Aristotle 4
Einstein 23, 163, 168, 174, 183, 193, 196, 211, 212, 215, 218, 222, 223
Newton 10–13, 22, 25, 61, 162, 169, 171, 192, 196, 199, 208, 213, 215, 218
Grossman, Marcel 18, 19, 168, 169, 171, 178

Hamilton's equations
(Hamiltonian) 72, 194, 241
heat:

kinetic theory 20, 56, 70, 71, 72, 76, 267, 291
production 65, 66, 67, 85, 91, 93
thermodynamics 45, 62, 94, 267, 282
transfer 33, 49, 50
Heisenberg, Werner 277, 279, 280
Helmholtz, Hermann von 18
Helmholtz free energy 72
Henry, Joseph 14
Heraclitus 57
Hertz, Gustav Ludwig 270 *see also*
Franck–Hertz experiment
Hertz, Heinrich 18, 20 *see also*
Maxwell–Hertz equations
heuristic 33, 39, 127, 290
hidden variable theory 280
Hilbert, David 222
homogeneous:
gravitational field 163
equations 176, 178, 179, 193
liquid (fluid) 62, 63
universe 219
Hubble, Edwin 220, 221, 292
hydrodynamic equations 64–65, 200
hydrodynamics 63, 64, 199
hydrogen 265, 266, 281
hydrostatic pressure 65
hypothesis:
atoms 56, 58, 59, 60, 77, 78
Bose–Einstein 273
Copernicus 6
cosmology 221
ether 269
general theory 169, 170, 185, 213, 214
kinetic theory 58
light quantum 47, 48, 49, 266, 292
Lorentz 111, 112, 113
Plato 2, 3
science 21, 22
spontaneous/induced transitions 271
hypothetico-deductive model of
science 21

ideal gas 43, 44, 45, 47, 69, 71
impetus 6, 9, 10, 11
index (indices):
Einstein convention 179
anti-symmetric tensor 182
contravariant tensor 181
contravariant vector 180
covariant tensor 181
covariant vector 180
Greek 179
Latin 179
symmetric tensor 181
indistinguishable:
energy elements 50, 274
particles 275
quanta 274

inertia:
Einstein 127, 129, 163, 219
  Galileo 10
inertial reference frames:
  Einstein 115, 116, 132, 168
  Maxwell–Lorentz 105, 113, 122
  Newton 105, 106, 107, 113, 116
inflexible boundary condition, *see*
    boundary condition, inflexible
invariant 178, 180, 183, 186, 187, 188,
    226, 229
irreversible 33
isotropy (isotropic) 115, 219

Jeans, James 37
Jew (Jewish) 15

Kaluza, Theodor 222, 223
Kaufmann, Walter 132, 133, 291,
    292
Kepler, Johannes 6, 7, 10, 13
Kepler's three laws 7–9, 12, 25, 26,
    250, 252
kinetic energy 41, 46, 48, 129, 130,
    131, 282
kinetic theory of atoms (gases) 19, 20,
    41, 56, 58, 59, 77, 267, 282
kinetic theory of heat 20, 56, 70, 71,
    72, 267, 291
Kirchhoff, Gustav Robert 18, 34, 78,
    79, 80
Kirchhoff's theorem 34
Kleiner, Alfred 19, 20
Koch, Pauline 15
Koch (family) 17, 19
Kuhn, Thomas 24, 40

Lagrange multipliers 42, 51
Lagrange's equations 195, 242
laser 271
Laue, Max von 47
Laue scalar 197
Lemaître, Georges 220, 221
Lenz, Heinrich Friedrich Emil 14
Leverrier, Urbain 170
levia (levity) 3, 4
Levi-Civita 168, 169, 171, 232
Lewis, Gilbert 49
light, bending of *see* bending of light
light, dualism (fusion) 269, 281
light, emission 18, 131, 269, 281
light medium 13, 14, 24, 40, 105, 114,
    138, 291
light, Michelson–Morley 111, 113,
    139, 140
light, Newton 11
light objects 3, 4
light quantum (photon, particle) 39,
    40, 45–49, 266, 269, 274, 275,
    281, 291

light ray (beam) 16, 40, 115, 116, 119,
    126, 129, 139, 141, 142, 142, 143,
    165, 166, 167, 170, 171, 174, 199,
    208, 210, 211, 212, 253
light spectra 164, 265, 266
light, speed of (velocity) 16, 25, 41,
    108, 113, 114, 116, 122, 125, 126,
    131, 132, 138, 139, 141, 142, 143,
    165, 168, 172, 174, 206, 207, 210,
    211, 212, 291
light wave 13, 14, 16, 40, 46, 47, 48,
    108, 111, 126, 127, 129, 130, 131,
    138, 211, 266, 269, 281
local time 110, 111, 112, 113, 123, 138,
    140, 168
Lorentz force 14, 128, 153, 291
Lorentz transformations 21, 108,
    112–115, 118–120, 123–125, 126,
    128, 132, 140, 145–148, 153, 172,
    291
Lorentz, Hendrik 13, 14, 20, 21, 25, 61,
    108–113, 123, 124, 131–133, 135,
    137, 140, 167, 169, 170, 215, 291
Lorentz–FitzGerald contraction 112
Lorentz's theory of electrons 14, 21, 61,
    62, 113, 128, 132, 281
Lucretius 58
Luitpold gymnasium 16
luminiferous aether, *see* aether,
    luminiferous

Mach, Ernst 18, 57, 59, 77, 78, 217, 291
Machian 173, 218
magnetic field 16, 108, 110, 111, 114,
    122–126, 132, 136, 149–153, 272,
    291
magnetic force 114
magnetic force vector 122
magnetism 13, 14, 19, 222
Marić, Mileva 18, 19, 290
Mars 3, 6, 7, 8, 9
mass 10–12, 61, 95, 131, 161–163, 172,
    173, 197, 203, 207, 209, 210, 217,
    250, 270
mass density (distribution) 68, 168,
    207, 218, 219
mass, gravitational 161, 162, 163, 165
mass, inertial 40, 161, 162, 163, 165,
    196, 219
mass, longitudinal 128
mass, Newton 162
mass, photon 271, 273
mass, point 209, 211, 212, 249
mass (rest, relativistic) 128, 132, 274,
    279, 291
mass, sun 169, 212, 215
mass, transverse 128
material composition 18, 161, 173
matrix formulation of quantum
    mechanics 277

matter 6, 34, 59, 61, 62, 77
  aether 14, 21
  Aristotle 57
  atoms 58, 59, 70, 77, 78
  cosmology 218, 219, 220
  general relativity 169, 170, 192, 193,
    195–199, 206, 207, 248
  quantum mechanics 277
  radiation 14, 34, 47, 49, 61, 266,
    267
matter-free gravitational fields 193,
    196, 197, 248
matter-free space, *see* space,
    matter-free
Maxwell distribution function 34, 59
Maxwell–Hertz equations 122, 128
Maxwell, James Clerk 1, 13, 14, 18, 20,
    21, 25, 59, 61, 62, 216
Maxwell–Lorentz theory of
    electrodynamics 105, 113
Maxwell's electromagnetic field
    equations 200
Maxwell's equations: Einstein 122, 124,
    128, 149, 200, 202, 216
Maxwell's equations (theory) 14, 17,
    18, 21, 39–41, 48, 61, 108–111, 113,
    122–124, 131–133, 138, 146, 147,
    149–151, 171, 192, 199, 200–202,
    222
Maxwell's kinetic theory 20
Maxwell's relations
    (thermodynamics) 52
mechanics (mechanical) 11, 13, 15,
    18–21, 25, 26, 33, 59–62, 85, 107,
    113–115, 131, 165, 172–174, 217,
    277
mechanical world view *see* world view,
    mechanical
Michelson–Morley experiment 105,
    111, 112, 113, 138, 139, 140
Michelson, Albert 111, 113, 140
microcanonical ensemble 20, 34
migration velocity 68, 69
Millikan, Robert 48
Minkowski, Herman 18, 167
Minkowski space 167, 168, 200
molecule 33, 34, 40, 41, 44, 47, 59, 62,
    63, 67, 68, 71, 72, 77, 93, 95, 207,
    273, 274
molecular:
  forces 19, 20
  dimensions 20, 56, 62, 63, 67, 68,
    69, 291
  kinetic theory of heat 20, 56, 70, 71,
    72, 267, 291
  kinetic theory of gases 39
  investigation of entropy 43
  movement (Brownian motion,
    diffusion) 56, 60, 70, 75, 77, 291
  energetics worldview 62

theory (thermodynamics) 33, 39, 40, 41, 44, 68
Morley, Edward 111, 140
momentum, angular 25, 26, 270
moon 3, 7, 8, 10, 11, 12
moons of Jupiter 10
Mössbauer, Rudolf Ludwig 216
Mössbauer effect 216
motive force 11, 106
moving reference frame, *see* reference frame, moving
Munich 15, 17, 290

natural motion 3, 4, 5
natural philosophy 61
Navier–Stokes equation 65, 78, 79
Neptune 9, 13, 213
Newton, Isaac 1, 6, 11, 12, 13, 15, 21, 22, 24, 25, 58, 60, 61, 107, 108, 162, 217, 254, 255
Newton's addition of velocities 115, 116, 132
Newton's bucket 217, 254, 255
Newtonian mass, *see* mass, Newton
Newtonian universe 217, 219
Newton's inertial reference frame, *see* reference frames, Newton
Newton's absolute space *see* absolute space, Newton
Newton's law of gravitation 12, 13, 22, 25, 61, 162, 168, 169, 170, 171, 192, 194, 196, 199, 205, 208, 213, 215, 218, 252
Newton's principle of relativity 11, 20, 21, 105, 107, 113, 123, 131, 132, 172
Newton's second law 107, 128, 129, 153, 154, 155, 162, 207, 250
Newton's third law 12
Newton's three laws (mechanics) 11, 12, 13, 15, 21, 24, 25, 26, 33, 59, 60, 61, 106, 115, 173, 217
Nobel prize:
    1909 (Ostwald) 62
    1926 (Perrin) 77
    1961 (Mössbauer) 216
non-accelerated reference system 163
non-locality 275
non-uniform motion 161, 163, 172, 177

observable 20, 59, 75, 163, 172, 173
optics (optical) 11, 21, 40, 113
Ørsted, Hans Christian 14
osmotic pressure 68, 69, 71, 72, 74
Ostwald, Wilhelm 20, 60, 62, 77, 78, 291

Pais, Abraham 20, 25, 214, 222, 269, 273

paradigm 24, 25
Paris Terminists 6, 9, 11
Parmenides 57
patent office 17, 18, 19, 290
Perrin, Jean 77, 291
phosphorescence 45
photoelectric effect 33, 45, 46, 48, 271, 291
photo-ionization 45, 291
photoluminescence 45, 291
photon 49, 271, 272, 274, 275, 276, 278, 279, 281
Planck–Einstein relation 270, 276
Planck, Max 18, 33, 38–42, 48–50, 133, 266–268, 272, 273, 291, 292
Planck's constant 45, 280
Planck's distribution function 267, 269, 271–275
plane waves 124, 125, 126
Plato 2–6, 10
Pluto 9, 13
Podolsky, Boris 275, 280
Poincaré, Henri 77, 112, 113, 140
pollen 60
Popper, Karl Raimund 23, 24
positivistic philosophical attitude 78
positivists 62
Pound–Rebka experiment 215, 216
precession of the perihelion of Mercury 22, 169, 171, 192, 199, 213, 214, 215, 221, 249, 292
preferred orbits 270, 276
Principe Island 214, 215
Principia 11, 12, 162
probability:
    blackbody radiation 45
    complexions 34, 40, 59, 272, 273, 274
    density $\left(|\Psi|^2\right)$ 277
    distribution law 75, 98
    electron 278, 280
    Kaufmann 132, 292
    spontaneous, induced transitions 271
    thermodynamics 34, 44, 47, 94, 96
    time ensemble 50, 273
projectile 105
projectile motion 9, 10
Ptolemeic system 5, 6, 7
Ptolemy 5, 6, 10
Pythagoras 2

quanta (quantum):
    energy 39, 46, 274, 275
    light 39, 45–49, 267, 274, 275, 281, 291
    radiation 13, 33, 39–41, 44, 45, 47–49, 164, 167, 265–269, 271, 272, 275, 281
quantum entanglement, *see* entanglement, quantum

quantum gas 273, 276
quantum mechanics 1, 13, 21, 216, 265, 266, 267, 275, 277, 278, 280, 282
quantum theory 21, 49, 216, 223, 268, 273, 275, 279, 280, 281, 282, 292
quasi-static field 206
quintessence 4, 24, 57, 217

radiation 33–35, 38–45, 47–51, 131, 164, 222, 266–274, 276, 279, 281, 292
radiation, blackbody, *see* blackbody radiation
radiation, cavity, *see* cavity radiation
radiation energy 34, 126
radiation energy density 37–39, 41–43, 50, 51, 269, 271, 272
radiation pressure 126
radiation, remnant (big bang) 221
radiation, thermal 34, 35
Rayleigh, Lord 37
Rayleigh–Jeans law (formula, theory) 37–42, 269
rectangular coordinates (Cartesian) 95, 223
rectilinear motion 12, 106, 107
reference frames 106–110, 115–120, 122–127, 130–133, 135, 169, 172–175, 179, 184, 185, 187, 192, 193
    absolute space *see* absolute space reference frames
    accelerated (non-uniform motion) 163, 164, 172, 173, 175, 193, 194
    aether 139
    inertial *see* inertial reference frames
    moving 108, 110–113, 117, 118, 121–123, 125, 128–131, 133, 135, 138–141, 143
    non-accelerated (uniform translational motion, constant velocity) 11, 163, 217
    privileged 173
    rotating 174
    special theory 193, 202, 240
    stationary (at rest) 11, 12, 14, 21, 107–110, 113, 116–118, 121–123, 126–128, 130, 131, 133, 138, 130, 141, 143, 145, 153, 163–165, 173, 217
relative motion
    absolute space 11, 14, 21, 107, 108, 113, 123, 131, 153, 217, 254
    accelerated 163, 172, 173, 177, 193, 218
    aether 105, 108, 109, 113, 114, 123, 131, 291
    earth 17
    light 131, 139, 143

relative motion (*cont.*)
  other bodies 10, 63, 64, 106, 107, 111, 217
  rotational 165, 172, 174, 217, 254, 255
  two reference frames 107, 111, 112, 114–119, 122, 123, 130, 131, 144, 145, 152, 153, 172, 174, 175, 217, 254, 255
relative velocity, *see* velocity, relative
relativity:
  Galileo 10, 105, 106
  Einstein 15, 23, 24, 25
  Lorentz 111, 113, 281
  Newton 11, 20, 21, 105, 107, 113, 123, 131, 132, 172
  principle of 113, 114, 116, 123, 124, 128, 130, 132, 149, 153, 163, 291, 292
relativity, general theory, *see* Chapter 5 (161–262)
  acceleration 164, 165, 166, 173, 217, 292
  measurement 162, 165–167, 172–176, 187, 210, 211, 215, 216, 222, 292
  non-uniform motion 161, 163, 172, 177
  time (clock) 168, 172, 174, 175, 208, 210, 213
relativity, special theory, *see* Chapter 4 (105–160)
  electromotive force (emf) 114, 122
  rigid body (sphere) 120, 121
  rigid rod 116, 174
  time (clock) 16, 114–120, 132, 140–142, 168, 172, 176, 177
  two postulates 113–116, 120, 132, 168, 169, 171, 172
  velocity addition 121, 122
resonator 38, 39, 41, 42, 48, 50
rest frame 21, 108, 110, 117, 120, 121, 123, 138, 143
retrograde motion 5
Ricci, Gregorio 168, 169, 171
Riemann, Bernhard 168, 169
  Riemann–Christoffel tensor 179, 191, 239, 240
  Riemann tensor 168, 192, 193, 194, 240
  Riemannian geometry 168, 222
  Riemannian space – curvature 222, 223
  Riemannian space – torsion 222, 223
rigidly rotating disc 165–167
Rosen, Nathan 275, 280
Rosenthal-Schneider, Ilse 215
rotation (rotational) 22, 34, 63, 64, 106, 170, 172, 174, 213, 214, 217, 255

Royal Astronomical Society 214, 215
Royal Society 215
Rubens, Heinrich 38
Rutherford, Ernest 270

$S = k \log W$ 34, 44, 45, 50
save the phenomena 2, 4, 5, 25
Schrödinger equation 277
Schrödinger, Erwin 266, 275, 277
scientific evolution 24, 25
scientific materialism 60
scientific revolution 24, 25
semi-permeable membrane 71
simultaneity 114–117, 132
Solvay conference:
  fifth 278, 282
  sixth 279, 282
Smith, R. W. 221
Sobral, Brazil 214, 215
solar eclipse 24, 168, 214
solute 62, 63, 67, 68, 71
solvent 62, 67, 68, 71
sources/sinks 108, 123, 126, 146
space, absolute:
  de Sitter 219
  Einstein 163, 173, 218, 221
  Lorentz 14, 21
  matter-free 192, 195, 196, 198, 219, 248
  Newton 11, 107, 108, 109, 113, 123, 131, 163, 217, 254, 255
space-time (spacetime) 168, 170, 175, 178, 200, 214, 221, 223, 279
space-time coincidences 175
space-time continuum 174, 184
space-time interval 176
specific heat 18, 19, 59, 267, 268, 281, 282, 292
specific heat capacity 267
specific volume 67
spectra:
  continuous 265
  discrete 265, 266, 270, 271
  starlight 170
spectroscopist 265
spherical surface 126
spherical wave 120
speed of light (velocity) 16, 25, 41, 108, 116, 125, 126, 132, 138, 168, 206, 210, 212
  c 114
  constant 113, 131, 168, 172, 291
  in aether 122, 131, 138
  variable 168, 207, 211
Spinoza, Baruch 290
Stachel, John 19, 20, 49, 67, 105, 115, 116, 131, 168, 274, 280, 291, 292
starlight, bending of *see* bending of starlight
stationary state 94, 270

statistical mechanics 20, 33, 40, 41, 59, 77, 273
Stokes' law (rule) 45, 46, 68, 291
superfluous 24, 40, 114
Swiss Federal Polytechnic School (ETH) 17, 18, *see also* ETH
synchronize (synchronous) 115–119, 140, 142

Taylor series 63, 95, 97
tensors 178–203, 223–240
  fundamental 183–186
  Riemann–Christoffel *see* Riemann–Christoffel tensor
Thales 2
thermodynamic equilibrium 73
thermodynamics, first law 93, 282
thermodynamics, second law 20, 50, 59, 60, 62
Thomson, Joseph John 215
time:
  averages 20, 40, 273
  cosmology 219
  dilation 172
  Einstein 16, 40, 114, 115, 132, 140
  general relativity 170, 174–177, 213
  local 110–113, 123, 137, 138, 140
  Lorentz 110–112, 123, 125, 133, 140, 142
  Michelson–Morley 111, 112, 139, 140
  Newton 60
  Ostwald 60
  proper 168
  special relativity 114–19, 121, 132, 140, 142, 145, 150, 167, 174, 176, 293
timeline to general relativity 167
transformations:
  electromagnetic field 110, 124, 146, 149, 150, 153
  energy 15, 61, 62
  Galilean 107, 108, 111, 112, 113, 122, 133, 140
  general relativity 169, 177, 178, 193
  light 33, 39, 40, 127, 290
  local time 110
  Lorentz 21, 108, 112–115, 118–120, 123–126, 128, 132, 140, 145–148, 153, 172, 291
transitions (induced, spontaneous) 271, 281
translation (translational motion) 11, 34, 63, 64, 77, 106, 107, 116–118, 129, 164

Ulm 15
ultraviolet catastrophe 37
uncertainty principle 277, 279, 280
unified field theory 178, 216, 221–223, 292

uniform motion 2, 3, 5, 17, 161, 163, 173, 177, 193
unit cross section 69, 73, 74

van't Hoff law 69, 71, 72
velocity:
  absolute 163
  relative 107, 117, 119, 217
velocity of light *see* speed of light
verification (verify):
  atoms 56, 70
  Brownian motion 77
  Einstein 1, 290, 291
  Galileo 10
  inertial and gravitational mass 161, 162
  general relativity 213, 221, 292
  Newton 13
  Popper 23, 24
  photoelectric effect 48
  $\rho(\nu, T)$ 35
  science 22
  special relativity 113, 120
  worldview 56

viscosity (viscous) 62, 63, 65–69, 74, 77, 86, 93, 291 *see also* coefficient of viscosity
viscous force 63, 68
volume scalar 184, 224

waves:
  de Broglie 275, 276
  electromagnetic *see* electromagnetic waves
  electron 276–278
  equation 108
  light 13, 40, 46, 48, 111, 127, 129, 130, 138, 211, 269, 281
  plane 124, 125, 126, 129, 130
  particle duality (fusion) 269, 275, 276, 278, 281
  radiation 35, 36, 47, 48, 269
  Schrödinger 277
  spherical 120
wavelength 35, 37, 47, 265, 266, 272, 276
wave–particle 275
Weber, Heinrich Friedrich:
  ETH 17–20, 40, 290

light emission 18, 40
specific heat 18, 19, 267, 268, 292
Weber's law 40
weight 3, 10, 11, 71, 93, 162, 163, 164
Wien, Wilhelm 35, 38, 39, 40, 42, 43
  displacement law 18, 35, 37–41, 43, 45, 47, 49, 50, 266, 269, 273, 274, 291
  entropy 42, 51
  principle 42
  relation 50
work 47, 65, 94, 129, 282
work function 46, 47
worldview (world-view):
  Aristotelian 4, 10
  electromagnetic 15, 20, 21, 56, 57, 62
  Einstein 1, 16
  energetics 15, 20, 56, 57, 61, 62, 78
  mechanical 15, 20, 21, 56, 57, 61, 62

Young, Thomas 13

Zeitschrift für Physik 273